AF324869

ELECTROPHYSICAL AND ELECTROCHEMICAL PHENOMENA IN FRICTION, CUTTING, AND LUBRICATION

ELECTROPHYSICAL AND ELECTROCHEMICAL PHENOMENA IN FRICTION, CUTTING, AND LUBRICATION

S. N. Postnikov

Gorky Polytechnic Institute, USSR

Translated by Ben Teague

 **VAN NOSTRAND REINHOLD COMPANY**

NEW YORK   CINCINNATI   ATLANTA   DALLAS   SAN FRANCISCO

LONDON   TORONTO   MELBOURNE

Van Nostrand Reinhold Company Regional Offices:
New York   Cincinnati   Atlanta   Dallas   San Francisco

Van Nostrand Reinhold Company International Offices:
London   Toronto   Melbourne

Library of Congress Catalog Card Number: 77-22249
ISBN: 0-442-26624-3

Manufactured in the United States of America

Published by Van Nostrand Reinhold Company
135 West 50th Street, New York, N.Y. 10020

Published simultaneously in Canada by Van Nostrand Reinhold Ltd.

15 14 13 12 11 10 9 8 7 6 5 4 3 2 1

Library of Congress Cataloging in Publication Data

Postnikov, S   N
   Electrophysical and electrochemical phenomena
in friction, cutting, and lubrication.

   Bibliography: p. 257.
   Includes indexes.
   1. Metal-cutting. 2. Friction.
3. Lubrication. 4. Thermoelectricity. I. Title.
TJ1185.P655   671.5'3   77-22249
ISBN 0-442-26624-3

to MY WIFE

Preface

This book deals with the problem of improving the wear-resistance of frictional couples and cutting tools, i.e., with a key problem in modern mechanical engineering. From the viewpoint of physics, it can be reduced to considering specific interactions of solids having their own chemical composition and structure, and controlling such interactions under various conditions.

In fact, both kinematic friction and metal-cutting processes can be considered to cause continuous changes in the energy of atomic and molecular interactions in a three-phase boundary system (including the most widespread version, metal I–lubricant–metal II). A whole complex of interconnected physicochemical phenomena takes place: thermal, sorptional, electrical, etc. The part of the mechanical energy imparted to a frictional system and transformed into other forms of energy depends upon the type of bonds between particles at an interface, in intermediate layers of lubricant, and in surface layers of solids (metals) subjected to elastoplastic strains. In turn, the phenomena peculiar to processes of friction and cutting can influence both the force of interaction between atoms and atomic collections and the very character of bonds. These phenomena continually disturb the conditions of the system and apply corrections to the path along which it would like to regain an equilibrium state. This sort of disturbance, in the end, results in changes in the macroscopic mechanical parameters: frictional force, work of cutting, etc. It is evident, therefore, that great applied value attaches to the study of specific features of the phenomena mentioned and to the simultaneous search for ways of controlling the mechanisms of their influence on friction and cutting processes. The study is aimed at decreasing friction, wear, resistance to plastic strains, work to overcome molecular forces in the formation of new surfaces, and so on.

It is pleasant to note that, in the devising of new friction theories, the emphasis is not on the mechanical models of interaction of solids, but on the little-studied category of electric and electromagnetic processes at the atomic level. These studies are especially facilitated by the parallel development of concepts about the role of an electrostatic component in the adhesion bond (B. V. Deryagin, N. A. Krotova, V. P. Smilga) and the remarkable ideas of P. N. Lebedev about electromagnetic radiation and absorption by interacting systems of atoms (E. M. Lifshitz, I. E. Dzyaloshinsky, L. P. Pitaevsky). Proceeding from these concepts, a trinomial friction formula was presented and experimentally confirmed (A. S. Akhmatov, G. N. Uchuvatkin). In addition, an attempt was made to substantiate the resonance electromagnetic mechanism of energy dissipation in external friction (V. A. Bufeyev). The elementary friction theory suggested by the author is also based on the forced vibrations of similar (also called *equivalent*) oscillators in rubbing bodies.

In the past two decades, careful study has been given to both physicochemical and engineering aspects of the macroscopic electrical phenomena peculiar to processes of friction and especially to those of cutting.

The attention of specialists had already been concentrated for a long time on the fact that the interaction zone of metals and alloys in relative sliding represents the hot junction of a natural thermocouple. If the external circuit of such a thermocouple is closed, a thermoelectric current will flow through it. The temperature in the area of actual contact can be evaluated from the magnitude of the electromotive force of the seat. However, an unexpected result has been the ability of the thermocurrent to interfere actively with the actual process of sliding.

In 1952, P. L. and S. L. Gordienko were the first to demonstrate the possibility of reducing the wear of a frictional couple by breaking its external thermoelectric circuit. In the field of metal cutting, an analogous procedure, based on the desire to decrease the detrimental effect of thermocurrent upon the cutting-tool life, found its application for the first time in Germany, in investigations by H. Axer and H. Opitz (1953-1954) and, somewhat later, in the work of T. H. G. Hehenkamp.

In 1958, V. I. Levin and I. V. Gogolev introduced drills fitted with tapered plastic shanks (USSR Inventor's Certificate No. 128,261). The inventors suggested that the improved wear resistance of the new drills was due above all to the damping, vibration-absorbing action of the plastic shank; meanwhile, this shank played a role somewhat like that of a circuit breaker in the thermoelectric circuit, tool-workpiece-machine.

M. T. Galey and A. A. Avakov (1962-1963) were the initiators in checking the effect discovered for the purpose of wide practical application. V. A. Bobrovsky made a great contribution to experimental approbation of this new procedure under laboratory and plant conditions. The essence of Bobrovsky's invention (Inventor's Certificate No. 206,972) is to use "bonded" holding devices for breaking the thermoelectric circuits, whose hot junction is the cutting zone. In appearance, bonded devices are like ordinary arbors, bushings, and so forth, but they consist of two parts glued together, the layer of glue being intended for insulating the tool from the machine. Bonded holding devices are simple to make and have a high stiffness as compared with the devices used previously (for example, Textolit bushings). According to Bobrovsky, their use extends tool life by approximately 20 percent, and the very act of breaking the thermoelectric circuit can make the tool life twice as long or even longer under the same working conditions.

The new procedure became more and more popular due to work by a number of other investigators, who showed the considerable influence of electric phenomena upon cutting-tool life, surface finish, and wear of the rubbing materials.

Simultaneously, beginning in 1965, investigations were carried out in an allied area, that of the phenomena connected with influence of the magnetic state of a tool upon metal-cutting processes (G. I. Yakunin, M. T. Balabekov, and others). Quite recently, research by the author and his colleagues has demonstrated the possibility of magnetostrictive strengthening of high-speed steel by magnetic treatment; this is another way of improving tool wear resistance.

In 1967, the first scientific-technical meeting devoted to the problem of electrical phenomena in friction and metal cutting and their practical application took place in Moscow.[1] It was sponsored by the Scientific Council on the Theory of Machines and Working Processes of the USSR Academy of Sciences, together with the State Scientific Research Institute of Mechanical Engineering and the Commission on the Technology of Machine Building. The meeting gave an impetus to the development of investigations with the final aim of effectively using electrical phenomena to obtain detailed information about specific processes in the interaction of solids and to control such processes.

In 1969, a second meeting was held, on Professor M. M. Krushchov's initiative (as the previous one had been). The considerable role of electrophysical processes under the particular conditions of the interaction of solids was confirmed in the proceedings of that meeting.[2] However, both the first and the second meetings clearly showed the extreme complexity of the phenomena under study and the necessity of more profound and systematic research.

Against the background of contradictory experimental data, which cannot be systematized on the basis of existing qualitative interpretations, it has become evident that some recommendations for utilizing the electrical phenomena were premature. Unfortunately, we have to state, for example, that the tendency to introduce bonded holding devices in practically all cases of metal cutting resulted in unnecessary waste of materials. Only when the use of such holding devices is scientifically justified can the electrical insulation of the tool be of much economic value.

Because of the rapid development of tribophysics and tribochemistry, the number of organizations and specialists studying electrical phenomena in friction and cutting has increased greatly. The range of questions on this subject has expanded considerably, especially since D. N. Garkunov and I. V. Kragelsky discovered the phenomenon of selective transfer. The use of this phenomenon, which is essentially electrochemical, has opened up broad opportunities for the creation of more advanced frictional systems and more effective lubricants.

The results of recent investigations into electrophysical and electrochemical phenomena in friction, cutting, and lubrication were discussed at the All-Union Scientific-Technical Conferences, held in Odessa in 1973[3] and in Tashkent in 1975.[4] Stress was laid on the need for work on such problems as the mechanism of formation and the properties of boundary layers with domain structure (M. M. Snitkovsky, V. N. Yuriev, and others); triboelectric phenomena in systems containing new plastics-based materials (G. A. Georgiyevsky, L. A. Lebedev, and others); the effect of polarization on the kinetics of electrochemical processes in friction and on the properties of surface structures (A. I. Porter, G. A. Preis, and others); and other problems.

In 1975, the author's book *Electrical Phenomena in Friction and Cutting* appeared.* It included the results of investigations carried out by the author at the Cavendish Laboratory of Cambridge University (1963–1964), and at the A. A. Zhdanov Gorky Polytechnic Institute (1965–1975) in cooperation with a number of the industrial enterprises at Gorky. The author naturally intended the Russian book as a monograph dealing with the then state-of-the-art. However, the book did not claim to have used all available literature on the subject, but concentrated rather on problems that interested the author and on allied topics.

The present book is in fact an English-language edition of the same monograph. In 1976–1977, the author visited the United States as a participant in the Senior Scholar Exchange between the Ministry of Higher Education, USSR, and the U.S. International Research and Exchange Board. His report at the 1976 ASME-ASLE Lubrication Conference in Boston, his participation at the 1977 AIME annual meeting in Atlanta, and his meetings with American scientists during his research at the Friction and Wear Laboratory of The University of Texas at Austin indicated general interest in the work described previously. That is why the author is grateful to Van Nostrand Reinhold Company for making this book available to his English-speaking colleagues.

S. N. Postnikov

---

*Gorky: Volgo-Vyatskoe Izdatelstvo.

List of Symbols

a Constant; period of identity of position of oscillators; activity of ion; coefficient

A Work of adhesion; amplitude; emission constant; adhesion component of friction force

b Constant; angular coefficient

B Magnetic induction; coefficient of electrical mass transfer; coefficient

c Electrodynamic constant

C Capacitance; concentration; constant; calibration

d Distance; diameter; clearance

D Drill diameter; variance of random function

e Electromotive force; charge of electron

E Electric field strength

$\mathscr{E}$ Electromotive force

f Frequency; specific force; exciting force

F Friction force; faraday

g Conductance

G Contact conductance

h Drilling depth; workpiece thickness

$\hbar$ Planck's constant

H Distance; magnetic field strength; thickness of boundary layer; microhardness

i Current

I Current; intensity of exoelectron emission

j Current density

k Constant; coefficient of elasticity; coefficient; Boltzmann constant; angular coefficient; component of wave vector

K Index of machinability; criterion of efficiency

L Thickness of oxide film; length; inductance; Debye length

m Mass

M Molecular weight

n Concentration; dislocation density; charge of ion

N Normal load; life

p Pressure; momentum; specific attractive force between condensed phases; mean pressure of molecular adherence

P Power

q Charge

Q Heat; Q factor; wear value

r Resistance coefficient; electrical resistance

R Resultant force; electrical resistance; constant; force; atomic radius

s Feed

S Area; spectrum

t Time; temperature; depth of cut

T Temperature; period; tool life

U Voltage; height of potential barrier

v Relative sliding speed; cutting speed

V Contact potential difference; electric potential; volume; molar volume; volume of transferred matter

W Energy; chemical potential; measure of adhesion

x Exponent in conductance-load relation

Z Distance; number of carbon atoms in fatty-acid molecule

GREEK LETTER SYMBOLS

α (alpha) Phase shift; temperature coefficient; specific emf; Seebeck coefficient

β (beta) Damping coefficient; coefficient

γ (gamma) Surface tension; stacking fault energy; electrical conductivity; rake angle of tool

δ (delta) Load-carrying capacity of boundary layer; criterion of absorbability; optimal criterion of dullness

Δ (delta) Change

ϵ (epsilon) Permittivity; strain; specific thermal emf at transition temperature

η (eta) Viscosity

θ (theta) Contact angle

Θ (theta) Temperature

λ (lambda) Thermal conductivity

μ (mu) Coefficient of friction; mobility of charge carriers

ν (nu) Frequency

ρ (rho) Resistivity

σ (sigma) Surface density of charge; surface energy; yield point; ultimate strength; strain hardening

Σ (sigma) Coordinate (sum of elastic and viscous resistance forces)

τ (tau)	Time; tangential shear stress
υ (upsilon)	Temperature
φ (phi)	Work function; potential; contact potential difference; electrode potential; oxidation potential; phase shift; surface potential; dimensionless potential energy
χ (chi)	Electronegativity; electron affinity energy of oxide
ψ (psi)	Height of potential barrier at metal-oxide interface; wave function; Lifshitz function
ω (omega)	Angular frequency

SUBSCRIPTS

a	Acoustic
c	Clearance; contact; closed circuit
e	Electric; efficiency; electrode
f	Integral emf
g	Galvanic
i	Inductive
m	Magnetic; metallic; maximum
n	Nominal; normal
o	Optimal; open circuit
s	Self-induction
t	Thermal
τ	Tangential
tr	Transition
x	Direction
y	Direction
z	Direction
0	Initial value

Introduction

The genesis of intermolecular forces involves electrostatic attraction or repulsion between electrons and nuclei and the presence of electrodynamic, magnetic, and valence (exchange) forces between atoms. In this sense, nearly all the phenomena that take place in, for example, the system metal I–lubricant–metal II are electrical. However, statistical summing of the atomic (electronic-ionic) and molecular interactions does not currently seem possible when the interactions differ greatly in physical and chemical nature. At the same time, with the development and generalization of ideas about the donor-acceptor character of bonds between atoms, molecules, or functional groups,[5] and also of ideas about the electromagnetic nature of van der Waals interactions between condensed phases,[6] the physical theory of friction could not remain in place. In separate calculations of molecular coherence in boundary layers and of noncontact adhesion between solids as a result of interaction between the fluctuations of their electromagnetic fields, it turned out to be possible to make the binomial law of friction[7] more precise and, for the first time, to undertake experiments aimed at measuring the specific forces of attraction between metals separated by a layer of boundary lubricant.[8] The notion that the near-surface force fields of condensed phases are electromagnetic is the basis for the quite recently advanced idea of a resonance-selective mechanism for the dissipation of energy, implemented by the equivalent oscillators of the rubbing solids.[9] It was just this idea (although a fundamentally different meaning was given to the mechanism of resonance losses itself) that provided the author with one starting point in the development of a new theory of friction,[10] which is constructed on analogy with the classical theory of light scattering. Chapter 1 presents this theory.

The actual form of the near-surface field of a solid depends on the actual crystal structure, which is defined by the lattice defects. In turn, the defect states of conjugate crystal lattices in friction depend on the surface energy at the interface. By using the model of a metal as an ionic lattice immersed in an electron gas, it is possible to evaluate the change in energy when the electron gases simultaneously press on the interface, each from the side of its own metal. It has been shown that the plasticization of the surface layer should be more pronounced in the metal with the lower chemical potential.[11]

The character of the contact interaction between surfaces involves their energy states directly. This fact determines the value of information obtained in work-function measurements, or, more precisely, measurements of the contact potential difference against a standard electrode. Measurements of this type have established, for example, that correlational links exist between the work function and the parameters of boundary friction.[12] The first part of Chapter 2 is devoted to the question of how and why the work function changes as a result of physical and chemical modification of a solid surface. A special analysis of the case where the electronic energy spectrum of a metal is distorted upon plastic deformation has made it possible to go into detail on the results obtained in an investigation of exoelectron emission due to friction.[13]

For the sake of the greatest possible clarity with respect to features of electrophysical and electrochemical phenomena in friction (cutting) of metals, Chapter 2 characterizes the principal seats of emf and some of the factors predetermining the electrical state of the interface. The hysteresis in the contact-conductivity curves in cyclic loading, which reflects the predisposition of the elements of the couple to adhesion and mutual dissolution,[14] is noted.

In order to study macroscopic triboelectric phenomena, whose development and practical significance is becoming more and more obvious, it is convenient (as the author has proposed)[15] to synthesize nonlinear electrical circuits with variable parameters, corresponding to any one of the states of the frictional contact. Indeed, electrical modeling covers the most varied cases (Chapters 2-4) and has proved almost universally applicable. At the same time, the choice of a specific version of the circuit and the possibility of calculating this circuit depend on the type and properties of the frictional system and the parameters and the trend of the process. For example, with the help of an equivalent electrical circuit for a three-phase frictional system, of which a simplified version is simple to calculate, it is possible to evaluate quantitatively how much the real area of contact and the average thickness of the lubricant layer change in friction between metals. This evaluation makes it possible to forecast the performance of sliding contacts in various lubricants and contributes to the realization of conditions favoring selective transfer.[16] Analysis of the circuit was also used in another way: in the development of a method for evaluating the load-carrying capacities of lubricant films by automatic statistical monitoring of the integral emf.[17-20] In particular, this method allowed the difference in lubricant properties between liquids containing salts of dicarboxylic acids to be explained. It turned out that the abilities of these salts to function as lubricants increase in proportion to the molecular weight in the homologous series. According to the results obtained, the size of the carbon chain can have quite a significant effect on the viscosity of an adsorbed boundary layer. This also provides indirect experimental confirmation that the Deryagin disjoining pressure depends on the length of polar chain molecules (Chapter 3).

The effect of electrically insulating the tool in the machining of a range of high-temperature stainless alloys has impelled researchers to begin studying all aspects of the thermoelectric phenomena taking place in the cutting of metals. Experimental data now available (although some of them are relative in value) enable us to assess the parameters of the tool-workpiece-machine thermoelectric circuit, local thermoelectric currents in the cutting zone, the predominant type of conduction in alloys, the specific thermal emf's of alloys, and so on. Work that the author started on the amplitude-frequency spectrum of the alternating-current component of the potential difference between tool and workpiece has been continued;[21] Chapter 4 reflects this development. At the same time, one of the principal tasks of the electrophysics of friction and cutting is to discover the mechanism by which thermoelectric phenomena influence the interaction of solids under specific conditions of sliding. As before, this problem has not been completely solved, not even in the first approximation. All the available explanations for the predominant destructive function of thermoelectric current—erosion of the rubbing surfaces,[22,23] variation in the kinetics of oxidation processes at points of contact between tool and chip and between tool and workpiece,[24,25] stimulation of the strengthening of an adhesion-cohesion joint,[26,27] acceleration of the diffusional degradation of the tool[28,29]—all are hypothetical. For this reason, Chapter 5 gives a brief analysis of the hypotheses just mentioned and discusses the matter of what factors predetermine the effectiveness of electrically insulating the tool.[30] If these factors are allowed for, then experimental situations under which breaking the thermoelectric circuit gives the expected results can be set up artificially. This has been demonstrated for the drilling of titanium alloys with high-speed steel drills. Justification has been found for the dependence of temperature fluctuations in the contact-interaction zone on the variation in electrical resistance of the thermoelectric circuit. This resistance determines the Peltier heating and Joule losses; in the ultimate analysis, this behavior affects the strength of an adhesion-cohesion joint.

Naturally it makes no sense to use bonded devices in the cutting of dielectric materials, since a dielectric is itself an insulator. As the author and coworkers have shown,[31] however, the use of the new holding devices may prove desirable in the machining of dielectrics, evidently because of insulation of the tool from the earth (from the grounded machine). The effect of "disconnecting the earth" is reported in Chapter 6, which deals with triboelectric phenomena in metal-dielectric couples.

The results of research into the effect of magnetic treatment on the life of a high-speed tool form the subject of Chapter 7. The author and coworkers have remarked that magnetostrictive strengthening is the most probable physical cause for the improvement in the properties of a high-speed steel under the action of an impulse magnetic field.[32] The mechanism of this process is examined. A comparison of the dynamic magnetic characteristics of several tool steels shows that it is useful to rate the materials by the quantity of heat energy evolved in them by eddy currents in cutting with a magnetized tool.

One aim of this brief introduction has been that the reader who has only a little acquaintance with the development of this new scientific field should not remain in the slightest doubt about the promise held out by the study of electro-physical and electrochemical phenomena in friction and cutting. The author hopes he has achieved this aim.

Contents

1

Some Theoretical Concepts on the Role of Electrical Phenomena in Contact and Friction of Solids

1.1. ELECTROSTATIC COMPONENT OF ADHESION

The resistance to motion that arises when condensed phases slide one past another depends on the adhesion between the surfaces. Because the literature is arbitrary in terminology when it defines *adhesion*, we will explain the meanings assigned to this and a number of other terms encountered in this book.

First of all, we speak of adhesion between two bodies if, when the bodies are brought into contact, the interface between them remains sharply defined. The interface is practically always preserved when two unlike bodies interact under conditions that preclude seizing and strengthening of the contact by diffusion.

According to Deryagin and coworkers, if the attraction between solids, an unavoidable attribute of adhesion, is due to heteropolar forces of chemical affinity, then the interface takes on the properties of an electrical double layer.[5] The more pronounced the donor-acceptor character of the bond, the more clearly the electrostatic component of adhesion manifests itself. A special case is the attraction between charges in the electrical double layer (microcapacitor) formed when crystalline solids, which in principle can be regarded as giant molecules, come into contact.

It must, however, be stated that in the case of a homopolar bond, which is marked by preservation of symmetry in the electron density, the nature of the adhesion between solids remains electrical, as before; but now the reason for the interatomic attraction is not a Coulomb interaction but a special quantum-mechanical effect that results in the transition of electrons from atomic orbits to the more energetically favorable molecular "orbits."

An adhesion joint is an extended two-dimensional defect belonging to both the boundary structures. If the conditions of these structures are such that the interface between them becomes smeared out, it is probably more correct to speak not of an attractive interaction between the two solids but of the formation between them of a third, "coupling" body with its own chemical composition, structure, and properties. An extended three-dimensional defect has not an adhesional but rather a cohesional nature; we will call it a *cohesion joint*. The most important processes resulting in the formation of cohesion joints are diffusional intergrowth of copolymers, diffusional welding of glasses, and, especially important to us in what follows, diffusional sintering of metals (from the moment when the interface disappears between the conjugate crystalline lattices).

Deryagin and coworkers use the terms *adherence* and *adhesion* to denote, respectively, the establishment of a bond between two solids and the quantitative measure of the result of this process.[5] Developing their idea, we can give the names *coherence* to the strengthening of an adhesional contact and *cohesion* to the quantitative measure of the result of this process. As we see, the formation of a cohesion joint presupposes the existence of an adhesional contact and the possibility that electrical double layers can take part in speeding up diffusional coherence processes. Clearly, the transition from an adhesion to a cohesion joint is irreversible: the appearance of a new adhesion joint inevitably involves the disruption of the system; this occurs in either of the bodies, on the surface characterized by the lowest bond energy.

The use of the terminology just presented should not be considered an attempt to make an absolute distinction between the phenomena of adherence (adhesion) and coherence (cohesion). The rupture of boundary lubricant layers, for example, is considered "purely cohesional" only when there is sufficient thickness in the part of the layer to which the effects of the lateral fields of the solid phases do not penetrate.[33] But if, under the action of these fields, changes in the molecular structure of the boundary layer extend over its whole volume, then the mechanism of the intermolecular cohesional interactions (in the form of van der Waals, hydrogen, or chemical bonds) becomes more complex precisely because of "purely adhesional" effects. An example of such effects is the adsorptive attraction between a metal and the active end groups of polar molecules adapted to the metal surface.

Thus, in the most general case, that of a contact interaction between rough solids with lubricant in between, what appears is a mixed adhesion-cohesion joint; the relative contributions of adherence and coherence processes to the formation of this joint are determined by the physical and chemical natures of the elements constituting the system and by the conditions in the system.

Leaving aside for the moment processes of diffusional strengthening of bonds, we must emphasize that, of the forces of chemical affinity, heteropolar forces are seen considerably more often in friction than are purely homopolar ones. Therefore, it seems to us that the electronic theory of adhesion,[5,34] which reveals the special role of the donor-acceptor interaction in the contact of two surfaces and has its roots in the electrical theory of adhesion,[35] has great heuristic potential in tribophysics.

A fundamental point in this theory is the possibility of a phenomenological approach to the use of statistical methods to describe the donor-acceptor bond. In the most general aspect, the idea in such an approach is extremely simple: instead of finding the wave function of a heteropolar molecule and the electron-density distribution in it, we solve an analogous problem, that of the redistribution of the electron-gas density when solid bodies are in contact. In practice, this means replacing highly complex quantum-mechanical calculations with relatively simple calculations of electrical double layers, using the macroscopic approximation of a parallel-plate capacitor. We will now dwell on this in more detail.

Macroscopic averaging of the electric field reduces, in the present case, to assuming that charges (which, strictly speaking, always occupy a known volume) are concentrated on surfaces of constant density σ that are "thin" in comparison to their separation from the points of the field under study. This assumption makes it possible to define the interaction force between two unlike bodies as the attractive force between the capacitor plates per unit area

$$f_e = 2\pi\sigma^2 \tag{1-1}$$

(for simplicity, the multiplier expressing some averaged value of the relative permittivity is not included).

Of course, the actual structure of the double layer is far more complex. The donor-acceptor reaction results in the formation of a system of dipoles oriented perpendicular to the interface. That is, on each of the contacting surfaces there appears a discrete structure consisting of like charges. Assuming the surfaces are homogeneous and ideal (this is equivalent to requiring the charges to form regular parquet structures), Deryagin and coworkers obtain the following formula for the specific adhesional force:

$$f_e = e \sum_{i,k} E_z(x_i, y_k) \tag{1-2}$$

where x_i and y_k are the coordinates of a charge in one of the charged planes and $E_z(x_i, y_k)$ is the field generated at this point by all charges of opposite sign in the other plane. Since the same field acts on a charge with any coordinates x_i and y_k,

$$f_e = \sigma E_z(x_i, y_k). \tag{1-3}$$

If a_1 and a_2 are the constants of the plane charge lattices and z is the separation between the lattices, then for $z \gg a_1 \sim a_2$ the true field, a periodic function with periods a_1 and a_2, will approximate the homogeneous field of a uniformly charged sheet, $\langle E \rangle = 2\pi\sigma$. But when $z \ll a_1 \sim a_2$, we may neglect the effect on a given charge due to all charges in the opposite lattice, except the nearest charge. Then

$$f_e = \frac{\sigma e}{z^2} = \frac{ne^2}{z^2} \tag{1-4}$$

where n is the number of charge pairs per unit area.

Because of the assumptions stated in the preceding, approximate estimates of f_e give too low results, especially with macroscopic averaging of the field. Nevertheless, the formulas presented make it possible to discover the qualitative laws governing adhesion; this is the value of the theory.

The theoretical study of the electrostatic component of adhesion presupposes that the charge density can be calculated for the double layer that appears when unlike solids are in contact. Because quantitative estimates of the adhesion forces based on an averaged charge distribution can be more or less correct only for $z \sim 10$–100 Å or more, and because effective thicknesses of this magnitude for the near-surface double layer are typical of a semiconductor, Deryagin and coworkers choose a semiconductor as the main object of investigation. The calculations are performed in the framework of the self-consistent field method. If the space-charge density and the charge of the surface states in a semiconductor are assumed to be functions only of the electrostatic potential, then the Poisson-Boltzmann equation can be reduced to quadratures.

For example, in considering the contact between a metal and a semiconductor with an arbitrary spectrum of surface states, when the semiconductor is separated from the metal by a microscopic clearance d with permittivity ϵ_c, Deryagin and coworkers obtain the following expression for the field in the clearance:

$$E_c = \frac{\epsilon_s kT}{\epsilon_c e L_D} \sqrt{\frac{2}{en_i} \int_0^{\tilde{\varphi}_p} \sigma(\tilde{\varphi}_p)\, d\tilde{\varphi}_p} + \frac{4\pi}{\epsilon_c}\sigma(\tilde{\varphi}_p) \tag{1-5}$$

where $\tilde{\varphi}_p = -\left(\dfrac{e\varphi_s}{kT}\right)$ is the dimensionless electron potential energy (φ_s is the surface potential);

$\sigma(\tilde{\varphi}_p)$ is the surface charge;

n_i is the mean concentration of charge carriers, characteristic of a given semiconductor;

$L_D = \sqrt{\dfrac{4\pi e^2 n_i}{\epsilon_s kT}}$ is the Debye length;

ϵ_s is the permittivity of the semiconductor.

It is obvious that the electrostatic term in the adhesion force is

$$f_e = \epsilon_c E_c^2 / 8\pi. \tag{1-6}$$

Concretely modeling the spectrum of surface states and actually narrowing the model to allow for only two types of surface levels, donor and acceptor, Deryagin and coworkers have shown that electric fields of the same order of magnitude as those at a metal-metal contact (10^7 V/cm) may arise at the metal-semiconductor boundary. The appearance of intense fields, which assure significant adhesion, is typical of semiconductors with a large number of surface states; it is as if the surface of such semiconductors became metallized. The surface turns out to be a good reservoir for negative (positive) charges if acceptor (donor) centers predominated on the surface before contact with the metal; when the situation is reversed, the surface changes from electrophilic to electrophobic.* Therefore, in order to weaken the adhesion, it is necessary to use a metal-semiconductor combination in which the work function of a semiconductor with a donor- (acceptor-) saturated surface is greater (smaller) than that of the metal.

We will not dwell on other results of the theoretical investigation, in particular on the laws governing the behavior of the specific adhesion force for the case of a semiconducting layer inserted between metals (although this case, as it will be easy to see later on, is of special interest). Nor will we present here the numerous experimental proofs that the adhesion-cohesion joint is electrical in nature; the reader is referred to the monographs of Deryagin and coworkers[5] and of Mambetov.[36] We remark only that one of the most convincing proofs that electrical double layers play a fundamental role in adhesion phenomena is the emission of fast electrons upon the disruption of a contact in a vacuum, observed by Karasev, Krotova, and Deryagin as early as 1952.

1.2. ELECTROMAGNETIC COMPONENT OF ADHESION

Let us now turn to the question of the origin of noncontact adhesion. By comparison with the short-range forces in regions of real contact, the forces of noncontact adhesion are long-range; they characterize the attraction of condensed phases on surface regions associated with the difference between the nominal S_n and real S areas of contact. According to Lebedev, as interpreted by Lifshitz, the nature of these molecular forces can be discovered through notions about the radiation and absorption of electromagnetic waves by interacting systems of atoms.

Spontaneous radiation from the surface of a solid is due to density fluctuations in the electron clouds of atoms in the superposed state. For any such atom, provided it is isolated, the mean value of the time-varying electric moment equals zero. When the electron clouds approach, their density fluctuations become correlated; as a result, an induced electric moment appears in each atom.

*We extend to semiconductors the terminology proposed by W. R. Harper for dielectrics (Section 6.2).

The attraction of induced dipoles controls the whole subsequent collectivization of the atoms. As a result of the association of atoms, which gives the fluctuations a directed character, the condensed solid as a whole becomes the source of a fluctuating electromagnetic field.

It is not difficult to see that the attraction between two condensed phases can again (as in the case of the donor-acceptor bond) be regarded as an interaction of two giant molecules. The correlation of fluctuations in the electromagnetic fields radiated by such molecules is the key to understanding the nature of the ponderomotive van der Waals forces responsible for noncontact adhesion.

The theory of an interfacial field shows that, if the separation between the surface of similar solids is great enough,* the magnitude of the electromagnetic component of the adhesion force, as a function of this separation, depends only on the permittivity of the medium in a steady-state (electrostatic) field:[37]

$$f_m = \frac{\pi^2}{240} \frac{\hbar c}{H^4} \left[\frac{\epsilon(0) - 1}{\epsilon(0) + 1} \right]^2 \psi\left[\epsilon(0)\right] \tag{1-7}$$

where the function $\psi\left[\epsilon(0)\right]$ can be given by a graph obtained by numerical integration (Fig. 1-1).

*By comparison with the principal wavelengths in the absorption (emission) spectrum of the substance.

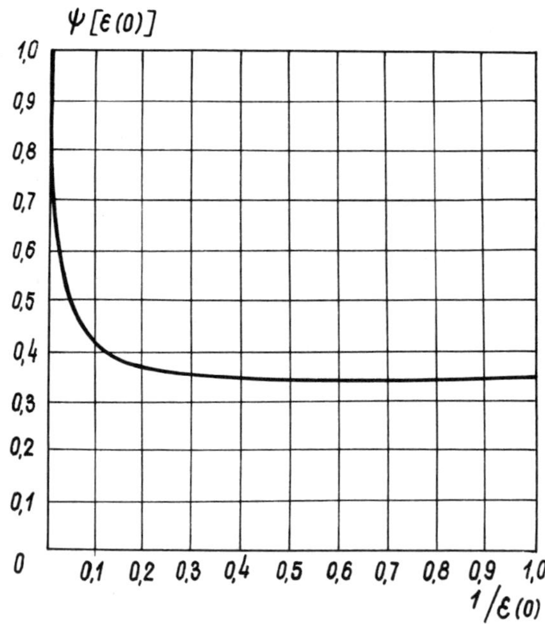

Fig. 1-1. Graph of $\psi\left[\epsilon(0)\right]$ versus $1/\epsilon(0)$, for use in the theory of interaction between condensed phases (after E. M. Lifshitz[37]).

For metals $\epsilon(0) = \infty$, so that

$$f_m = \frac{\pi^2}{240}\frac{\hbar c}{H^4} = \frac{0.013}{H^4} \ (\text{dyn/cm}^2) \tag{1-8}$$

where H is expressed in micrometers.*

The macroscopic theory of Lifshitz, which avoided the artificial assumption that the van der Waals forces are additive, has received direct experimental confirmation in the work of Deryagin and Abrikosova (Fig. 1-2).

It is entirely possible, however, that even by using radioactive or ultraviolet radiation to ionize air, Deryagin and Abrikosova did not completely eliminate charging of the quartz surfaces. Fused quartz is an electrophilic dielectric; therefore, experimental data obtained with it may express the effects of ponderomotive forces of a mixed character, both molecular and electrical.

Now it becomes clear that both adhesion components under consideration, the

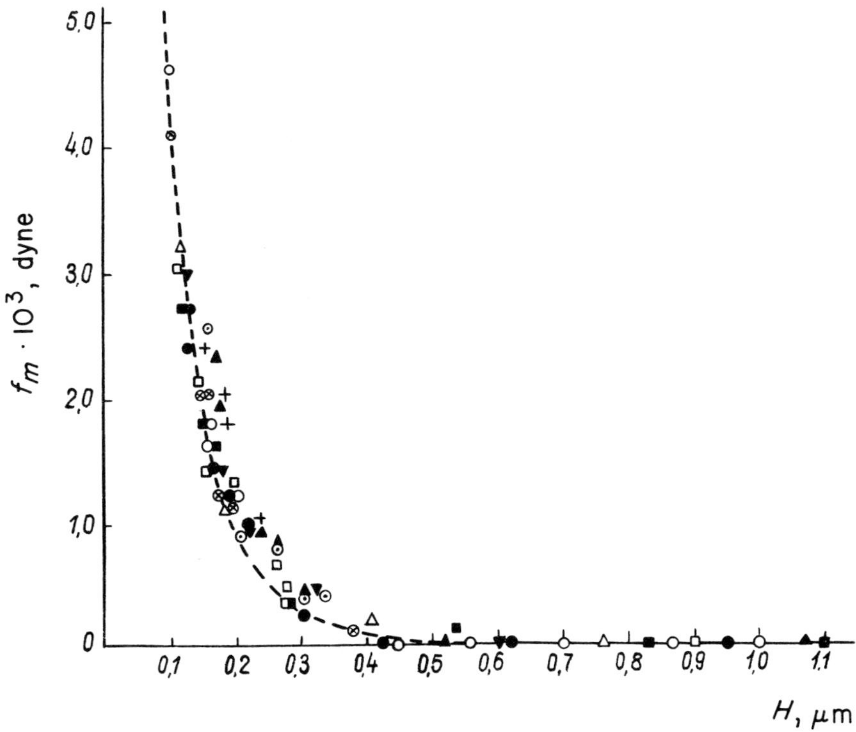

Fig. 1-2. Force of molecular attraction between a lens with a radius of 26 cm and a fused-quartz flat, versus the distance between them (after B. V. Deryagin and coworkers[6]). The dashed curve is the theoretical curve (after E. M. Lifshitz[33]).

*This partial result (for the case of conducting solids only) was obtained earlier by Casimir.[38]

electrostatic and the electromagnetic, should be taken into account as components of the overall attractive force in a binomial law of friction. We will proceed to do this, after giving a brief treatment of the case of rolling friction, which is secondary for us.

1.3. ROLE OF ELECTROADHESION FORCES IN ROLLING FRICTION

We must not ignore electroadhesion forces in rolling friction, because in those cases where energy losses on elastoplastic deformation and slipping of the contacting bodies are negligibly small, friction (the resistance to the rolling of one body over the surface of the other) is due above all to the long-range forces mentioned above. If the adhesion joint comes apart at a high rate, then the specific amount of work done to overcome the electrostatic attractive force between the sheets of the double layer may be large. The reason for this is that when the charge leakage is slight, and the more so when there is none at all, the separation of the sheets of the double layer is a thermodynamically nonequilibrium process.[5] The potential difference between the sheets increases many thousandfold over the equilibrium value (on the order of a volt), making disruption of the contact more difficult. According to the theory of the electrostatic component of rolling friction, treated in detail by Deryagin and coworkers,[5] an increase in the braking moment due to this term involves an asymmetric charge distribution relative to the midline of the contact, which is the instantaneous axis of rotation of the body (for example, cylindrical in shape); the charge of the body remains constant under steady-state conditions. This asymmetry appears because both the recombination of the sheets of the electrical double layer and the appearance of sheets in the newly formed contact zone lag somewhat behind the kinetic processes of separation and approach of the surfaces. Quite obviously, the type and rate of charge recombination depend on the surface conductivities of the contacting solids and the rate at which the sheets of the double layer are separated. Low conductivity and a high rolling speed are necessary preconditions for the appearance of these "overvoltages," at which a gas discharge develops. In the contrary situation (high conductivity, low speed), there is practically no chance of gas-discharge phenomena, and recombination (like charging) occurs either in a region of direct contact or, less important, as a result of the tunnel effect. There is support for the physical correctness of this fundamentally new mechanism of rolling friction: the same phenomena characterize the new mechanism as occur in ordinary adhesion separation (gas-discharge luminescence, emission of highly energetic electrons, X-ray radiation due to braking of fast electrons in the fields of the atoms of the material, and so on). Moreover, an experimental investigation has shown that the analogy cited extends not only to the electroadhesion phenomena themselves but also to the functional relations that describe them.[39]

1.4. ELECTROSTATIC AND ELECTROMAGNETIC COMPONENTS IN SLIDING FRICTION

The donor-acceptor interaction of solids, in the form seen in the purely adhe-sional type of separation, also appears in external sliding friction. We have in mind a process accompanied by vibrational rebounding of an indenter. The sig-nificance of electrical phenomena caused by frictional oscillations will be dealt with later (Section 5.1); here, a generalized way of writing the binomial law of friction is suggested, with allowance for several of the points in Sections 1.1 and 1.2.

In the molecular theory of Deryagin, which makes concrete the nature of the link between friction and adhesion forces,[40] friction between polycrystalline solids is examined through the statistics of pair interactions of single crystals. Bornian repulsive forces between electron shells of atoms play a decisive role in the elementary event of translation. When atomic systems slide past one another, whether external friction or internal plastic displacements are involved, the state of the frictional contact, with allowance for molecular roughness of the surfaces, is expressed in the force polygon (Fig. 1-3), where $\mathbf{N}$ is the normal load, $\mathbf{P}$ is the resultant of the molecular attractive forces, $\mathbf{R}$ is the resultant of the repulsive forces, and $\mathbf{F}$ is the friction force. On the assumption that atoms and molecules are bodies of constant shape, the true coefficient of friction can be represented as the ratio of tangential and normal components of the reaction of the lower surface (crystallographic plane):

$$\mu = R_\tau / R_n. \tag{1-9}$$

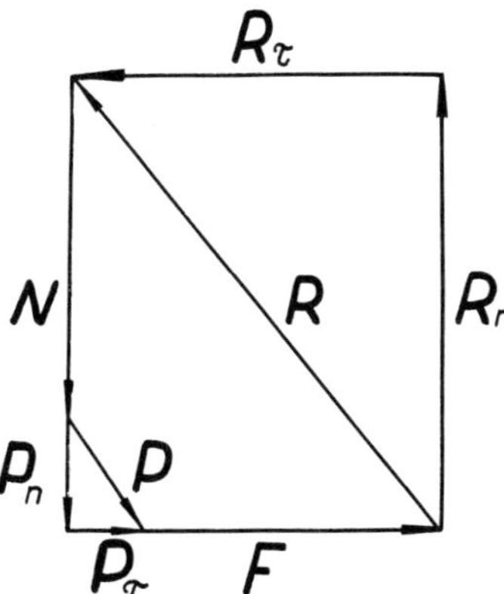

Fig. 1-3. Polygon of forces acting at a frictional contact (after B. V. Deryagin).

Neglecting the tangential components of the attractive forces, Deryagin takes the friction force as equal to R_τ:

$$F = \mu R_n = \mu(N + P_n) \tag{1-10}$$

or

$$F = \mu(N + p_s S) \tag{1-11}$$

where S is the area of real contact and p_s is the mean pressure of molecular adherence acting on this area.*

It should be emphasized that the binomial formula, Equation 1-11, the prototype for which is the Amontons-Coulomb law, $F = \mu N + A$, allowing for adhesion of the surfaces (A), includes any interpretation of the monomial Amontons formula $F = \mu N$ as a special case. For example, the Terzaghi-Bowden-Tabor theory[41,42] is based on the assumption that, because of the negligibly small losses on elastoplastic displacement of the material, coherence of the surfaces can be by itself the main cause of external friction. This theory leads to the formula $F = \tau N/\sigma_f$, which is valid only for the case of plastic contact, when the component $\mu p_s S$ (which depends on the reaction of Bornian forces to molecular attractive forces) varies in proportion to the load $(S = N/\sigma_f)$. Here, the coefficient of friction is given by the ratio of the tangential shear stress to the yield point of the less solid material $(\mu = \tau/\sigma_f)$.

As the area of real contact S approaches the area of nominal contact S_n, the effect of contact adhesion on the external friction force becomes more and more perceptible. Since the theory of Deryagin does not take account of forces of noncontact adhesion, the binomial law of friction given in Equation 1-11 corresponds, strictly speaking, to the frictional contact shown in Fig. 1-4(a); that is, to $S = S_n$.

With allowance for noncontact adhesion forces, which act between the portions of the surface associated with the area $S_n - S$ (see Fig. 1-4[b]), we have

$$F = \mu[N + p_s S + p_m(S_n - S)] \tag{1-12}$$

where p_m is the mean specific attractive force between the condensed phases.

The interaction of sliding bodies through the fluctuating electromagnetic fields that they radiate may play an important role; it was mentioned previously that Akhmatov first noted this role. The following formula[7] corresponds to frictional contact between smooth surfaces separated by a lubricant layer of a completely boundary nature (Fig. 1-4[c]):

$$F = \mu[N + (p_0 + p'_m)S_n] \tag{1-13}$$

---

*Here and below, the mean value p of any pressure p_i acting on microareas ΔS_{pi} is defined by the relation $pS_p = \Sigma p_i S_{pi}$.

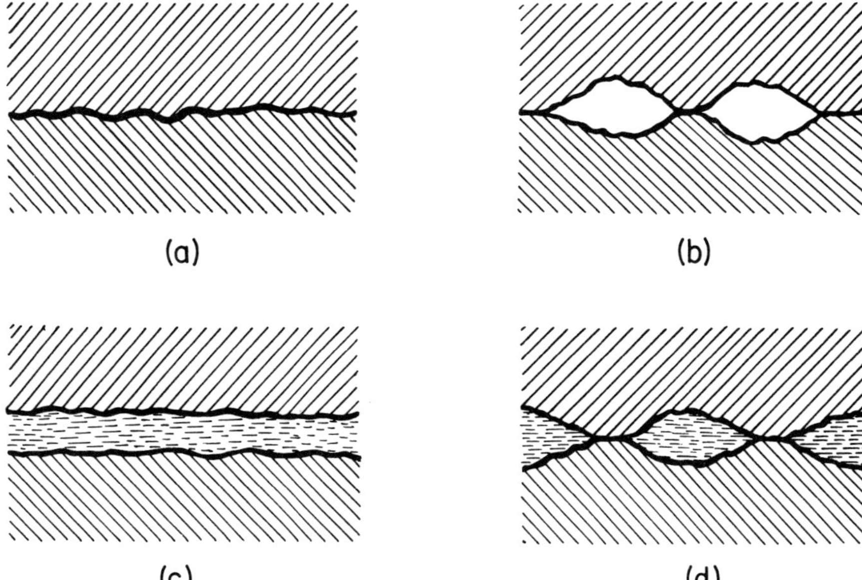

Fig. 1-4. Varieties of frictional contacts: (a) $S = S_n$; (b) $S \neq S_n$ (without lubricant); (c) $S = 0$; (d) $S \neq S_n$ (with lubricant).

where p_0 and p'_m are the mean specific attractive forces of the boundary layers and the condensed phases.

According to the theory of Lifshitz, extended with the help of Dzyaloshinsky and Pitaevsky to the case of the interaction of solid bodies when there is an intermediate substance between them, the value of p'_m depends on the dielectric properties of the lubricating layer, so that $p'_m \neq p_m$. In the first approximation, however, it can be assumed that the electromagnetic field of the condensed phases "penetrates" the boundary layers without substantial absorption of energy and with no distortion. Then, taking $p'_m = p_m$ and combining Equations 1-12 and 1-13, we obtain

$$F = \mu[N + p_s S + (p_0 + p_m)(S_n - S)]. \qquad (1\text{-}14)$$

This expression for the external friction force corresponds to the most general and, naturally, the most widely encountered variety of frictional contact, that shown in Fig. 1-4(d). All other expressions for F follow directly from Equation 1-14 as special cases.

Because relative sliding, as a translational displacement in atom-electron systems, requires that an interface exist between the rubbing solids, a frictional contact resembles nothing more (in its physical nature) than an adhesion joint. At the same time, under conditions favoring the appearance of a diffusional component of cohesion, when friction becomes markedly discontinuous (Bowden's "stick-slip" process), the strengthened regions of the adhesional contact are transformed into portions of a cohesion joint. Distinctive features of the state of these portions, such as weld junctions formed in the friction of metals, have to do with the fact that, from the moment of formation, they experience the combined effects of normal and tangential stresses.[43]

If the donor-acceptor bond predominates over the whole area of real contact between the rubbing solids, then the pressure p_s in Equation 1-14 has the same physical meaning as an electrostatic component f_e.

The quantity p_0 reflects the contribution of interaction between boundary layers to the total attractive force. This interaction depends directly on the degree to which molecular forces (orientational, inductional, dispersional) and the hydrogen bond are apparent.

Noncontact adhesion in friction is characterized by the quantity p_m, the specific force of attraction. Akhmatov and Uchuvatkin have shown that p_m, which is, like p_0, a van der Waals force, has practically the same value as the electromagnetic component f_m.[8] The generality of this result can hardly be doubted, although experiments to determine the forces of the attractive electromagnetic interaction between metals have been conducted under conditions of static (but not kinematic!) boundary friction. We will now deal with these experiments in more detail.

The dependence $F = F(N)$ was studied for various metals in a special setup comprising a tribometer with two sensitive systems, one for measuring the normal and one for measuring the tangential force. The lubricant was pure stearic acid; a boundary layer of given thickness H was formed by the Rayleigh-Pockels titrated-solution method. The value of H was approximately 10 times the arithmetic mean microgeometrical profile of the surfaces; sliding was therefore considered to take place only within the lubricant layer, over an area equal to S_n.

The experimental results, presented in Fig. 1-5, indicate that the static boundary friction force of the stearic acid is a linear function of pressure in the range of small and negative loads. A line $F = F_0 + kN$ corresponds to each of the like pairs under study, where k is the slope, F_0 is the ordinate intercept, and $N_0 = -F_0/k$ is the abscissa intercept. As the boundary-layer thickness gradually increased, the nature of the metal had less effect on the frictional force, and at some critical value $H \approx 1$ μm, F as a function of the normal load was given by a single line $F = F_0' + k'N$ for all the pairs under study (Fig. 1-6). This result yielded a very important conclusion for the development of a theory of friction: "The force field of metals extends beyond their surfaces to a distance not greater than 1 μm and can be completely screened by a boundary lubricant layer."[8]

Theoretical treatment of the results obtained using Equations 1-11 and 1-13 enabled the authors to easily determine the specific molecular attractive force be-

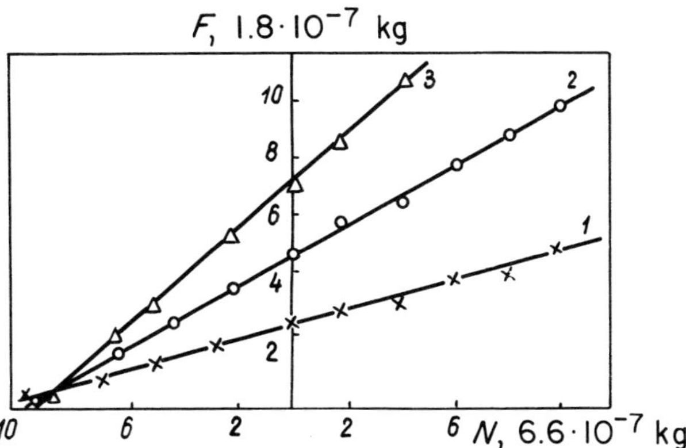

Fig. 1-5. Frictional force F versus load N at small and negative loads (thickness of lubricant layer 7.5 · 10^{-8} meter), for different metal couples:[8]   (1) Cr-Cr; (2) steel-steel (type St3); (3) Ni-Ni.

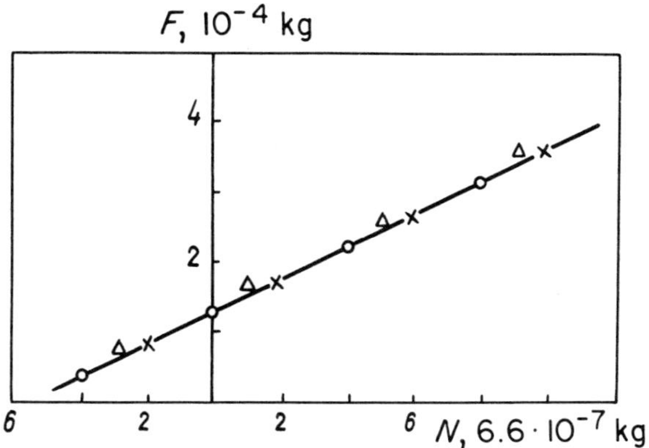

Fig. 1-6. Frictional force F versus load N for various metals, for large thickness of the boundary layer of stearic acid ($H = 10^{-6}$ meter).[8]

tween condensed phases. Indeed, for $F = 0$, having an experimental value N'_0 and using Equation 1-11, one can find the specific attractive force between boundary layers $p_0 = N'_0/S_n$ ($p_m = 0$); an analogous procedure, using Equation 1-13 instead, yields

$$p_m = \frac{N_0 - N'_0}{S_n}. \qquad (1\text{-}15)$$

Experiments with steel and stearic acid have shown that p_m decreases rapidly with increasing thickness of the lubricant layer; the behavior of the corresponding experimental curves (Fig. 1-7) is expressed by the power law $p_m = CH^{-4}$, where $C = 1.32 \cdot 10^{22}$ kg/cm^2.

The fact that C nearly matches the theoretical value of the constant in the Casimir-Lifshitz formula once again indicates the usefulness of assuming p_m and f_m identical.

Now we proceed to conclude that the role of electroadhesion forces in forming the macroscopic friction force must become quite obvious as soon as the reaction of the repulsive forces to them becomes substantial.

The degree to which the electrostatic and electromagnetic components of friction are manifest (this follows from *a priori* considerations) is determined by the fine structure of the metallic surfaces: the presence or absence of adsorbed gas films or extremely thin films of physicochemical compounds, and the surface structure of the metals, with atoms of the chemical elements formerly in the environment that have diffused into the surface.[44] In the light of the concepts just characterized, we should now direct our attention to the significance of extremely quick-forming oxides, which are highly diverse in structure and properties (Section 2.1). By screening the electromagnetic fields radiated by the solids, oxide films reduce noncontact adhesion and in this sense act as lubricants.[33] These same oxides, however, may be semiconductors with a large number of surface states, of such a type that contact adhesion between the film and the metal (or between films) will be rather strong. That is, the oxidation of the rubbing surfaces may be a reason that the electrostatic and electromagnetic components of friction vary in opposite senses.

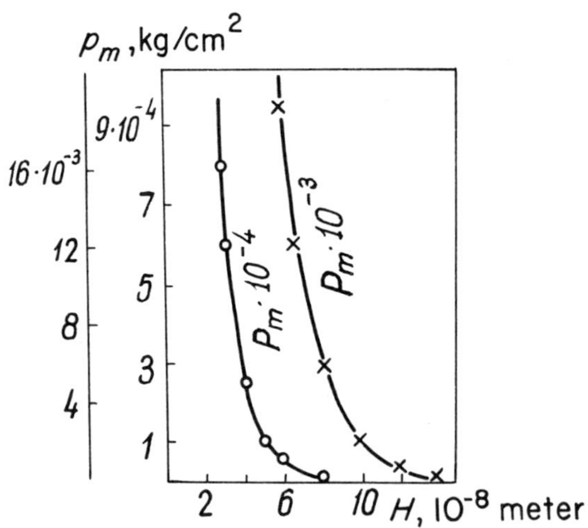

Fig. 1-7. Specific force of molecular attraction between condensed phases as a function of the distance between them (two scales on the vertical axis).[8]

1.5. RESONANCE ELECTROMAGNETIC MECHANISM OF ENERGY DISSIPATION IN EXTERNAL FRICTION

Ideas about the character and interaction of near-surface fields of solid phases in friction have been treated many times. Brillouin,[45] Woog,[46] and others have tried to use them in constructing an electromagnetic theory of friction. These attempts, however, have not succeeded, although Brillouin demonstrated the dissipation of energy through the swinging of a magnetic needle moving in an inhomogeneous magnetic field.

As has been made clear, the idea of structuredness (discreteness, periodicity) of the molecular fields associated with crystalline planes of equal reticular density is also the basis for Deryagin's theory of friction. In explaining the physical nature of the electromagnetic friction term, Akhmatov considered the external fields of solids (as giant molecules) to be qualitatively different.* We think, however, that the further development of notions about the dominant role of the force fields of solids in frictional interaction will take the route of perfecting the oscillator model of friction.

According to the first model of that type, proposed by Bufeyev,[9] and his theoretical generalizations from it, the dissipation of energy in external friction takes place mainly through the electromagnetic interaction of the condensed phases, with the help of a selective-resonance mechanism involving their equivalent oscillators. This mechanism is considered to relate equally to all the known types of interaction (electromagnetic, exchange, etc.) and thus to each of the terms appearing in the total energy of interaction of rubbing solids in the general case:

$$W = W_1 + W_2 + \cdots + W_n. \tag{1-16}$$

As Bufeyev emphasized,[9] by taking account of only the first harmonic of the energy of an atomic oscillator moving in the near-surface field of the substrate, he succeeded in showing that the classical laws of Amontons, Coulomb, Deryagin, and Akhmatov are linear approximations to "the true law of friction, which is in reality defined by the actual shape of the near-surface field of a solid phase in the immediate vicinity of the surface."[9]

We note the exceptional fruitfulness of this energetic approach to the construction of a completely rigorous physical theory of friction, an approach that is, in a sense, the logical continuation of earlier work.[47] But at the same time we must not fail to point out an important fact: Bufeyev in his theory thinks of the resonance-selective mechanism as requiring a certain closeness between the eigenfrequencies of the interacting oscillators of the rubbing solids. Thus, regardless of which physically equivalent oscillators are responsible for friction (Einstein

---

*The change in character of the field of a condensed phase with increasing distance from its surface is expressed, for example, in the transition from a system of elementary spherical waves to a plane wave.[33]

oscillators, Born-Kármán lattice oscillations, oscillations of atomic groups, or whatever), they have been in resonance before the relative sliding begins. The sliding only creates the probability of an encounter between different equivalent oscillators and of an additional resonance buildup of the oscillators. Using the elementary Einstein model of a solid, Bufeyev calculated the resonance component of the friction force. At the same time, he showed that, if the sliding-speed-dependent frequency of the periodic perturbations of the principal oscillators is far less than their eigenfrequency, then other losses are small. These other electromagnetic losses are treated as nonresonance losses, although it is not difficult to see that, in the present case, the resonance phenomenon itself takes on a meaning different from that in Bufeyev's basic model. Instead of natural, "static" resonance, which involves periodicity in time only, we should of course speak of two close frequencies, one of which, as before, reflects only the natural periodicity in time, whereas the other reflects forced periodicity in time depending on the character of the ordered positioning of the oscillators (periodicity in space). This kind of "dynamic" resonance results directly from the relative motion of the solids. The proof that this resonance can occur is precisely the existence of a complex spectrum of coherent oscillations of atoms and of whole structural groups. This spectrum contains, along with the Einstein characteristic frequency, a set of other components, right down to the frequencies of acoustical vibration of the solid itself. As regards the nonresonance (after Bufeyev) component of the friction force, the specific value of this component (referred to unit area of contact) was determined from the relation $f_{fr} = s_n/v$, where s_n is the normal component of the Poynting vector and v is the speed of relative sliding.* Maximum numerical estimates were found through model representations (Smolukhovsky, Herring, and Guyot) of the surface electrical double layer (homogeneous pairs), through a model of an instantaneously charging capacitor (heterogeneous pairs), and so on. These estimates turned out comparatively small. Thus, the conclusion that radiative energy losses are small[33] can be generalized; that is, it can be considered valid not only for elastic but also for electromagnetic waves. It must be noted, however, that the dissipation of wave energy through the generation of waves inside the sliding bodies cannot be the main dissipative process when it occurs by "pure" radiation. From the physical point of view, such a process should be associated with other relaxation phenomena (but above all with dynamic resonance), and also with elastic and electromagnetic hysteresis of the materials, which depends on the actual crystalline structures of the materials, which are defined by the lattice defects.

Taking into account everything that has been said, the author has attempted to use the Maxwell-Lorentz microscopic electrodynamics as the basis for a new the-

---

*In the case, considered by E. I. Adirovich and D. I. Blokhintsev, of the generation of elastic waves, $f_{fr} = j_n/v$ where j_n is the normal component of the Umov vector.

ory of friction. In constructing this theory, he has been guided by the following points:

(1) A theory of friction should clearly reflect both the periodicity of the processes in time and the periodicity of the states in space.

(2) The mechanism of energy dissipation in friction involves chiefly a transformation of the kinetic energy of the rubbing solids into energy of oscillatory motion of the oscillators in their periodic encounters; that is, ultimately, transformation to heat.

1.6. ELEMENTARY THEORY OF FRICTION

In the first approximation, let us consider the case of mutual translation of two adjacent single crystals, as, for instance, Deryagin's theory does.* Assuming these crystals to be in the metallic state and using the model of a metal as an ionic lattice immersed in an electron gas, we will say that the role of the equivalent oscillators is played by positively charged atomic residues executing harmonic oscillations with some characteristic cyclic frequency ω_0. Each such oscillator, associated with one of the sliding surfaces, is in an advancing electric field that is distributed over the adjacent crystalline plane, with a period equal to the lattice constant. It is not hard to see that the very fact of relative motion between the solids makes it possible to convert from spatial periodicity to temporal periodicity and to suppose that each of the surface ions is in an external electric field that varies periodically in time. It is useful to consider this field a simple sinusoidal function with frequency

$$\omega = 2\pi \frac{v}{a} \tag{1-17}$$

where v is the speed of relative sliding and a is the lattice constant.† If the field actually present is of any other form, it can always be represented as a collection of such functions (by Fourier's theorem); therefore, solution of the general problem would reduce to the same steps.

Through the use of the superposition principle it is easy to conclude that the preferred direction of oscillation of the ions in the applied electric field is along the normal to the interface (z-axis). The same result follows, incidentally, from the character of the interaction of neutral atoms, since the Bornian electrostatic repulsive forces vary much more rapidly with interatomic distance than do the attractive forces.

---

*We deliberately do not call these single crystals ideal, in order to emphasize that they possess two-dimensional defects, namely external surfaces.

†In the general case, a is the period of identity of position of the oscillators.

Now, by analogy with the classical theory of light scattering, we can write a Newtonian equation of motion for any surface ion:

$$m\ddot{z} = -kz - r\dot{z} + qE_0 \cos \omega t \tag{1-18}$$

where m and q are the mass and charge of the ion; k is the constant of the quasi-elastic bond; r is the resistance coefficient; and E_0 is the amplitude of the field strength E. We put this equation into the familiar form of the differential equation of forced oscillation,

$$\ddot{z} + 2\beta\dot{z} + \omega_0^2 z = \frac{q}{m} E_0 \cos \omega t \tag{1-19}$$

where $\beta = \dfrac{r}{2m}$ is the damping coefficient and $\omega_0 = \sqrt{k/m}$.

The distance between the rubbing atomic planes, with allowance for the approach of the planes under the normal loading N, can be expressed approximately by the relation

$$h = z_0 - bN + z(t) \tag{1-20}$$

which simplifies to

$$h = z_0 - bN$$

since $z(t) \ll a$. Here b is a constant.

Then,

$$E_0 = \frac{q}{4\pi\epsilon_0 (z_0 - bN)^2}, \tag{1-21}$$

and, according to Equation 1-19, the amplitude of the forced oscillations of an ion is

$$A = \frac{q^2}{4\pi\epsilon_0 m(z_0 - bN)^2 \sqrt{(\omega_0^2 - \omega^2)^2 + 4\beta^2 \omega^2}}. \tag{1-22}$$

The decrease in energy of the ion over a period $(T = a/v)$ is made up by the forcing work $(f = qE)$ and equals

$$\Delta W = \int_0^T f(t)\dot{z}dt = \pi A q E_0 \sin \varphi \tag{1-23}$$

where φ is the phase angle by which the forced oscillation lags the external force.* Since

*In the case $\omega \neq \omega_0$, the energy ΔW dissipated over a period corresponds precisely to the area of the dynamic hysteresis loop in the Σ-z coordinate system, where Σ is the sum of elastic kz and viscous rz resistance forces $(s = \pi r \omega A^2)$.[48]

$$\sin \varphi = \frac{2\beta\omega}{\sqrt{(\omega_0^2 - \omega^2)^2 + 4\beta^2\omega^2}},$$ (1-24)

by substituting Equations 1-21, 1-22, and 1-24 into Equation 1-23, we obtain

$$\Delta W = \frac{2\pi\beta\omega q^4}{m(4\pi\epsilon_0)^2 (z_0 - bN)^4 [(\omega_0^2 - \omega^2)^2 + 4\beta^2\omega^2]}.$$ (1-25)

Because the dissipation of energy in friction begins from the monoatomic boundary layers, the total fraction of energy losses taking place over a period in the two symmetrical systems of oscillators (symmetric with respect to the interface) can be represented in the form

$$\Sigma\Delta W = 2\Delta WnS$$ (1-26)

where n is the number of ions per unit area in each of the conjugate surfaces and S is the real area of contact. If F is the friction force, then in physical meaning

$$F = \frac{\Sigma\Delta W}{vT} = \frac{\Sigma\Delta W}{a}$$

or, taking into account Equations 1-25 and 1-26,

$$F = \frac{\beta\omega q^4 nS}{4\pi\epsilon_0^2 ma(z_0 - bN)^4 [(\omega_0^2 - \omega^2)^2 + 4\beta^2\omega^2]}.$$ (1-27)

The least definite value occurring in Equation 1-27 is the damping coefficient. On the assumption that the principal losses of oscillational energy are due to the radiation of electromagnetic waves, this coefficient can be calculated by the formula[49]

$$\beta = \frac{q^2\omega_0^2}{12\pi\epsilon_0 mc^3}$$ (1-28)

where c is the electrodynamic constant. As one would expect, the value of β found in this way (e.g., for copper ions) leads to implausibly low values for the friction force.

Turning to the other processes by which oscillational energy is dissipated, we must emphasize the following very important circumstance: all these processes, including damping due to radiation, are equally characteristic of both the atoms forming the rubbing surfaces and the atoms in internal layers. In other words, the mechanisms by which oscillational energy is converted to heat energy comprise phenomena identical in their physical essence, regardless of what kind of friction is under consideration, external or internal. This fact, incidentally, enabled the author to remark that the similarity between Deryagin's and Kragelsky's formulas for the binomial law of friction is not accidental. The similarity indicates that energy dissipation in the mechanical impinging of asperities also results from the translational shift of periodic molecular fields.[26]

The resemblance between the models of external friction and of frequency-dependent internal friction of the relaxation-resonance type makes it possible to introduce the concept of a frictional absorption spectrum, $Q_{ext}^{-1}(\nu) \sim \Delta W/W$, similar to the mechanical spectrum of a solid $Q_{int}^{-1}(\nu)$. Obviously,

$$F \sim (Q_{ext}^{-1})_\nu. \tag{1-29}$$

Such an approach makes it possible to estimate the Q factor of the external oscillators from experimental data presented earlier.[48] It is therefore useful to cast Equation 1-27 in a form containing Q and not β.* To this end, supposing that the phase shift φ is sufficiently small, we write the energy of oscillation of an atomic residue in the form

$$W = \tfrac{1}{2}AqE_0. \tag{1-30}$$

Then, from Equations 1-23 and 1-30, it follows that

$$\sin \varphi = \frac{1}{2\pi} \cdot \frac{\Delta W}{W} = \frac{1}{Q}, \tag{1-31}$$

and, using Equation 1-24, we obtain

$$F = \frac{q^4 nS}{8\pi\epsilon_0^2 maQ(z_0 - bN)^4 \sqrt{(\omega_0^2 - \omega^2)^2 + 4\beta^2\omega^2}}. \tag{1-32}$$

The surface concentration of ions is

$$n = \frac{\delta}{a^2}$$

where δ is a multiplier depending on the type of conjugate lattices. For body-centered cubic structures, for example, $\delta = 1$; for face-centered cubic structures, $\delta = 2$. Substituting the expression for n into Equation 1-32, we have

$$F = \frac{\delta q^4 S}{8\pi\epsilon_0^2 ma^3 Q(z_0 - bN)^4 \sqrt{(\omega_0^2 - \omega^2)^2 + 4\beta^2\omega^2}}. \tag{1-33}$$

Atoms of the elements in the middle of the periodic table (copper, iron, germanium, etc.) oscillate at eigenfrequencies of $\omega_0 \sim 3 \cdot 10^{13}$ sec^{-1} ($\nu_0 \sim 5 \cdot 10^{12}$ Hz), which correspond to an Einstein temperature of $\Theta_E \sim 300$ K. Consequently, in the range of relative sliding speeds most often encountered in practice, $\omega \ll \omega_0$ (relaxation-type friction model), and

---

*Only the β in the numerator of Equation 1-27 is referred to.

$$F = \frac{\delta q^4 S}{8\pi\epsilon_0^2 \omega_0^2 m a^3 Q (z_0 - bN)^4}.$$ (1-34)

Let us try to evaluate the force of friction between juvenile surfaces formed by divalent copper ions sliding at a speed of 1 m/sec. According to Schulze,[50] the Debye boundary frequency for copper is $\nu_0 = 6.5 \cdot 10^{12}$ Hz (characteristic Debye temperature for copper, $\Theta_D = 310$ K). That is, $\omega_0 \approx 4 \cdot 10^{13}$ sec^{-1}. Since $a = 3.61$ Å, $\nu \approx 2.8 \cdot 10^9$ Hz ($\omega \approx 1.7 \cdot 10^{10}$ sec^{-1}). For this frequency, we take $Q \sim 10^4\text{-}10^5$.* Assuming $(z_0 - bN) \approx 3.4$ Å, $S \sim 10^{-6}$ m^2, and taking $\delta = 2$, $q \approx 3.2 \cdot 10^{-19}$ C, and $m \approx 10^{-25}$ kg, we obtain $F \sim 1\text{-}10$ N. This order of magnitude for the frictional force makes sense, and its plausibility has been confirmed by elementary experiments.

1.7. DEPENDENCE OF FRICTIONAL FORCE ON SLIDING SPEED

We now represent the external force as a trigonometric series

$$f(t) = \Sigma q E_{0k} \cos(\omega_k t - \alpha_k).$$ (1-35)

Then, because several kinds of oscillators† may be present at the rubbing surfaces, with parameters q_i, m_i, ω_{0i}, β_i, Q_i, we replace Equation 1-32 by the expression

$$F = \frac{S}{8\pi\epsilon_0^2} \sum_i \sum_k \frac{n_i q_i^4}{m_i a_i Q_{i(k)} (z_{0i} - bN)^4 \sqrt{(\omega_{0i}^2 - \omega_k^2)^2 + 4\beta_i^2 \omega_k^2}}$$ (1-36)

where $i(k)$ denotes the functional relation between Q_i and ω_k.

This expression makes possible a clear statement about the physical nature of the relation between friction force and speed. From the equation, it follows that the frictional absorption spectrum, and thus the function $F(v)$, can have a number of alternating maxima and minima (similarly to the light-scattering curve when there are several absorption bands). Accordingly, as the author and Navrotskaya emphasized, any formal models based on the assumption of a sliding-speed-independent friction force are acceptable only for steady-state regimes.[51] They do not reflect the periodicity of the pair interactions of particles forming the condensed phases.

The energy absorption attains maxima and the Q factor attains minima at sliding speeds such that

$$\omega_k = \sqrt{\omega_{0i}^2 - 2\beta_i^2},$$ (1-37)

*See V. S. Postnikov,[48] p. 44, Fig. 25.

†Including linear chains of ions oriented along the z axis, whose lengths—and thus, to some extent, the friction force (we mention the analogy with a vibrating string)—are determined by the dimensions of the bodies ("scale factor").

that is, at dynamic resonance of one of the principal groups of equivalent oscillators. In our judgment, it is just this fact that forms the basis for the resonance-selective mechanism of energy dissipation in friction.

A clear instance of the appearance of dynamic resonance at frequencies near the characteristic frequency of the atomic oscillators is furnished by experiments in which the sliding speed was taken up to 1,000 meter/sec ($\omega \sim 2 \cdot 10^{13}$ sec^{-1}).[52] Energy absorption at these speeds is so intense that, because the metal has a finite thermal conductivity, a very thin molten layer appears at the surface and acts as a lubricant (like the layer of water when a skate moves over ice). We remark, incidentally, that at this frequency of external excitation, there also is a maximum of the relaxational internal friction ($\omega\tau \approx 1$, where τ is the temperature relaxation time).[48] The excited region in which the heat flux relaxes is pressed against the interface; this effect also facilitates the transition of the frictional couple from the athermal sliding regime to the thermal (quasi-adiabatic) regime.

The positions of the extrema on the $F(v)$ curve depend on the initial physical and chemical state of the surface layers and on the way the state changes as the frictional interaction develops. Thermal transition from mainly elastic deformations to plastic displacements, formation of screening oxide films with lubricant functions, various secondary effects due to the presence of polar lubricants, and so on—all these factors can drastically influence the way the friction force depends on speed. This is why the force-speed relation is one of the most complicated problems in tribology.[33]

1.8. RELATIONSHIP BETWEEN FRICTION FORCE AND ENERGY ABSORPTION

To prove the validity of the elementary theory of friction in the absence of direct experimental data connecting F and Q^{-1}, the functional dependence expressed by Equation 1-29 can be verified only if a third parameter can be found on which F and Q^{-1} depend simultaneously.*

One convenient way to vary Q^{-1} at constant frequency is nuclear irradiation of the material. Experimental results show that the energy absorption decreases with increasing time of irradiation (Fig. 1-8).[53] At the same time, the friction force is known to decrease drastically when the surface is irradiated with accelerated particles.[54]

Since both the energy absorption and the friction force depend in the same way on the time of irradiation, it can be tentatively concluded that the functional dependence $F(Q^{-1})$ of Equation 1-29 is valid, at least in this case.

In the opinion of the author and of Eliezer, the disturbance of the periodicity

---

*The relationship between F and Q^{-1} was discussed in terms of available experimental data in a paper entitled "On the Elementary Theory of Friction," by the author and Z. Eliezer, submitted for publication.

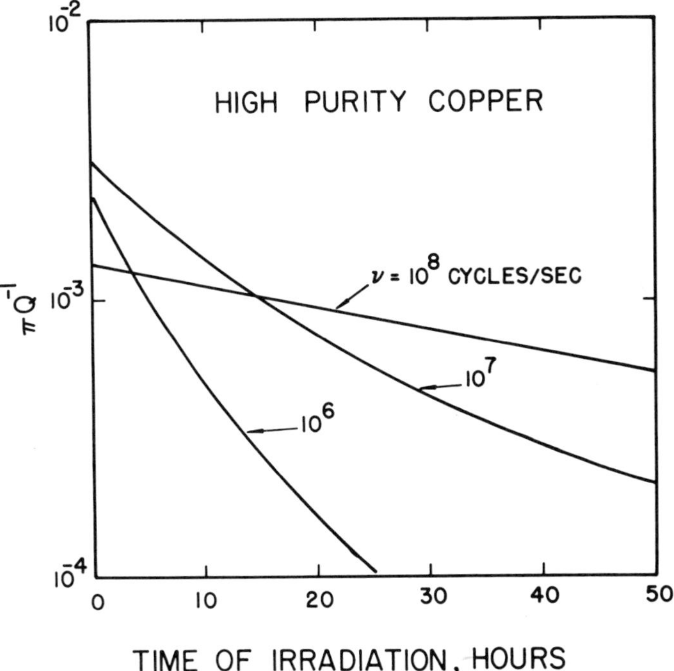

Fig. 1-8. Variation of Q^{-1} (at constant frequency) with time of irradiation of a copper specimen with Cobalt Gamma. Data were taken from Fig. 3 of reference 53.

of the lattice by irradiation is the physical basis for the variation in the same sense of F and Q^{-1}. This lattice disturbance impedes the free movement of dislocations, resulting in hardening of the material and a corresponding decrease in F.[42] Furthermore, Q^{-1} should also decrease for the same reason, as Granato pointed out.[53]

The mechanism of external friction in real materials has much in common with that of hysteretic internal friction. The existence of many unlike defects in the rubbing solids of course leads to irreversible changes in their states. Nevertheless, in the case of nonlinear viscous and restoring forces, the force F can still be evaluated if the areas of the dynamic hysteresis loops in Σ-z coordinates are known. These areas represent the superposition of relaxation and hysteresis losses.

Resonance, hysteresis, and relaxation, coexisting in external friction processes, make different contributions to the amount of energy absorbed. A comparative evaluation of the relative dissipation of energy $\Delta W/W$ for metals shows that the dissipation amounts to hundredths or even thousandths of a percent ($Q \sim 10^4$-

10^5) for relaxation losses, tenths of a percent for static hysteresis, and several percent ($Q \sim 10^2$) for dynamic resonance. As for alloys, when they are used in frictional assemblies, the fraction of energy losses due to static hysteresis may be quite perceptible.

In the light of what has been said, we must not fail to emphasize the special significance attaching to the study of the properties of the oscillators that are defects of structure or directly linked with such defects (dislocation oscillators, Langmuir spots on the surface of a metallic polycrystal, etc.).

We must, however, remark that the variation in energy of the dislocation oscillators cannot be described within the framework of classical electrodynamics: the loss of energy by excited dislocations is considered as a collection of independent quantum transitions occurring at different times.[55] Therefore, further development of the theory presupposes that generalizations can be made on the basis of quantum mechanics.

1.9. ELECTRONIC MECHANISM FOR THE PLASTICIZATION OF SOLIDS IN CONTACT INTERACTION

Because the energy state of the rubbing surfaces plays a fundamental role, there is indisputable interest in an article by Sidorov and coworkers, which gives a qualitative, quasi-classical explanation of the variation in surface energy at the interface between metals as a result of their direct contact.[11]

It is well known that in the free-electron approximation, a three-dimensional metal specimen in the shape of a cube of edge L represents a rectangular potential well. The crystal faces that coincide with the faces of the well are a potential barrier to conduction electrons; in encountering it, the electrons impart a certain momentum to the surface. As for the variation in the energy state of the interface (metal-vacuum), the solution of the problem actually reduces to examining the way in which conduction electrons are scattered at the surface.

After remarking that this kind of process does not alter the Fermi-Dirac distribution at absolute zero, Sidorov and coworkers turn their attention to this fact: the mean time τ between collisions with opposing faces of the crystal by those electrons in states with wave-vector components (k_x, k_y, k_z) and $(-k_x, k_y, k_z)$ is finite in value, although very small:

$$\tau \sim \frac{mL}{p_x} \tag{1-38}$$

where m is the mass of an electron and p_x is the projection of the electron momentum on the x-axis. Thus, it is useful to replace the condition of elastic reflection of electrons by another condition, obtained for inelastic reflection of the particles, that is more probable. With this aim, let us examine the mechanism of inelastic scattering, using as a basis the considerations presented above, as Sidorov and coworkers did.[11]

The surface of the crystal, having received some energy W' from an electron colliding with it, goes into a metastable state for some time τ', after which it returns its excess energy to electrons. From the Heisenberg uncertainty relation, we have

$$\tau' \sim \frac{\hbar}{\Delta W'}. \tag{1-39}$$

We still do not know where the region of most probable localization of the electron is when the front of its wave function reaches the surface of the crystal. But localization of the electron in the boundary layer seems possible if we take into account that all the energy levels up to the Fermi level are filled at the moment when the electron encounters the potential barrier. Consequently, the inelastic scattering of electrons is, as it were, a product of their intermediate localization, which is expressed in the transition, over time τ', from the state (k_x, k_y, k_z) to a state with a zero quantum number $(0, k_y, k_z)$.* It follows from the condition of dynamic equilibrium that

$$\tau' \sim \tau \tag{1-40}$$

that is, the total number of localized electrons (or, what is the same thing, of electrons in states with zero quantum numbers) must remain invariant.

Since the localization of electrons in inelastic collisions is accompanied by the generation of phonons, the boundary conditions for the wave functions ψ do not correspond to the generally accepted Kármán-Born conditions, which, in application to solids of finite size, are written in the following way:

$$\psi(0, y, z) = \psi(L, y, z) = 0. \tag{1-41}$$

It is just in permitting zero quantum numbers that the approach set forth here differs from those proposed earlier.[57, 58]

The total energy of the electron gas can be represented in the form

$$W = W_v + W_s \tag{1-42}$$

where W_v is the energy of electrons within the volume of the specimen (L^3), and W_s is the energy of electrons localized near the surface.

Making reference to the results just cited,[57, 58] Sidorov and coworkers do not dwell on the determination of values for W_v and W_s, at the same time remarking that the energy received by the lattice from localized electrons is $W_l \sim W_s$; the electron contribution to the surface energy is proportional to the height of the potential barrier at which scattering takes place.

If we continue using macromolecular analogies (regarding a crystal as a giant molecule) and follow Becker,[59] then for the one-dimensional potential wells

---

*The equivalence of the wave-vector components to the quantum numbers n_x, n_y, n_z (the electron energy depends on $n^2 = n_x^2 + n_y^2 + n_z^2$) is obvious from the relation $k = \frac{2\pi}{L} \cdot n$.[56]

bounding the three-dimensional well, we can write the following formula for the Fermi energy:

$$W_F \approx \frac{26.07}{V^{2/3}} \tag{1-43}$$

where V is the molar volume of the metal.*

If we know the chemical potentials of the metals placed in contact (W_{F1} and W_{F2} in Fig. 1-9), it is not difficult to arrive at an idea of the surface energy at the interface. From Fig. 1-9, it follows that only those conduction electrons supplied by the metal with the higher chemical potential will be scattered at a potential barrier of height U. That is, it is in this metal that the surface energy accumulates, while the surface energy falls off sharply in the other metal. If the contacting metals have equal chemical potentials (Fig. 1-9[b]), then the surface energy of the interface equals zero; that is, this interface actually disappears. The absence of asymmetry in the electron density in the direct contact of solids is a necessary condition for the formation of cohesion bonds (see Section 1.1).

All further results considered by Sidorov and coworkers are valid only if they satisfy the condition that the chemical potential and the work function (φ) vary in the same sense from one metal to the other. The lack of justification for this like sense of variation, which in no way follows from *a priori* considerations, is unfortunately a serious shortcoming of the article.[11] Nevertheless, it is worthwhile looking at the consequences of the preceding ideas, which were obtained on the assumption that this condition is satisfied.

Among these consequences, in particular, the possibility of evaluating the stacking fault energy is of special interest. Besides the article already cited,[11] Sidorov and coworkers had mentioned this possibility in earlier publications.[61,62] Indeed, the disruption of correct stacking of slip planes results in the appearance of an

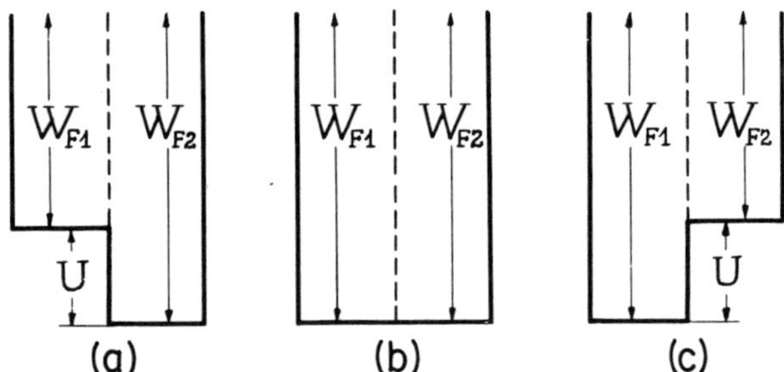

Fig. 1-9. Typical relations between the chemical potentials of contacting metals.[11]

*The validity of this formula, written in a somewhat different form, has received indirect confirmation, to an accuracy of about 5 percent, for 26 metals and semiconductors.[60]

interface between different crystallographic modifications within a single metal. If we knew the work function of each of these modifications, then there would be no problem. However, because this work function cannot be determined directly by experiment, Sidorov and coworkers attempted to establish a relation between the stacking fault energy and the work function of the metal in its initial crystallographic structure. Figure 1-10 shows a composite graph of this relation, which has a correlational character. Because the correlation coefficients for both body-centered cubic (bcc) and face-centered cubic (fcc) structures came out equal to approximately 0.8, Sidorov and coworkers suggested that the contribution to the stacking fault energy by electrons "evaporating" from the Fermi surface could be quite substantial. Thus, to the list of factors whose effects on the variation in stacking fault energy were linked with the electronic structure (electron concentration, difference in atomic radii, variation in density of states, variation in the topology of the Fermi surface),[63] they added yet another, the excess pressure of the electron gas. The physical nature of this pressure is probably the same whether the pressure acts on the surface of a stacking fault or manifests itself in a change in the energy state of the metal-metal interface (the interface is in fact a two-dimensional structural defect).

In accordance with the theoretical groundwork developed, Sidorov and coworkers concluded that plasticization of the surface layer in contacting metals takes place more freely in the metal with the lower work function. This view complements the conclusions of Novikov, who considers the increased mobility

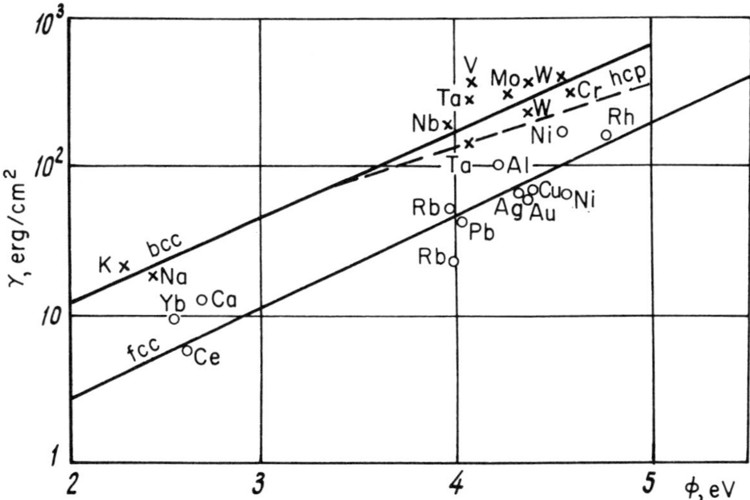

Fig. 1-10. Correlation relationship between work function and stacking-fault energy.[11] X = body-centered-cubic metals; O = face-centered-cubic metals (hexagonal-closest-packed metals are not shown).

of dislocations and the resulting acceleration of plastic deformation in the surface layer to be results of the decrease in surface energy.[64] This point of view is in complete agreement with the idea that the strength of solids is lowered by the adsorption of surfactants (Rebinder effect).

The assumption of an electronic mechanism for the plasticization of solids has made it possible to shed new light on the processes taking place in fretting corrosion.[11]

It is clear from what has been said that, in order to extend the life of a test piece, the counterface material should preferably be chosen such that $\varphi_c < \varphi_{sp}$. Then, as soon as a thin surface layer of the counterface begins to plasticize, it will act as a lubricant (up to a certain stage of strain hardening), by a mechanism very similar to that of "electron lubrication," which reduces frictional wear.[65]

Confirmation for these conclusions comes from data on how fretting corrosion affects the life of test pieces of type Kh18N10T steel and copper (work functions 3.6 and 4.4 eV, respectively) under cyclic loading (see Fig. 1-11). The trials were carried out by the method that Stepanov described;[66] the same materials were used for the counterfaces.

Sidorov and coworkers compare their results with those of Alyabiev and co-workers, who found the material of the test piece protected against fretting corrosion if it contacted a material having an electrode potential more negative than its own.[67] This comparison bears out the ideas developed by Sidorov and co-workers[11] and affirms that a functional interrelation exists between the work function and the standard electrochemical potential of a material.

In conclusion, we remark that the electronic theory of plasticization, although still sketchy, may become a sound basis for the development of practical recommendations. (See also Section 5.2.)

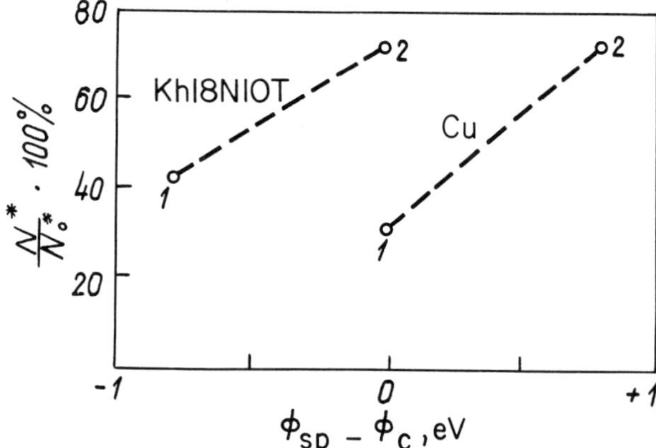

Fig. 1-11. Relative change in life of test pieces in cyclic loading, under the action of contact friction of counterfaces.[11] The life without contact friction (N_0^*) was taken as 100%. The materials of the contacting counterfaces were (1) copper ($\varphi = 4.4$ eV) and (2) type-Kh18N10T steel ($\varphi \approx 3.6$ eV).

2

Experimental Studies of Electrical Phenomena in the Friction of Metals

2.1. THE WORK FUNCTION AS A CRITERION OF THE PHYSICOCHEMICAL ACTIVITY OF SLIDING SURFACES

The strength of an adhesion or cohesion joint is determined by the energy state of the surfaces forming the frictional contact. Since some mechanical action on the surface layers of metals is unavoidable when friction takes place, the energy states of portions of the contact region will continually vary. Surfaces whose free energies have been raised by the interaction will tend to return to a state of thermodynamic equilibrium with the environment, that is, to the state of maximum passivity. Both the parameters of friction and the ratios between them depend on, first, how quickly the frictional mechanical activation can be lost and, second, what kinds of physical and chemical processes go on as this happens.

The most comprehensive physical characteristic reflecting the energy state of a solid surface is the work function. Specific indications of the desirability of using the work function to analyze friction and wear processes include the previously described electronic mechanism by which the surface layers adjacent to the metal-metal interface are plasticized, the possibility of monitoring the state of the sliding surfaces under conditions of selective transfer,[68,69] and so on. The value of the work function (φ) figures in many semiempirical formulas suggested for the determination of the surface energy (σ). Thus, for example, according to Demchenko and Homutov,[70]

$$\sigma = \varphi Z/1.885 R^2 \cdot 10^{-3}\,\mathrm{erg/cm^2}, \tag{2-1}$$

where Z is the number of free electrons per atom and R is the atomic radius in angstroms.

The strength of the contact electric fields (which assure a significant adhesion) and, therefore, the electrostatic component of friction depend on the difference between the work functions of the metals in contact.

The work function is determined by measurements of the contact potential difference (CPD or V_c) between the metal and a standard (usually gold):

$$\varphi_{\mathrm{Me}} = \varphi_{\mathrm{Au}} - eV_c \tag{2-2}$$

where e is the charge on an electron. Often it is possible only to compare the V_c values corresponding to the instrument readings.

For measurements of the CPD in vacuum or under atmospheric conditions, it is customary to use the vibrating-electrode (dynamic-capacitor) method. One version of a device of this kind was described by Avdentova and Starodubrov-skaya.[71] The device, whose fabrication followed a schematic diagram by Markov,[72] consists of the following principal elements (see Fig. 2-1): a capacitive sensor S, a tuning-fork oscillator O, an amplifier A, a compensating network C, and a null indicator I. The oscillations of the tuning fork, sustained by a low-power electromagnet, are transmitted to the standard specimen, which is rigidly attached to one prong of the fork. Thus, the direct-current signal proportional to the difference in work functions $\varphi_{\mathrm{Au}} - \varphi_{\mathrm{Me}}$ is converted to an alternating-current signal; after two amplification stages (first with an electrometer tube,

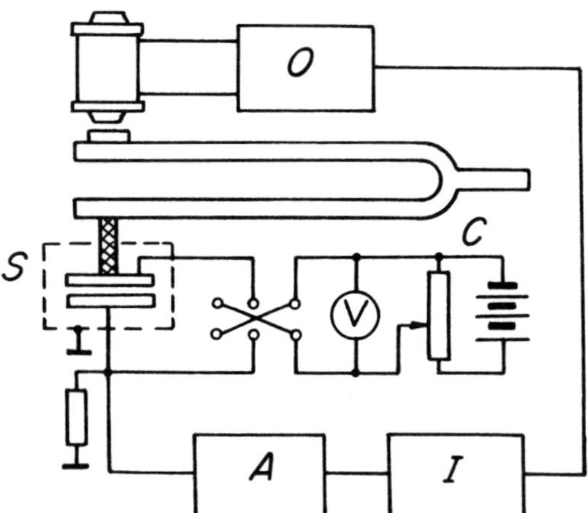

Fig. 2-1. Schematic diagram of setup for measuring contact potential difference.[71] O = tuning-fork oscillator; S = capacitive sensor; C = compensating network; A = amplifier; I = null indicator.

then with a transistor amplifier), this signal is fed to the null indicator (e.g., an oscillograph). The method of compensation is used to measure the signal strength.

Aligning the device reduces in practice to replacing the plate under study with another gold electrode. If a spurious signal appears on the indicator, a null correction is made.

This device, with its measuring chamber shielded from the effects of external electromagnetic fields by a detachable steel casing, has a sensitivity of 4–5 mV. The sensitivity depends on the modulation factor of the signal (the ratio between the amplitude of the oscillations of the standard capacitor plate and the separation of the plates with the tuning fork at rest). In order to prevent the specimens coming into contact (which would contaminate the surface of the standard) and at the same to obtain a signal of the requisite magnitude to feed to the amplifier, Avdentova and Starodubrovskaya recommend modulation factors between 0.55 and 0.80.[71] We note, by the way, that the sensitivity of the device may be markedly affected by frequency instability in the tuning-fork oscillator, which operates in a strict resonance regime. Even a slight departure of the oscillator frequency from that of the tuning fork, which has a high Q factor, involves a drastic reduction in the signal as a result of the lowered modulation factor.

Markov measured the CPD of metals against gold in two states: with the surface cleaned with emery paper, and in friction of like metals without lubrication.[72] He showed (see Table 2-1) that a sharp drop in the absolute value of the work function accompanies removal of the oxide films and destruction of the thin surface layers of a metal (see Equation 2-2). After mechanical activation, the initial state of the surface was restored only when the metal was held under atmospheric conditions for 10 to 15 days. When there was boundary lubrication and plastic deformation of the metal occurred on the friction track, the metal surface, saturated with various kinds of defects (dislocations, vacancies, extrusions, intrusions, etc.), became so active in relation to the environment that,

Table 2-1. Contact potential differences in mV for metals under mechanical action.[72]

Metal	Starting (Oxidized) State	Cleaning with Abrasive Paper		Dry Friction	
		After 0.1 min	After 1 min	After 0.1 min	After 1 min
Copper, type M2	50	250	250	230	225
Alloy AK4	1,000	1,300	1,280	1,250	1,220
Brass, type L62	200	580	560	540	520
Nitrided steel, type 38KhMYuA	170	340	335	–	–

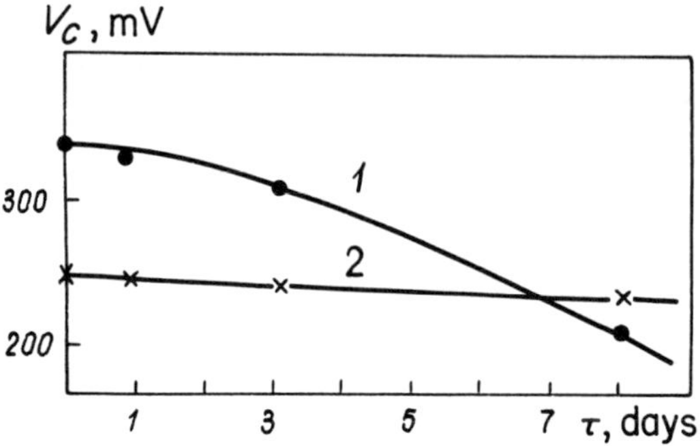

Fig. 2-2. Variation in work function of type-30KhGSA steel on holding under room conditions.[72] Curves taken (1) on friction track, (2) off friction track.

as time passed, the work function was greater on the friction track than off it (Fig. 2-2).

By measuring the work function, Markov and coworkers studied the effects of liquid lubricants on the energy state of the surface.[72, 73] They suggest using the adsorption curves to judge how metals interact with lubricants differing in composition and properties. These curves are constructed in the following way. Before the adsorbate-lubricant is applied to the specimen surfaces, the surfaces are carefully cleaned of contaminants by washing in petroleum ether, treatment with abrasive paper and activated carbon, and then a second washing with petroleum ether.* Then the specimens are used in measurements of the starting value $V_{c(0)}$ and of the corresponding work function. Finally, the adsorbate is applied to the surfaces of the plates, by immersion of the plates in the lubricant and holding at room temperature for selected times. After excess lubricant is removed from the specimen surfaces (with ash-free filter paper), $V_{c(ads)}$, the contact potential difference in the presence of an adsorbed layer that alters and weakens the field of the condensed phases, is measured. The curve of $\Delta V_c = V_{c(ads)} - V_{c(0)}$ versus holding time τ gives some idea of the kinetics of adsorption of the lubricant molecules and the possible rearrangement of the adsorbed layer on the metal surface. Whenever we use the somewhat imprecise terminology of Markov and coworkers,[72, 73] we must always keep in mind the effect of this rearrangement on the CPD.

From the adsorption isotherms—for example, those shown in Fig. 2-3—we can compare the adsorbability of two equally viscous oil bases (thickened oil and a compound made by mixing of mineral oils) containing equal concentrations of different antiwear additives. Because the CPD may either increase or decrease, depending on the physicochemical affinities of the liquid lubricants and the solid surfaces, we evaluate adsorption by the absolute value of ΔV_c and also by

*This treatment of the specimens is recommended by Markov and coworkers.[74]

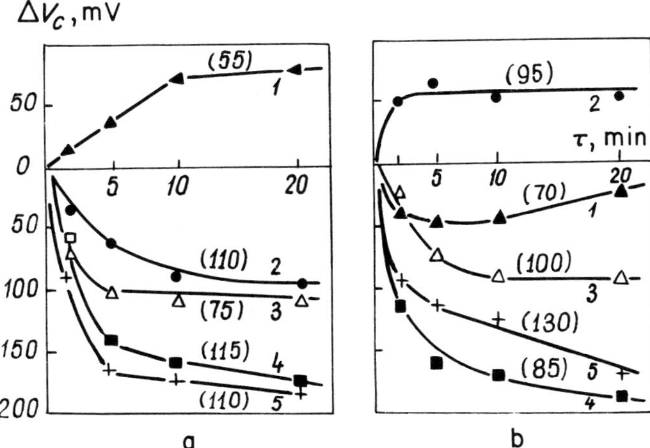

Fig. 2-3. Kinetics of variation of the contact potential difference in the adsorption on a metal of additives from oils.[73] Oil bases: (a) thickened oil; (b) type M-1 mineral compound. Additives: (1) ABS-2; (2) tricresyl phosphate; (3) LZ-318; (4) chloroprene; (5) "Sovol" chlorinated biphenyl. τ is the time of holding in the oil. The figure in parentheses on each curve is the critical seizing load for the oil, in kilograms, measured on a four-ball friction tester.

the rate of change of this parameter. From Fig. 2-3, it follows that adsorption of the additives from the oil solutions goes at different rates and to far different extents. Obviously, the determining factors here are the predominant type of adsorbate-adsorbent bond and the extent to which the bond is manifest. An analysis has been made of the adsorption curves and the results from tests of doped oils on a four-ball friction tester. The analysis showed that chlorine-containing compounds ("Sovol" chlorinated biphenyl, chloroprene) surpass all other additives tested in adsorbability and antiwear properties. A chlorine-containing additive forms a protective lubricant layer at a particularly high rate (almost instantaneously), while a sulfur-containing additive takes a rather long time to form such a film.[73]

The construction of adsorption isotherms makes it possible to evaluate the lubricating ability of fuels and take measures to limit the wear of rubbing parts in fuel systems. For example, according to the adsorption curves of Fig. 2-4, of the three fuels studied, type T-1 has the best lubricant properties (from the standpoint of the value and rate of change of the work function for steel) and type TS-1G has the lowest properties. These findings were confirmed in operating tests and also in trials where the coefficient of friction and the work function were measured at the same time.[72] The most valuable result of these trials is the correlation that was established for the first time between $\Delta\mu$ and

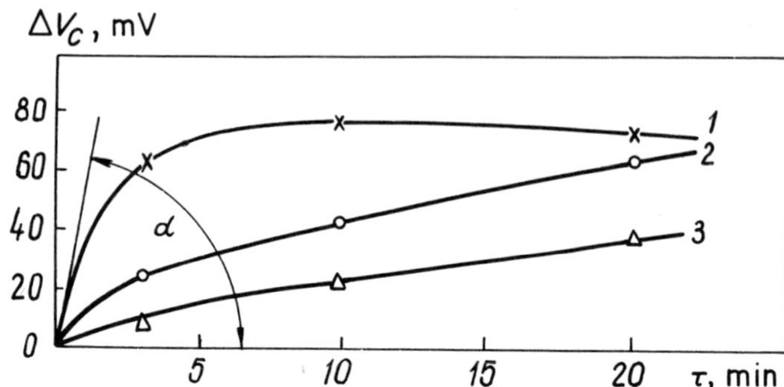

Fig. 2-4. Adsorption curves for type-St3 steel in interaction with various grades of fuel:[72] (1) T-1; (2) TS-1; (3) TS-1G.

$\Delta\varphi$: the more the work function changes, the more the coefficient of friction changes.

The value of ΔV_c determines the variation in the difference between the work functions of two metals, only one of which can be taken as the standard. Therefore, in a comparative evaluation of the adsorbabilities of lubricants using the $\Delta V_c = \psi(\tau)$ curves, it is necessary to be sure from the start that the curves were obtained for one and the same adsorbent metal. If the lubricant were applied to different metals, it would probably be more correct to use a criterion such as $\delta_{ads} = \Delta V_c / \Delta V_{c(0)}$, as indicated by Avdentova and Starodubrovskaya,[71] for the adsorbability of the lubricant. From Fig. 2-5 it follows, for example, that type A-66 gasoline with "Sigbol" antistatic additive affects the work function of steel more than that of an aluminum alloy (solid curves No. 2). At the same time, the values of ΔV_c were more than a factor of 2 lower for steel than for the aluminum alloy (dashed curves No. 1).

Antistatic additives are used in fuel to prevent explosive situations, which may develop if large electrical charges accumulate during the filling of fuel systems. In studying how the percent content of "Sigbol" additive in gasoline affects ΔV_c, Starodubrovskaya and Avdentova observed considerable deviations in the $V_{c(0)}$ values for type St3 steel used as the adsorbent. Therefore, in the statistical compilation of which the results appear in Fig. 2-6, data were used from the trials with the most frequently occurring values of $V_{c(0)}$, which lay in the range of 220–260 mV.

An analytical treatment of the experimental data enabled those authors to express the dependence $\Delta V_c = F(C_a)$ by an equation of the form

$$\Delta V_c = -1.032 \cdot 10^6 \, C_a^2 + 19.73 \cdot 10^3 \, C_a + 65.28. \tag{2-3}$$

It is not hard to see that the parabolic curve constructed from this equation (the dashed curve in Fig. 2-6) is within the range of deviation of ΔV_c from the statis-

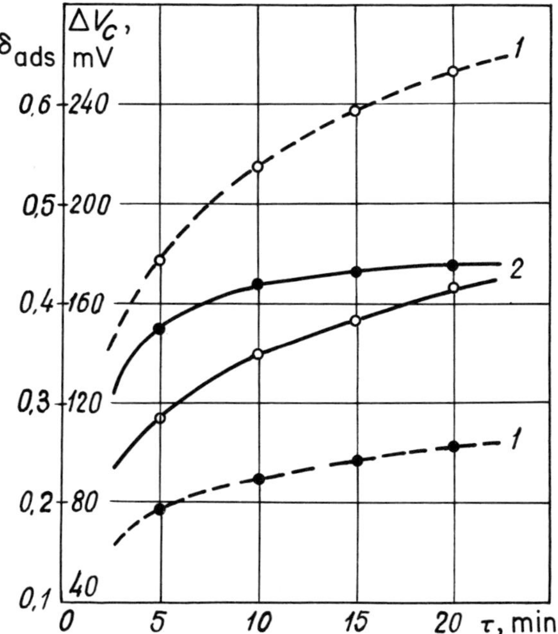

Fig. 2-5. Absolute (ΔV_c) and relative (δ) variations in contact potential difference versus the time of application of adsorbate to the surface of a steel ($\bullet$) and an aluminum ($\circ$) alloy:[71] (1) $\Delta V_c = \psi(\tau)$; (2) $\delta = f(\tau)$.

tical mean. The existence of a maximum on the curve indicates that there is some optimal concentration of the additive, at which the rate of formation of the adsorbed layer is greatest.

Starodubrovskaya and Avdentova explained the parabolic character of the relation $\Delta V_c = F(C_a)$ in the following way. The transport of molecules from the bulk phase to the metal surface means a drastic change in the conditions of the molecules. Since the activation energy of the additive molecules is far less than that of the particles making up the pure petroleum product, when the adsorbed layer is formed, the concentration of gasoline in it diminishes considerably faster than the concentration of additive rises. Whether the adsorbed layer is formed from gasoline molecules only or from additive molecules, we have $\Delta_g V_c = \alpha C_g$ or $\Delta_a V_c = \beta C_a$, where α and β are constants depending on the activation energy. If there is a two-phase system (gasoline + additive) and we suppose that $C_g \gg C_a$ and $\alpha \gg \beta$, we can put the expression for the total variation in contact potential difference into the form

$$\Delta V_c = \alpha(C_g - C_a^m) + \beta C_a = -\alpha C_a^m + \beta C_a + \alpha C_g. \tag{2-4}$$

The similarity between Equation 2-4 and the empirical formula of Equation 2-3

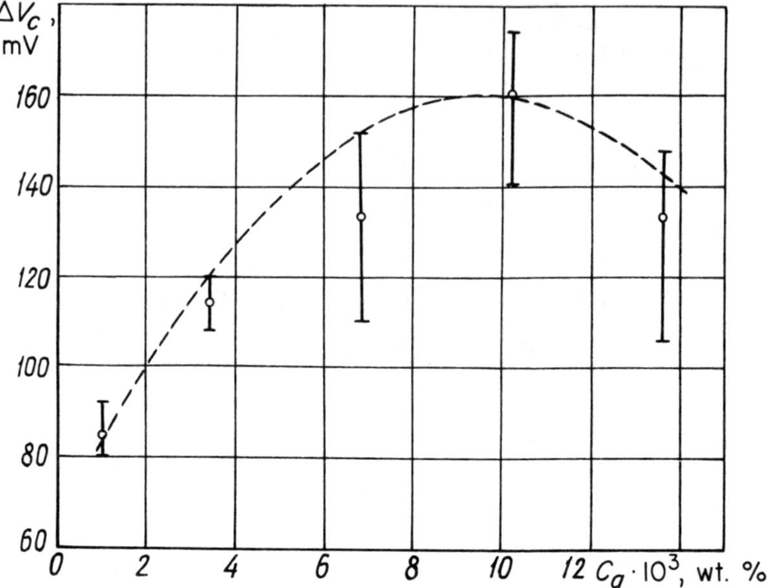

Fig. 2-6. Absolute variation in contact potential difference ΔV_c as a function of the concentration of antistatic additive in gasoline.[71]

probably indicates the correctness of the molecular-physical interpretation of Equation 2-3.

Uchuvatkin has reported on the accurately reproducible change in contact potential difference ΔV_c when a given boundary layer is applied to a given metal surface.[12] Simultaneously with the measurements of $V_{c(0)}$ and $V_{c(ads)}$, he determined the forces of the attractive electromagnetic interaction of metals in the boundary friction of homogeneous couples (see Section 1.2), and also the coefficient of linear extension of the frictional contact.* The experiments showed that a correlation exists between the work function and the parameters mentioned. For example, in a certain set of metals in the fourth row of the periodic table (Cr, Ni, Cu, Zn), the work function φ, the specific force of attraction between condensed phases p_m, and the coefficient of extension k are largest for chromium, somewhat smaller for nickel, and far smaller for copper and zinc; that is, the values decrease toward the metals of groups I and II.

Uchuvatkin studied the function $F = f(N)$ under identical conditions for various fatty acids, and also the CPD as a function of the number of carbon atoms in the molecules of these acids z. He concluded that the relation between $\Delta\mu$

---

*The method of resonance curves, the idea for which was advanced by L. I. Mandelshtam and realized by S. Eh. Haikin and coworkers,[33] was used to find the physical and chemical constants characterizing the state of the contact.

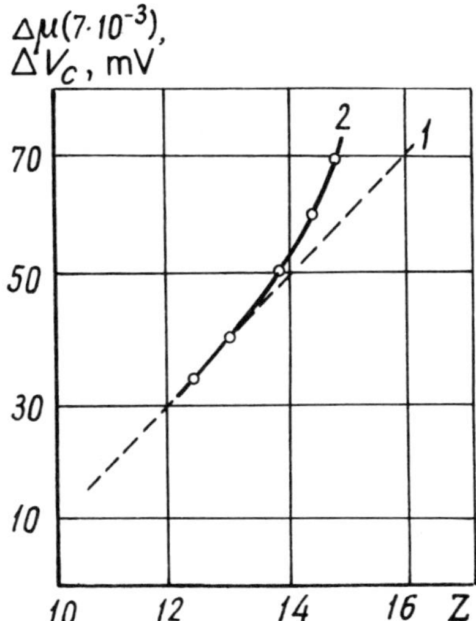

Fig. 2-7. Variations in the coefficient of friction (curve 1) and the contact potential difference (curve 2) of a steel, versus the number of carbon atoms in fatty-acid molecules (after
data of Uchuvatkin[12]).

and ΔV_c is linear.   From Fig. 2-7, however, it follows that $\Delta\mu$ and ΔV_c are
proportional only in the first approximation.   At the same time, there is every
reason to suppose that the law of variation in $\Delta\mu$ versus $\Delta\varphi$ (or ΔV_c), seen by
Markov in alloys when industrial lubricants were used, is general for all mate-
rials, both lubricants and materials forming frictional couples.

Garkunov, Markov, and Golikov studied the interrelation of oxidation-reduction
reactions in frictional couples formed by components differing in work function.[75]   Their aim was to explain the phenomenon of atomic transfer and learn
to control it.   By that time it was considered established that when copper alloys
rub against steel in glycerin or an alcohol-glycerin medium, copper atoms are
separated from the alloy solid solution and gradually transferred back and forth
from one surface to the other.   The mechanism by which copper was separated
later needed refinement, since the first stage of the process more closely resembles the preservation of a copper "sponge" at an alloy surface when the alloying
elements (Mn, Al, Zn, etc.) are transported selectively into the lubricant.   In
essence, what is going on is spontaneous anodic dissolution, where the seminoble
metal (Cu) acts as a cathode at some compromise alloy potential.   As for the
back-and-forth transfer of copper, ever more recent experiments have confirmed
its reality.   The very small values of friction force and the absence of damage to
the conjugate surfaces, which would have occurred in that kind of transfer
(now called *selective*; see Section 2.7), distinguish back-and-forth transfer in
principle from the well-known phenomenon of seizing of metals, as Garkunov

and Kragelsky have pointed out.[76]   Atomic transfer has also been observed in type TsIATIM lubricating grease, type AMG-10 oil, and other media.

Using the electronegativities (κ) of metals and alloying elements in alloys, one can attempt to predict what route the chemical modification of the friction track will take. However, if one considers this modification as a process of transforming the surface into a more stable thermodynamic state, then it is convenient to use the work function (φ) as a criterion of the degree of passivity— or conversely, activity—of a sliding surface.* In application to a frictional couple this means that it is expedient to measure the contact potential $V_c \sim \varphi_{Au} - \varphi_{Me}$ for each of the metals. The results obtained by Garkunov and coworkers indicate the same thing.[75]

After analyzing the data of Table 2-2, it is not difficult to conclude that a series of contact potentials can be constructed for every lubricant.

Such a series, in the judgment of Garkunov and coworkers, does make it possible to foresee on which of the paired metals the oxidant will be reduced, and which of the metals should be oxidized in friction. To this end, it should be sufficient to use the method of polarization curves to represent the spontaneous oxidation of the metals. This method was developed by Frumkin and coworkers.[78]

As one example, the authors consider the forecasting of processes in the frictional couple of bronze and constructional steel, which is intended for operation in a glycerin medium. They remark that as long as $V_{c(st)} > V_{c(br)}$ (both contact potentials being positive against gold; see Table 2-2), the bronze should be oxidized and pure copper should be reduced at the steel surface. After the steel surface becomes covered with a thin film of metallic copper, the reduction of copper will begin to take place at the bronze, since $V_{c(br)} > V_{c(co)}$ for the spontaneously formed copper-bronze couple. The transfer of copper between

*We note, by the way, that $\varphi = 0.75 + 2.5\kappa$ according to Conway and Bockris.[77]

Table 2-2. Contact potential differences against gold (in mV).[75]

Material	Cleaned Surface	After Application of Lubricant (oil)[a]			
		AMG-10	TsIATIM-201	MS-20	Glycerin
Electrolytic chromium	148	128/148	−120/−80	180/180	164/168
Steel, type 30KhGSA	192	156/172	−540/−368	232/232	360/368
Bronze, type BrAZhMts	276	280/296	284/292	356/360	264/264
Brass, type LS 59-1	244	224/240	100/144	260/264	312/316
Copper, type M-1	192	136/156	−8/60	216/216	212/216
Steel, type 1Kh18N9T	216	260/276	56/96	328/320	300/312

[a]Immediately after application/30 sec later.

the sliding surfaces will stop when both surfaces are plated with copper and the CPD of the Cu-Cu couples equals zero.

With regard to this kind of theoretical treatment of atomic-transfer results obtained by Garkunov and coworkers in tests on friction machines, it should at least be stated that such a treatment is not the only possibility, or else that the treatment needs more interpretation. Incidentally, later publications by the same authors have also pointed this out.[72] Indeed, if the frictional system copper alloy–glycerin–steel is considered as a short-circuited galvanic cell,* then it must be remembered that the emf of this cell includes, besides the contact potential difference, the potential difference appearing at the metal-electrolyte interfaces. In this case, the polarization diagram reflects the oxidation, simultaneously and by a single oxidant, of two contacting metals that are different until both are copper-plated. It is precisely the exceedingly slow dissolution of steel in glycerin that gives steel a positive potential against the faster-oxidizing bronze; atoms of the alloying elements, becoming ionized, escape from the bronze into the lubricant, and copper atoms associated in groups are transported to the steel.[79]

Of course, under conditions where glycerin acts like a weak acid, the contact potential difference, as one of the principal components of the emf, may bring about a drastic change in the character of selective electrochemical corrosion. The values of φ_{MeI} and φ_{MeII} will determine the direction the process takes: dissolution or passivation of the copper alloy. In this sense, the data of Garkunov and coworkers are of indisputable interest.[75]

Here the "passivation" of a copper alloy means the appearance of a "servovital" (self-regenerating) copper film at the alloy surface, preventing further dissolution. The formation of such a film, however, is often complicated by the alloy's going over to the passive state because an oxygen barrier has appeared on it right in the electrolyte solution. What is more, after an initial residence in air (in practice this nearly always occurs), the components of the frictional couple are already covered with an oxygen barrier consisting of a phase oxide, a quasicrystalline limiting oxide, and adsorbed oxygen. Realization of conditions favoring the initiation of selective transfer requires the removal of this barrier as the first step.

On the other hand, the oxide film is known to perform a lubricating function that depends on its thickness. Even the first faces of oxide crystal lattices have a powerful screening effect: the coherence of the metals is weakened severalfold. A thicker oxide film with a pronounced intrinsic structure changes the coefficient of friction in accordance with its volumetric characteristics.[44,80] The question arises, How does the oxidation of a metallic surface affect the work

---

*The author proposed the "metal I–electrolyte–metal II" model of a frictional system as a short-circuited galvanic cell.[15]

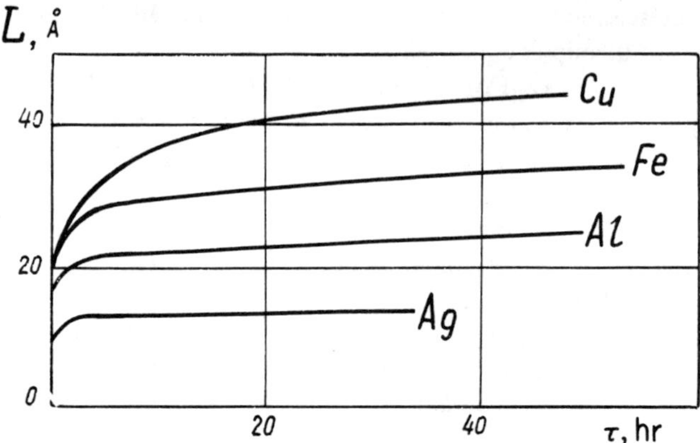

Fig. 2-8. Thicknesses of oxide films formed on various metals at room temperature.[81]

function? Although, as was mentioned above, Markov touched on this question,[72] Yevdokimov and Syomov have given the clearest and best-grounded answer to it.[13] We will dwell on their answer.

Characteristics of oxide films formed at room temperature include small thickness and a very high rate of growth upon initial contact of the juvenile metal surface with the atmosphere (Fig. 2-8).

According to current ideas, the rate of growth in the initial stages of oxidation is determined mainly by (1) drift of metal ions to the oxide-gas interface in the strong electric field generated in the thin oxide film when electrons tunnel to the external surface of the oxide and adsorbed oxygen atoms capture them;[82,83] (2) accelerated migration of ions from defect to defect in the oxide crystal lattice;[84] and (3) diffusion of ions when there is local overheating by the heat of the exothermal oxidation reaction.[85]

If we follow, for example, the Mott-Cabrera model, then the appearance of a negative surface charge on the oxide film as a result of tunneling by electrons should set up an additional potential barrier and increase the work function. As Table 2-3 shows, however, cleaning of the metal surface in air leads to φ_{om} values not only larger but also smaller than the corresponding values of φ.* Among the metals under study, especially large changes in work function upon cleaning occur in titanium (+0.56 eV) and aluminum (−0.6 eV). Yevdokimov and Syomov emphasize that such changes cannot result from just one deformation of the metals[13] (of course, this factor is taken into account),[86]† and that another cause of the changes is the formation of oxide layers and adsorbed layers. An

---

*The work function of the oxidized metal φ_{om} was determined from measurements of the CPD; the reference electrode was oxidized nickel (φ_{Ni} = 4.87 eV).
†The next section deals with the deformational distortion of the electron energy spectrum in solids.

Table 2-3. Work functions of metal surfaces in eV.[13]

Metal	Clean Surface in Vacuum (φ)	Surface in Air 1 min after Cleaning (φ_{om})	Long-Oxidized Surface (φ_{om})
Al	4.25	3.65	4.04
Mg	3.64	3.80	3.61
Zn	4.24	3.94	4.25
Cd	4.10	4.30	4.20
Sn	4.38	4.35	4.31
Nb	3.99	4.06	4.76
Ti	3.95	4.51	4.68

evaluation of the Debye screening length L_D for oxide films gives values on the order of 10^3 Å; ordinarily, the film thickness L is far less than this. Hence the conclusion suggests itself: that, upon oxidation under atmospheric conditions, the work function of a metal surface either should not change at all or else can only increase (to about 0.5 eV)[87] as a result of negative surface charging in the adsorption of oxygen and water vapor. The same result was seen in a study of the oxidized surfaces, but not in every case (see Table 2-3). With aluminum, for example, although the Mott-Cabrera model has been confirmed for $L \ll L_D$, the work function does not increase upon cleaning but decreases. Yevdokimov and Syomov explain this apparent contradiction by using a zone model of an oxide film containing Tamm zones.[88] Here the nature of the surface states is not an important element in the overall line of argument, although in real metal-oxide systems the principal role belongs not to Tamm levels (whose existence has been established for a highly idealized model)[5] but to levels due to defects in the metal substrate.

The surface states (S levels) are localized in the immediate vicinity of the metal-oxide interface, as the energy-level diagrams of Fig. 2-9 show. For clarity, the oxide film is first arbitrarily separated from the metal and its surface charge is taken equal to zero (Fig. 2-9[a]). Since some of the surface states below the Fermi level are occupied by electrons, when the metal-oxide contact comes into being, the surface states will discharge if $\varphi_m > \varphi_o$, or charge if $\varphi_m < \varphi_o$, until the Fermi levels of the metal and the oxide coincide. It is not hard to see that Fig. 2-9(b) corresponds to the case where the work function of the metal is larger than that of the oxide, so that the oxide film must take on a positive charge. From Fig. 2-9(b), it follows that the height of the potential barrier at the metal-oxide boundary is

$$\psi = \varphi_m - \chi - V_S \tag{2-5}$$

where χ is the electron affinity energy of the oxide and V_S is the positive potential to which the oxide film has become charged.

Let us now assume that the external surface of the oxide has taken on a nega-

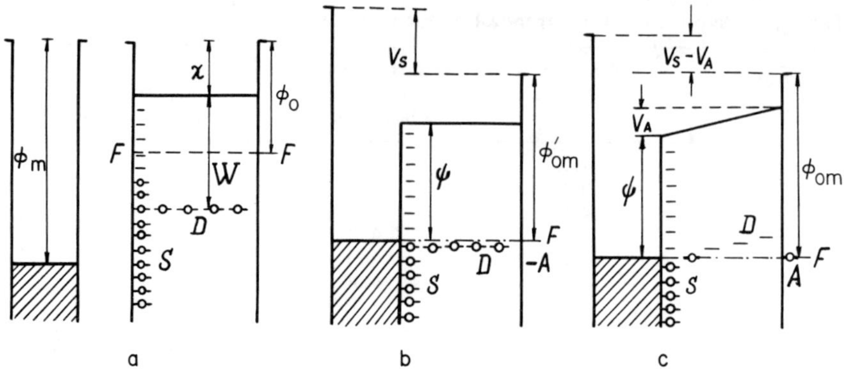

Fig. 2-9. Change in work function of a metal when it is covered with a thin oxide layer $(L << L_D)$:[13] (a) metal and oxide film arbitrarily separated; (b) positive charging of oxide film on the destruction of S levels of surface states; (c) negative charging of surface on filling of A levels of adsorbed oxygen.

tive charge through filling of the A levels of adsorbed oxygen by electrons (due to tunneling of the electrons from the metal or destruction of the donor centers D). Then the potential of the surface changes by an amount V_A, and, according to Fig. 2-9(c), the work function of the oxidized metal surface is

$$\varphi_{om} = \psi + \chi + V_A \tag{2-6}$$

or by Equation (2-5),

$$\varphi_{om} = \varphi_m - V_S + V_A. \tag{2-7}$$

Thus, for $\varphi_m > \varphi_o$ $(V_S > 0)$, the work function of a metal may either increase or decrease as the metal is oxidized. Which will occur depends on the relation between V_S and V_A. It is also obvious that, for $\varphi_m < \varphi_o (V_S < 0)$, the oxidation

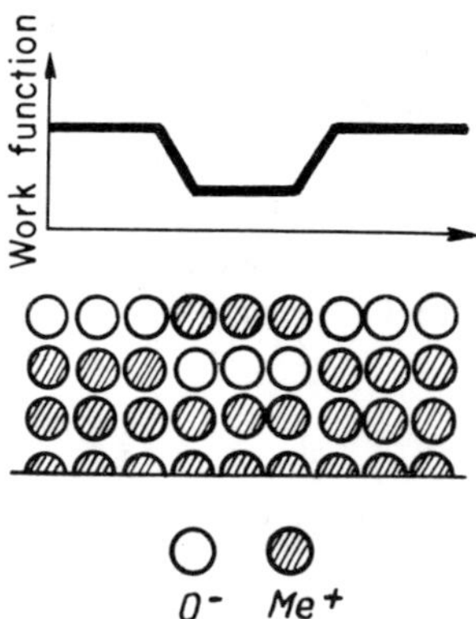

Fig. 2-10. Local depressions in work function of a metal on oxidation.[13]

of the metal should lead only to an increase in the work function. The results obtained in the study of titanium probably fit in here.

Another interesting point noted by Yevdokimov and Syomov is that, along with the overall increase or decrease in the work function of a surface being oxidized, local depressions of φ_{om} are possible. These depressions are located preferentially where defects pile up, making it easier for the positive metal ions to escape to the surface of the growing oxide (Fig. 2-10). We see how sensitive the work function is to even a slight variation in the electrophysical state of the surface.

2.2 ELECTRON EMISSION

Initiators and Stimulators

A phenomenon closely linked with a change in the work function is "exoelectron emission" (Kramer effect),* the emission of electrons from solid surfaces as a result of changes in the electron energy spectrum under various influences. It has been found, in particular, that the probability of an electron's surmounting a potential barrier may rise sharply upon phase transformations, oxidation, mechanical actions in air (deformation, scratching, spalling), an increase in the temperature of a previously excited body (thermal stimulation), surface treatment in a glow discharge, electromagnetic radiation (gamma, X-ray, ultraviolet), or bombardment with alpha particles.

The variety of forms of the Kramer effect has naturally given rise to a number of hypotheses concerning its mechanism (see Yevdokimov and Syomov's short review of these hypotheses).[13] It is thought, for example, that the exoelectrons can be excited by the latent heat of phase transformations;[90] the heat of chemosorption of gases[91] or the energy of the chemical reaction between an aggressive gas medium (chlorine, oxygen, fluorine) and a clean metal surface;[92] the energy of defects in the crystalline lattice of the growing oxide,† transmitted by the Auger mechanism to other surface states with low potential barriers against the escape of electrons;[94] the energy liberated on the annihilation of vacancies reaching the surface;[95] and the mechanical energy of deformation accumulated by active centers of low work function in the neighborhoods of structural defects.[96,97]

Now let us attempt to use these and other notions to write the equation of electron emission in the most general form.

Let $W_{F(0)}$ be the Fermi energy of electrons in a solid of ideal structure at $T = 0°$ K. In a real solid of the same type but having an imperfect structure, at ordinary temperatures and under atmospheric conditions, part of the electrons

---

*The term "exoelectron emission" was established in scientific usage by decision of a conference in Innsbruck,[89] but the term basically does not reflect the physical essence of the phenomenon, the systematic study of which was started by J. Kramer in 1950.[90]

†Such defects are considered special F' centers analogous to the color centers in alkali-halide crystals.[93]

are in an excited state. That is, the electron energies have been raised above the Fermi level by an amount W_0, different for each electron, which we will call the *initial excitation energy*. Depending on what kind of energy provided the excitation, we write W_M for mechanical, W_C for chemical, W_A for van der Waals adsorption, . . . , W_E for electromagnetic, and W_T for thermal. Then the equation of electron emission will have the form

$$W_0 + W_M + W_C + W_A + \cdots + W_E + W_T = \varphi_{\max} + \frac{mv^2}{2} \qquad (2\text{-}8)$$

where $\varphi_{\max}$ is the maximum possible work function.

Here we should emphasize that the Fermi level coincides with that energy level for which the occupation probability equals 1/2. The definition found in the literature for the Fermi level, as the highest energy level occupied by electrons at a temperature of absolute zero, does not suit us in the present case. In metals, for example, this definition is often applicable only because the energy of thermal motion kT ($\sim$0.025 eV at room temperature) is considerably less than $W_{F(0)}$, the Fermi energy at absolute zero ($\sim$5 eV). This fact makes it possible to suppose $W_{F(T)} \approx W_{F(0)}$ and go on to determine the work function experimentally, as if it did not depend on temperature. However, from what has been said, and especially from the last hypothesis concerning the mechanism of the Kramer effect, the actual (or effective) work function is a function of the initial excitation energy, which depends on the temperature and the degree of imperfection of the solid:*

$$\varphi = \varphi_{\max} - W_0 . \qquad (2\text{-}9)$$

Thus,

$$W_M + W_C + W_A + \cdots + W_E + W_T = \varphi + \frac{mv^2}{2} . \qquad (2\text{-}10)$$

The application of energy of any nature may be the proximate cause of emission; in this theoretical sense, no term on the left side of Equation 2-10 differs from any other term. However, the contributions of the several terms to a total energy sufficient for an electron to escape from the boundaries of the solid with a maximum velocity v are far from equal. If under given conditions some energy source is the principal factor responsible for the external excitation of an electron, then that energy source can be considered the initiator of electron emission. A source playing the role of an incidental factor facilitating the emission is usually considered a stimulator of emission.

Clearly, emission, even from different portions of the same solid, can be

---

*The degree of imperfection of a crystal lattice is determined mainly by the dislocational structure of the lattice, i.e., the distribution and density of dislocations.

caused by a number of principal and incidental factors. It is convenient to call the energies communicated to an electron by stimulating sources ΔW_E (electromagnetic stimulation, with photostimulation as one of its varieties), ΔW_T (thermal stimulation), ΔW_C (chemical stimulation), and so on. Then, using Equation 2-10, it is easy to write an expression for any special case of emission. For example, the equation for thermionic emission stimulated by oxidation and by electromagnetic radiation will have the form

$$W_T + \Delta W_C + \Delta W_E = \varphi + \frac{mv^2}{2}; \qquad (2\text{-}11)$$

for thermally stimulated photoelectron emission, we have

$$W_E + \Delta W_T = \varphi + \frac{mv^2}{2}. \qquad (2\text{-}12)$$

Now if we try to write the equation of exoelectron emission without knowing whether the major contribution to the electron excitation energy comes from structural defects appearing in deformation and irradiation or from oxidation and adsorption processes, then we will have to write

$$W_M + W_E + W_C + W_A = \varphi + \frac{mv^2}{2}. \qquad (2\text{-}13)$$

In applied studies, however, low-temperature emission is encountered more frequently, with plastic deformation as the initiator and with heating, oxidation, and illumination as the clearly defined stimulators. From this simple fact, it is obvious that Equation 2-13 should be written in a more concrete form:

$$W_M + \Delta W_T + \Delta W_C + hv = \varphi + \frac{mv^2}{2}. \qquad (2\text{-}14)$$

Incidentally, we should mention that in the great majority of cases the emission of exoelectrons is quite impossible without additional energy from light quanta. As Yevdokimov has shown, when friction between metals takes place in air, cutting off the exciting light stops the emission with no time lag, even though deformational and chemical processes continue.[98] Of course, the illumination is always of long wavelength ($hv < \varphi$); this is what distinguishes the photostimulation of exoelectron emission from the ordinary photoelectric effect ($hv > \varphi$).

The intensity of exoelectron emission increases drastically upon first-order and second-order phase transitions.[99] Somewhat further on, we will take up the question of why the Kramer effect is seen in second-order transitions occurring without latent heat. As for first-order transitions and generally for energy changes of any kind involving the sudden liberation of accumulated energy (thermal, mechanical, electrical, etc.), special provision can be made in the formula for exoelectron emission. How to do this—whether through additional terms or by introducing appropriate coefficients, regarding energy changes as emission multipliers—still depends on the personal tastes and interests of the investigator.

It is not difficult to see that the mechanism of low-temperature emission is just as many-sided as the extremely wide range of phenomena causing this emission. Thus, any unity of opinion that exists on the nature of exoelectron emission is just the recognition that all the mechanisms listed previously have a single physical basis, even though each concrete type of exoelectron emission has its own specific mechanism. But to be consistent we must keep in mind that the common features of the energy sources (from the standpoint of enabling electrons to surmount the potential barrier) extend to both types of electron emission, high-temperature (thermionic) and low-temperature (exoelectron).

The need for a distinct classification of the various types of emission, including those covered by the term *exoelectron emission*, becomes more and more obvious. It seems to the author that the physical basis for such a classification should be the changes in the electron energy spectrum of a solid due to either individual energy sources or a combination of such sources differing in physicochemical nature. Regardless of which factors govern the rearrangement of the electronic structure—thermal or mechanical influences, exothermal reactions of various kinds, electromagnetic radiation, bombardment with a stream of fast particles, and so on—the rearrangement alters the topology of the Fermi surface, with the inevitable result that the work function changes. A reduction in the value of φ signifies the conversion of electrons to a "prefiring" state. If the energy communicated to an electron is enough to surmount the potential barrier, then the electron will escape from the solid. It should be noted, however, that as several competing energetic processes go forward, their roles as stimulators or even initiators of exoelectron emission begin to weaken rapidly. What is more, a process such as oxidation, which reflects the tendency of the surface to go into a passive state with the formation of a protective screening film, should be accompanied by an increase in the work function. Evidently, it is only under conditions where there is no oxygen barrier (so that defects appear at the surface as a result of corrosion of the metal) that steady exoelectron emission can be seen over the whole period of chemical excitation. As for the emission peaks detected upon second-order phase transformations, they are probably due to a rearrangement of the electronic structure in which the statistical fraction of lower-work-function electrons rises so as to leave the internal energy (as opposed to the thermodynamic potential) of the system invariant when there is a simultaneous increase in the statistical fraction of higher-work-function electrons.

From these ideas, it is not hard to understand the interaction of processes taking place in the same microvolume of a solid subjected to energetic influences. For example, distortions of the electronic spectrum of a metal when it is plastically deformed affect the nature and kinetics of the adsorptive adaptation of lubricant molecules or those of direct chemical reactions. In turn, marked distortions in the electron-density distribution in the solid, which are due to the effects of van der Waals forces and especially chemical-bonding forces, make it easier for plastic deformation to develop and proceed.

The electronic structure of solids is a physical platform that has made possible important advances in the explanation of the mechanical (in particular, strength) properties of solids.[100] Recent publications have shown, for example, that conduction electrons are able to migrate from compressed regions of the crystal lattice to regions where the interatomic spacing is relatively large;[101] that the energy of formation of stacking faults is linked with the electronic structure of a metal;[102] and that the metallization of interatomic bonds is significant in the brittle-to-ductile fracture transition.[103]

According to the researches of Kryuk, who considered electron emission as a deformational effect, the shape of the exoelectron-emission curves can be used to judge (1) the degree of metallization of the bonds in a covalent crystal upon deformation (the concentration of free electrons increases) and, conversely, (2) the degree to which a homeopolar component appears upon significant deformation of a metallic crystal (the concentration of conduction electrons declines).[104] The conclusion is in complete agreement with the behavior of φ seen earlier upon a change in the character of the coherence forces in a crystal lattice.[105] With regard to the information content of the curves of exoelectron-emission intensity (I) versus time and temperature, Kryuk used this very fact, that the deformation of solids with the polar type of interatomic bonds (purely metallic and covalent) leads to opposite reactions: highly developed plastic deformation for typical metals, and brittle fracture for typical dielectrics.[104] Mints and Kortov assumed that the character of these processes is determined by the relative abilities of submicrovolumes of the lattice to dissipate and accumulate the energy of deformation.[106] On that assumption, Kryuk emphasizes that this energy, "within the limits of an association of several atoms (ions), is first taken up by electrons that participate in interparticle coherence in the lattice." In other words, the formation of defects in the crystal begins at distortions of its electronic structure. In metals, the energy of deformation is quickly dissipated because of the relative freedom and high concentration of collectivized electrons. In distinction to metals, dielectrics (but also semiconductors with a pronounced homeopolar component) have a tendency to a rapid buildup of mechanical energy, right up to a critical value. The directionality and rigidity of the covalent bonds cause brittle fracture. The exoelectron-emission curves presented by Kryuk have indeed confirmed that the level of energy dissipation depends on the electronic structure of solids.[101]

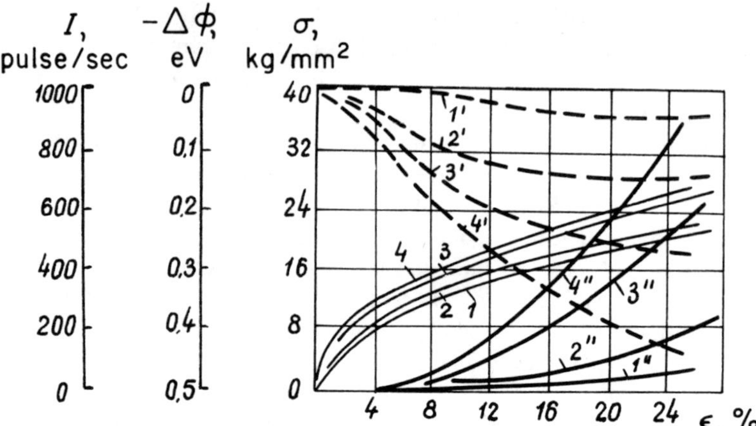

Fig. 2-11. Effects of plastic deformation on strain hardening σ (curves 1–4), variation in work function (curves 1′–4′), and exoelectron emission I (curves 1″–4″) of copper alloys: [96] (1, 1′, 1″) Cu; (2, 2′, 2″) Cu + 2.3% Al; (3, 3′, 3″) Cu + 4.5% Al; (4, 4′, 4″) Cu + 7.4% Al.

Analysis of theoretical and experimental results on the Kramer effect, in particular of the data of Figs. 2-11 and 2-12, favors the following assumption: Exoelectron emission, as an indicator of the thermodynamic instability of the metal surface, is caused by distortions in the electron energy spectrum of the solid and by the subsequent rearrangement of its electronic structure, which involves the appearance of defects. Plastic deformation, the most competent type of deformation from the standpoint of the formation and multiplication of defects, should be taken as the principal type of energy source responsible for the emission of exoelectrons.*

*The dislocation density increases by an average of four orders of magnitude (e.g., from 10^7 cm^{-2} to 10^{11} cm^{-2}) upon plastic deformation.

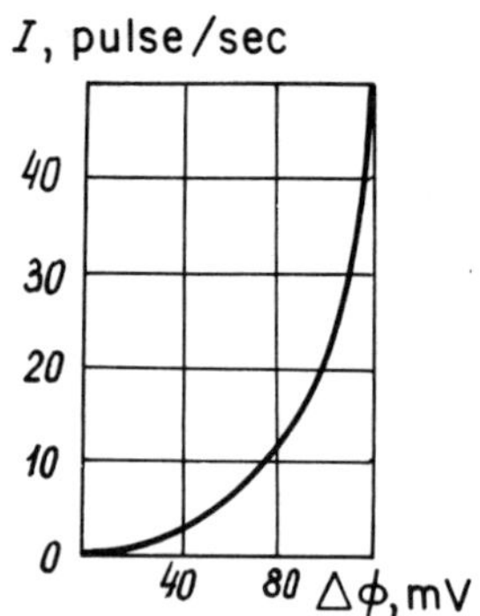

Fig. 2-12. Exoelectron emission by silver as a function of the change in work function upon extension.[97]

Mints writes,

The approach to the deformation phenomenon as excitation has great value in clarifying the role of nonmechanical interactions between interparticle structural elements. This fact makes possible a quantitative evaluation of a controlled influence on the hardening of metals aided by mechanical energy, electric and magnetic fields, and radioactive irradiation.[96]

The correctness of this forecast is confirmed by the development of new ways to strengthen materials through laser treatment, magnetic treatment (see Chapter 7), and so on.

Exoelectron Emission in Sliding Friction

A complex set of interrelated phenomena takes place in the friction of metals: mechanical, thermal, chemical, electromagnetic, and other phenomena. But friction is above all a deformational process; as such, it is accompanied by a considerable increase in the dislocation density in the surface layers of the sliding bodies, and thus by exoelectron emission upon stimulation (or multiplication) by external light, artificial heating, spontaneous oxidation, phase transformations, and so on. From the standpoint of dislocations, if we suppose that the active centers of emission are sites where the defects of the crystal structure are concentrated, then we can explain why deformational excitation may prevail over all other types of excitation, which are mainly reflected in the boosting of weakly bound electrons to higher energy levels. It is precisely in this area that Yevdokimov and coworkers are carrying on systematic research on exoelectron emission in sliding friction.[13]

In sign-alternating shear deformations, there is constant disorientation of blocks and grains of the metal; as a result, defects and new surfaces form rapidly. For this reason, it was expected that sign-alternating friction would increase the free surface energy and lower the work function more than unidirectional friction. The observed dependence of electron emission on the sign alternation of friction gave complete confirmation of this supposition.[107] The $I = f(\tau)$ curves obtained in static tests with excitation by ultraviolet light (Fig. 2-13) show that the level of exoelectron emission after sign-alternating friction is invariably higher than that after unidirectional sliding under the same conditions. According to Yevdokimov and Syomov, one probable reason for the initial peaks on the curves is the chemosorption of oxygen and the effect of metal-oxygen compounds on the work function.[13] In trials with steel or cast-iron test pieces, for example, the increase in I with time can be linked with the decrease in work function upon the formation of iron(II) oxide. Afterward, the emission intensity declines because of absorption of the electrons in the growing oxide layer.

Of particular interest are the studies of Yevdokimov and coworkers concerning

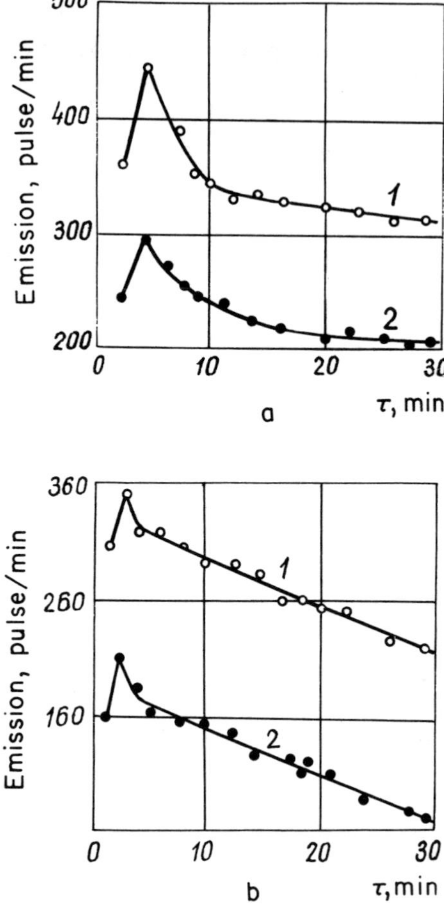

Fig. 2-13. Time dependences of exoelectron emission from the surfaces of (a) type-25 steel, (b) type-45 steel, (c) type-SCh15-32 cast iron, and (d) aluminum, after (1) sign-alternating and (2) sign-constant friction without lubrication.[13]

the general laws of exoelectron emission in friction under dynamic test conditions.* Continuous recording of exoelectron emission during the actual sliding process certainly yields an incomparably large amount of information on the energy states of the sliding surfaces. The accessible portions of the surface, which have just been the arena for frictional effects, at times bear traces of the violent dislocation processes that have occurred on them (multiplication, migra-

*Yevdokimov and Syomov[13] discuss the results of these investigations, which were published at various times.

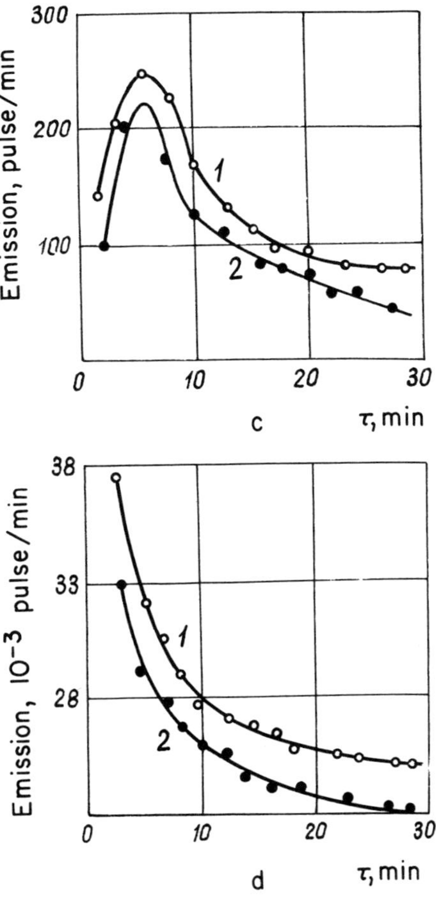

Fig. 2-13. (*continued*)

tion, and interaction of dislocations; dislocation reactions). The process of oxidation begins immediately on these sections as they become denuded. Oxidation is speeded up by the rise in surface temperature resulting from the intrinsic thermal effect and the heat of deformation, which depends on the sliding conditions. Here, according to the ideas just set forth, is where both an overall change (over the whole oxidized surface) and local depressions in the work function should be expected. These sections might result from photostimulated and thermally stimulated mechanical-chemical emission, if it is allowed that the additional energy of a quantum of light is needed for the excited electrons to escape.

The factors causing the decline in exoelectron emission include: (a) the screening effect of the oxide film, an obvious reflection of which is that, when its donor centers are destroyed, the film turns into something like a trap for electrons escaping from the metal; (b) the disappearance of local depressions of the work function in the concluding stage of the oxidation reaction;[13] and (c) rearrangement of the surface dislocational structure of the metal, in the direction of increased thermodynamic stability.

The last factor, it seems to us, is decisive. The degree of imperfection of a solid subjected to elastoplastic deformation and then left to itself is characterized by annihilation of dislocations of unlike signs, reduction in the vacancy concentration, and other similar processes that lower the surface energy. Figuratively speaking, the electron-emitting "wounds," inflicted on the surface when asperities of the counterface rubbed against the sections of real contact, now "heal" and "scar." The shorter the time between frictional deformation and the detection of the resulting emission, the more reliably the energy state of the rubbing portions of the surface can be assessed from the exoelectron-emission curves.

Figures 2-14 through 2-16 present such curves for two steels, reflecting how the load, speed, and sign alternation of motion affect the exoelectron emission.

The identical shapes of the exoelectron-emission and microhardness curves in Fig. 2-14 made it possible to assume[13] that the shape of the $I = f(\tau)$ curve can be used to assess the degree of cold work to which the surface layers are subjected in friction. In fact, section I corresponds to running-in of the surfaces, a process in which the multiplication and development of structural defects exceed the annihilation of defects from the start. Of course, the increase in dislocation density and the appearance of new dislocation networks (which are known to be very stable) result in an increase in the microhardness of the

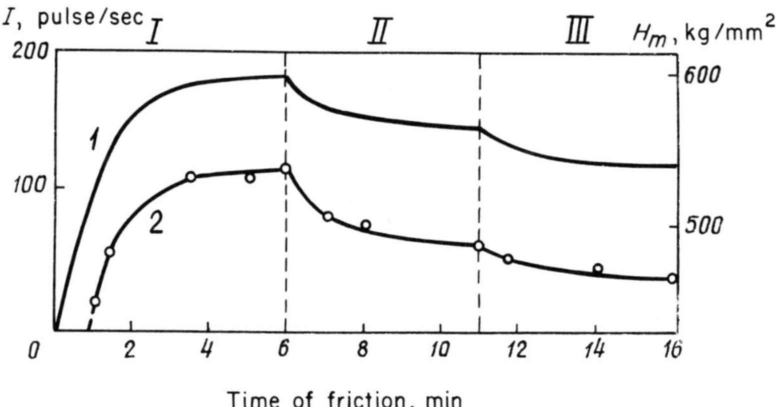

Fig. 2-14. Effects of varying normal load in friction on exoelectron emission (curve 1) and microhardness (curve 2), for type-45 steel.[13] The load equals 13.2 kg on section I, 26.4 kg on section II, and 52.8 kg on section III. The speed is 1.5 meter/sec.

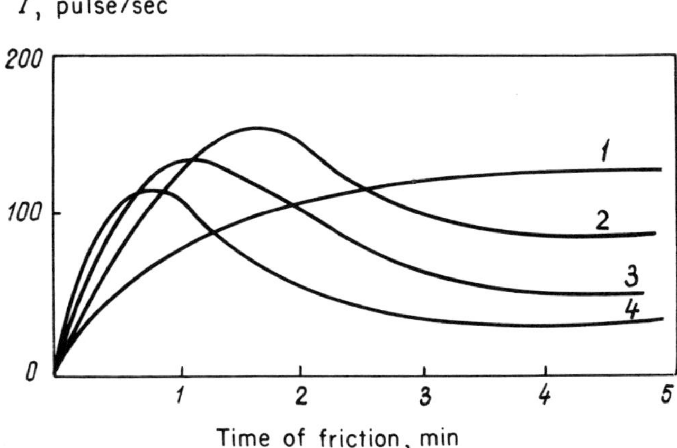

Fig. 2-15. Effect of sliding speed on exoelectron emission by type-25 steel, in dry friction under a load of 26.4 kg, at speeds of 1.5 meter/sec (curve 1), 2.7 meter/sec (curve 2), 4.5 meter/sec (curve 3), and 6.3 meter/sec (curve 4).[13]

friction surface and a rise in the emission level. This continues until the competing processes of strain hardening and recovery come into equilibrium. With an increase in the normal load, and therefore a rise in the surface temperature (sections II and III), the maximally strain-hardened zone of plastic deformation begins to soften much more rapidly, in spite of the lower-lying metal layers that have been pulled into the zone.*   This change manifests itself in a reduction in microhardess and a drop in the level of exoelectron emission.

The experimental curves in Fig. 2-15 are explained in the following way.[13] As the sliding speed increases, the surface running-in time (that is, the time during which transverse deformations develop most rapidly, structural defects are generated, chemosorption sites appear, and the temperature of the active emission centers rises) becomes shorter. For this reason, the steepness increases from curve 1 to curve 4; these curves reflect the ability of thermally stimulated exoelectrons to pass through the very thin oxide film. At the same time, the larger the value of v, the more quickly the $I = f(\tau)$ curve attains its maximum and the more steeply it begins to fall off. This decline has to do with the continuing oxidation of the steel, which results in the formation of iron(II, III) oxide (Fe_3O_4) and iron(III) oxide (Fe_2O_3) along with iron(II) oxide (FeO). The Fe_3O_4 and Fe_2O_3 create an additional potential barrier, which prevents the escape of electrons. Yevdokimov considers the associated rise in work function a manifestation of the "screening-like" action of the chemical compounds formed on the

---

*In steels with good cold-workability, short-term strain hardening of these layers upon a discrete increase in load causes a peak to appear on the exoelectron-emission curve.[13]

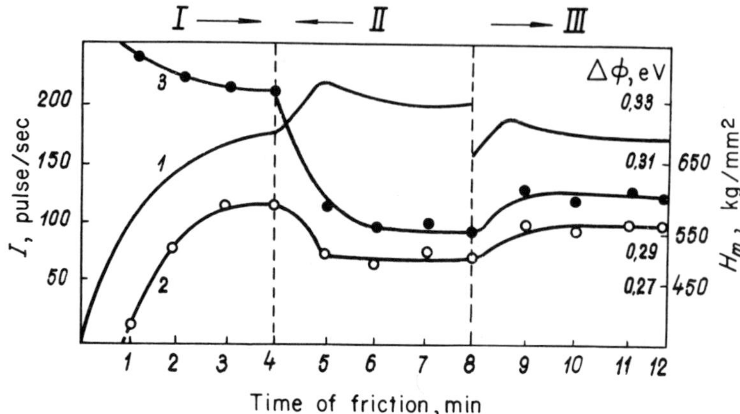

Fig. 2-16. Effects of sign alternation in sliding on the properties of frictional surfaces, for specimens of type-45 steel.[13] The load is 13 kg, and the sliding speed 1.5 meter/sec. (1) exoelectron emission; (2) microhardness; (3) contact potential difference.

surface. He writes, "As the sliding speed, and thus the temperature and the oxidation rate, rise, this oxide effect on the emission process comes to dominate the thermal stimulation of electrons that is occurring at the same time."[108] We can agree with this view; however, the treatment of the character of curves 2 through 4 would probably be more complete and even more precise if it took account of competition between the processes of strain hardening and recovery, as was done in the preceding case.

Yevdokimov once again finds it necessary to allow for this competition.[109] He analyzes, first, the behavior of the exoelectron-emission curve taken in sign-alternating sliding, and, second, the discrete microhardness and CPD values corresponding to the curve (Fig. 2-16). His conclusion is presented here in full:

Sign-alternating friction leads to softening of the contacting surfaces, after an intermediate and rather latent stage in which they are maximally strain-hardened. Because of the increased degree of deformation and imperfection of the structure, sign-alternating friction lowers the work function and intensifies exoelectron emission.

In concluding our examination of the nature and features of the Kramer effect, we should look at one more result of practical importance.

According to Yevdokimov and Syomov, exoelectron emission from frictionally deformed surfaces of metal specimens increases when the frictional interaction takes place in an inactive oil with a surface-active agent (such as oleic acid) added.[13] The reason for the increased emission level is thought to be the increased dislocation density, which is due to the Rebinder adsorption effect. The same effect causes plasticization of the surface layer and localization of shear deformations in it. The sign-alternation factor of the friction strengthens the effect. For this reason, when a surface-active lubricant with antiwear effectiveness is used, the ratio of the exoelectron-emission intensity upon a change in sign to the intensity in unidirectional friction, I_r/I_u, approaches unity. Experiments* have shown that the same law governs the ratios of dislocation densities n_r/n_u, of internal stresses σ_r/σ_u, and of wear values Q_r/Q_u. On this basis, the condition $I_r/I_u \longrightarrow 1$ is recommended as a guideline in the selection of lubricant additives to eliminate the detrimental effects of sign alternation.

ADDITIONAL REMARKS

The emission of high-energy electrons[5] should not be confused with the emission of low-energy electrons seen in the Kramer effect. High-energy emission occurs in a high vacuum, when an adhesion joint is destroyed and a strong electric field appears between the plates of the arbitrary capacitor. The emission of fast electrons can thus be viewed as a special variety of spontaneous field emission, which does not require special stimulation.

In the friction of metals in air, emission of high-energy electrons can evidently take place only upon vibrational rebounds of the sliding indenter. At these moments, despite the partial neutralization of the charges in the sheets of the double layer by an electrical-conduction mechanism,† the field in the clearance (where a vacuum forms instantaneously and disappears immediately) remains rather strong. Quite obviously, the electron flux arising under these conditions, the head part of which (resembling a streamer) consists of particles moving at very high velocities, causes collisional ionization of the air pulled into the clearance between sheets by the suction of the vacuum. From this moment on, neutralization of the charges of the electrical double layer goes by the heavily studied gas-discharge mechanism. In the present case, we find an analogy between discharge phenomena caused by the destruction of an adhesion joint and the electrical processes in a gap between electrodes when an external field is applied.

This analogy does not, however, validate in any way Dubinin's idea that the friction of metals in air involves spark discharges;[110] as we will see later on, this idea has its proponents. There is no need to dwell here on Dubinin's incorrect estimate of the voltage between sheets of the double layer, which by his reckoning attains implausibly large values, sufficient for the occurrence of spark discharges (*between metals*). The essential point here, and one that Dubinin did not keep in mind, is that the friction is between *metals*; presumably, the whole dispute springs from this point.

---

*See Yevdokimov and Syomov,[13] Chap. 6.

†This neutralization (ohmic leakage) can be made far more difficult by the presence of oxide films.

The rate and duration of the transition of a perturbed system into the equilibrium state are defined by the relaxation time τ:

$$v_D(t) = v_D(0)e^{-t/\tau} \qquad (2\text{-}15)$$

where $v_D(0)$ is the mean or drift velocity of conduction electrons corresponding to the starting, nonequilibrium distribution. According to the rough calculations of Kittel, $\tau \sim 10^{-14}$ sec;[56] confirmation of this value comes from a study of the damping of short electromagnetic waves propagating in a copper waveguide ($\tau < 10^{-11}$ sec). The adherence phase in the stick-slip process lasts about 10^{-6}–10^{-5} sec. As we see, the redistribution of charges in the friction of metals goes so fast that no significant potential difference can result from the retention of charge on individual asperities, as Dubinin and his followers proposed. The author has written about all of this;[26] Bobrovsky made just such a critical comment several years later.[111]

We by no means rule out electrical microerosion of the sliding surfaces being oxidized, but the only apparent reason for such erosion is "silent" discharges. Furthermore, other theoretical prerequisites can sometimes be used (see Section 5.1) to explain the mechanism by which fields with "breakdown" field strengths are formed.

It should be added that thermionic emission is the most intense kind under severe conditions of sliding. Thermionic emission is characterized by a saturation current density computed by the Richardson-Dushman formula:

$$j_s = AT^2 e^{-\varphi/kT} \qquad (2\text{-}16)$$

where A is the emission constant.

The instability of thermionic emission in friction (which is due to the variation in the number and distribution of the points of real contact, where the temperature flashes are localized) makes itself seen when the alternating-current component of the integral emf of a frictional system is recorded. We now take up the question of which seats of emf are most significant in the macroscopic aspect of these phenomena.

2.3. PRINCIPAL SEATS OF EMF AND EQUIVALENT CIRCUITS FOR FRICTIONAL SYSTEMS

Thermoelectric Component

It is rather widely known that, because of the intensive heating of sliding metal specimens, the region of direct contact between them can be regarded as the hot junction of a natural thermocouple. If the electromotive force generated in a thermocouple of this type is recorded (Fig. 2-17), then it is possible to obtain information on the mean temperature at the rubbing surfaces and the behavior of the deformed contacting areas in the relative sliding of metals. Special

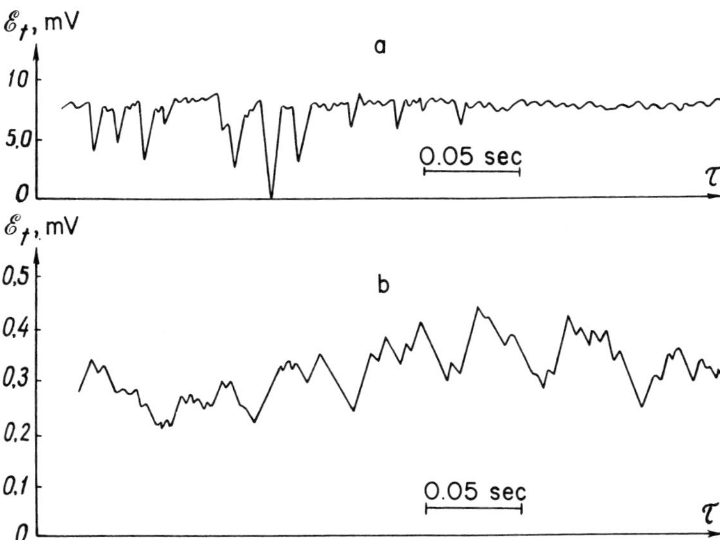

Fig. 2-17. Oscillograms of thermal emf's of frictional couples: (a) constantan and copper; (b) high-speed steel and mild steel.

methods have been worked out for calibrating natural thermocouples; some make provisions for eliminating parasitic thermoelectric currents in the closed measuring circuits.

Thus, an electrical parameter becomes a sensitive indicator. In fact, it was clear *a priori* that electrical phenomena had some importance beyond determining the intensity of heating of the rubbing metals in the region of contact.[26] Along with the auxiliary role of thermoelectric phenomena, the intrinsic, independent effect of these phenomena on the process of friction (cutting) has commanded and continues to command an enormous amount of interest.

Now, the first component of the integral emf is the thermoelectric component.

As the author has shown, the macroscopic study of electrical phenomena in friction is made more highly systematized through the use of diagrams of non-linear electrical circuits with variable parameters, synthesized to correspond to some elementary event in a frictional interaction.[15,27]

Suppose four asperities belonging to physically and chemically clean surfaces come into direct, pairwise metallic contact in the relative sliding of solid bodies. Figure 2-18 shows the internal thermoelectric circuit they form. A highly probable case in friction is the case where the temperatures are different in the two regions of real contact ($t_1 \neq t_2$), so that a thermal emf e_t acts in the closed electrical circuit made up of conductors I and II (r_1 and r_2 are the resistances of the constrictions).

From the quantum theory, we have no reason to expect much of a thermal

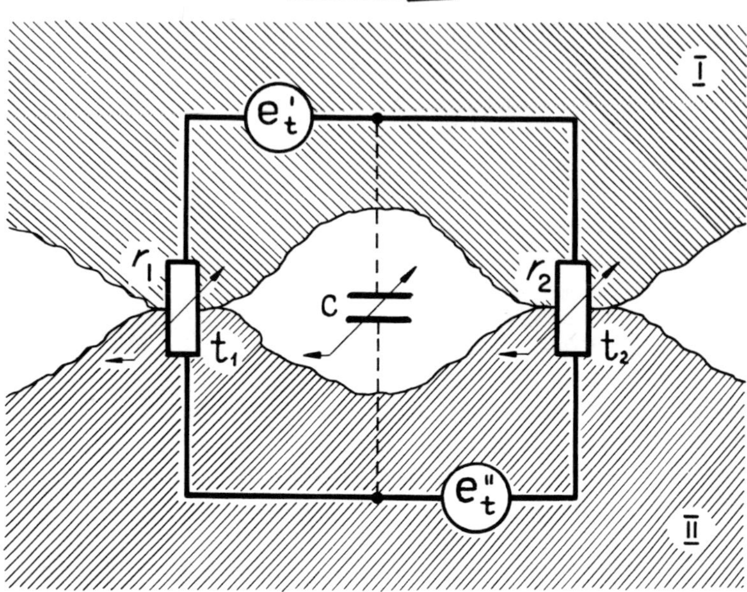

Fig. 2-18. Elementary contact cell and schematic diagram of its thermoelectric circuit in the friction of metals.

emf to appear in the case of juvenile surfaces, since the "contribution" of each of the seats created by the surfaces is very small or nonexistent. The seats are:[112]

1. The difference between the contact potential differences at the "micro-junctions."
2. The temperature dependence of the kinetic energy (mobility) of the charge carriers.
3. The difference between the free-electron concentrations.

But if the contacting metals are oxidized, then because of the semiconducting properties of oxide films, we should allow for the ability of these films to act rather like amplifiers of the thermoelectric effects at the points of closest approach of the condensed phases. Then, classical Boltzmann statistics is applicable to the analysis of thermoelectric phenomena in friction. The low thermal conductivities of the oxides (high temperature gradients on the areas of contact) mean that rather large eddy currents will appear in friction. These currents may in turn bring about the localized evolution of heat.[113]

The electromotive forces set up by the difference in temperature between the two branches of the circuit may act in either opposite senses (the same conduction mechanism in metals I and II) or in the same sense (the circuit comprises electron and hole conductors). Therefore, the total thermal emf of the circuit is

$$e_t = e'_t \mp e''_t. \tag{2-17}$$

The diagram of Fig. 2-18 includes a nonlinear parametric capacitance C. The value of the capacitance depends on the separation between the rubbing surfaces and the permittivity of the medium filling the crevice between the phases. The differential capacitance and the "single-turn" inductivity of the circuit characterize the state of the electromagnetic field in the frictional capacitor.

The real contact carries a set of looped thermoelectric currents $i_1, i_2, \ldots, i_n$ (Fig. 2-19). Local microcurrents may also flow on the surfaces of the polycrystalline metals, because of the differences in work function on the crystal faces.

Lebedev[114] turned his attention to the fact that, in speaking of real contact between solids in friction, it is necessary to allow for an important feature, pointed out by Frenkel,[115] that must be reflected in the equivalent circuits. This is that electric current may flow both through portions of the surfaces that have approached to interatomic distances on the order of a few angstroms, and through portions separated by distances up to 30 Å. If only the first portions are considered seats of emf, then load resistances should be indicated in the internal circuit of the contact, corresponding to tunnel conduction (on the second portions). In Figs. 2-18 and 2-19, these resistances are not shown; the reason is just that we are considering the sliding region as not only a thermoelectric but also a thermionic converter (that is, as a special thermoelectric converter with a gas cell).[27] In other words, the thermionic component is included in the thermoelectric component of the integral emf. What is more, it would be to the point here to mention the role of mixed electron emission, remembering that the same portion of the surface can serve as a source of thermionic electrons, exoelectrons, field-emission electrons, and so on.

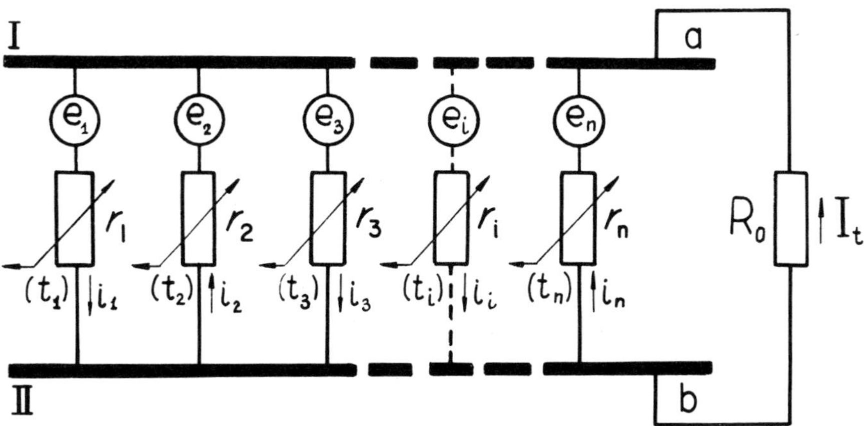

Fig. 2-19. Internal and external thermoelectric circuits of a frictional couple.

The total resistance of the discrete contact is

$$R_m = 1/\Sigma \frac{1}{r_i} \tag{2-18}$$

where the r_i ($i = 1, 2, \ldots, n$) are the resistances of single constrictions. The hot junction of the natural thermocouple represents several thermocouples connected in parallel. If the external circuit of the frictional couple is shorted across the transfer resistance, R_0, and e_i ($i = 1, 2, \ldots, n$) is the emf of the external circuit r_i-R_0, then the instantaneous mean value of the resultant thermal emf is

$$\mathcal{E}_t = \sum_{i=1}^{n} e_i g_i / \sum_{i=1}^{n} g_i \tag{2-19}$$

where g_i ($i = 1, 2, \ldots, n$) is the conductance of the i-th constriction. The overall thermoelectric current is

$$I_t = \mathcal{E}_t/(R_m + R_0). \tag{2-20}$$

If we take the absolute Seebeck coefficients α_I and α_{II} as constants (it is not nearly always valid to do so), then, depending on the types of conduction of the materials, the specific thermal emf of the couple is

$$\alpha = \alpha_I \pm \alpha_{II} \tag{2-21}$$

$$(\mathcal{E}_t = \alpha \, \Delta T).$$

It is generally known that n-type conduction is due to valence electrons; in the excited state, these electrons can move freely into unoccupied energy levels situated somewhat above the Fermi level. At the same time, metals with almost

Table 2-4. Valences and conduction mechanisms for elements contained in the alloys of Table 2-5.

Element	Valence(s)	Intrinsic Conduction Mechanism
C	2, 4	Dielectric
Al	3	*n*
Si	4	Semiconductor
Ti	2, 3, 4	*n*
V	5	*n*
Cr	3, 6	*p*
Mn	2, 3, 4, 6, 7	*n*
Fe	2, 3	*n*
Ni	2	*n*
Cu	1, 2	*n*
Mo	6	*p*
W	6	*p*

Table 2-5. Compositions and conduction mechanisms of alloys containing the elements of Table 2-4.

Alloy	Alloy Chemical Composition (%)												Alloy Conduction Mechanism	
	C	Al	Si	Ti	V	Cr	Mn	Fe	Ni	Cu	Mo	W	Predicted	Found
Tool steels														
R-18	≤0.7–0.8		≤0.4		1.0–1.4	3.8–4.4	≤0.4	Base	≤0.4		≤0.3	17.5–19.0	*p*	*p*
R18F2	≤0.85–0.95		≤0.4		1.8–2.4	3.8–4.4	≤0.4	Base	≤0.4		≤0.5	17.5–19.0	*p*	–
Constructional steels														
30KhGSA	0.28–0.35		0.9–1.2			0.8–1.1	0.8–1.1	Base	≤0.4				*n*	*n*
EI643	0.36–0.43	≤0.1	0.7–1.0	≤0.1		0.8–1.1	0.5–0.8	Base	2.5–3.0			0.8–1.2	*n*	*n*
Stainless steels														
VNS-2	≤0.08		≤0.7	0.1–0.3		14.1–15.5	≤1.0	Base	4.5–5.5	1.75–2.50			*p*	*p*
VNS-5	0.11–0.14		≤0.7			14.0–15.5	≤1.0	Base	4.0–5.0		2.3–2.8		*p*	*p*
Kh18N9T	≤0.12		≤0.8	≤0.8		17.0–20.0	≤2.0	Base	4.0–5.0				*n*	*n*
EI703	0.06–0.12	≤0.5	≤0.8			20.0–23.0	≤0.7	Base	35.0–40.0			2.5–3.5	*n*	*n*
Titanium alloys														
OT4-1	≤0.1	1.0–2.5	≤0.15	Base			0.8–2.0	≤0.4					*n*	*n*
VT5	≤0.05	4.0–5.5	≤0.15	Base				≤0.3					*n*	*n*

filled valence bands have p-type conduction; that is, the principal charge carriers in these metals are vacant energy levels in the Fermi distribution ("holes"). The pure hole conduction mechanism is seen in tungsten, molybdenum, zinc, and some other metals. As for the conduction mechanism in alloys, it obviously depends on the percent contents and intrinsic types of conduction of the chemical elements making up the alloy. As a rule, alloys have a mixed type of conduction; however, either n-type or p-type conduction predominates in any alloy (see Tables 2-4 and 2-5). Which type it is can easily be determined from the sign and magnitude of the thermal emf for different combinations of materials (the author has used, for example, the nearness to zero of the value of α_{Pb}).

Acoustoelectric Component

The action of another seat of emf, whose role is usually not taken into account in the determination of surface-layer temperature by the natural-thermocouple method, is a rather interesting question. This is the "acoustoelectric effect," that is, the appearance of an emf resulting from the entrainment of charge carriers by mechanical stress waves. In the quantum theory, of course, phonons are associated with these waves.

Lebedev, the first to examine this question in application to the friction and treatment of solids,[114] starts from the fact that deformed portions of the rubbing surfaces are sources of elastic and plastic waves.[116] He performs a qualitative analysis of the problem for the example of the propagation of plane waves from an interface, without allowance for reflection and interference. Having made use of the phenomenological formula of Weinreich[117] for the strength of the acoustoelectric field,

$$E = \mu\beta_c I/\gamma v, \tag{2-22}$$

and assuming that the intensity of the plane waves decreases in the direction of propagation in accordance with

$$I(x) = I_0 e^{-\beta x}, \tag{2-23}$$

he obtains the following expression for the potential difference between points with coordinates x_1 and x_2:

$$U_{12} = \frac{\mu\beta_c}{\gamma v\beta} \cdot I(x_1)[1 - e^{-\beta(x_2 - x_1)}]. \tag{2-24}$$

Here μ is the mobility of the charge carriers; β_c is the damping coefficient of waves interacting with charge carriers; γ is the electrical conductivity of the material; v is the velocity of the waves; and β is the damping coefficient of the waves in the material.

Since oscillations and waves of different frequencies are generated in friction, the potential differences determined for all components of the spectrum, with allowance for variations in the acoustoelectrical properties of the material, should finally be summed algebraically. A similar approach is suggested for use in the calculation of potential difference between conjugate bodies, that is, in fact, of the acoustic emf $\mathcal{E}_a$.

Clearly, the intensity of stress waves, and thus the value of $\mathcal{E}_a$, depend on the factors that determine the character and the substance of frictional interaction processes. The same factors determine the frequencies of mechanical waves in the audio range; this follows from the results of a harmonic analysis of the vibrational spectrum of the frictional force.[118]

Galvanic Component

Now let us examine the interaction of metals in the presence of an electrically conductive liquid lubricant. Participation of the electrolyte (as a crevice-shaped capillary interlayer) leads to the formation of potential discontinuities at metal-solution boundaries (Fig. 2-20). As a rule, the values of $_I\varphi_{III}$ and $_{II}\varphi_{III}$ are such that their algebraic sum is not equal and opposite to the contact potential difference $_I\varphi_{II}$. In other words, we have a short-circuited galvanic microcell. This means that it is potentially possible for oxidation-reduction reactions to take place at the electrodes of the cell. The emf of the cell is

$$e_g = {}_I\varphi_{III} - {}_{II}\varphi_{III} - {}_I\varphi_{II}. \tag{2-25}$$

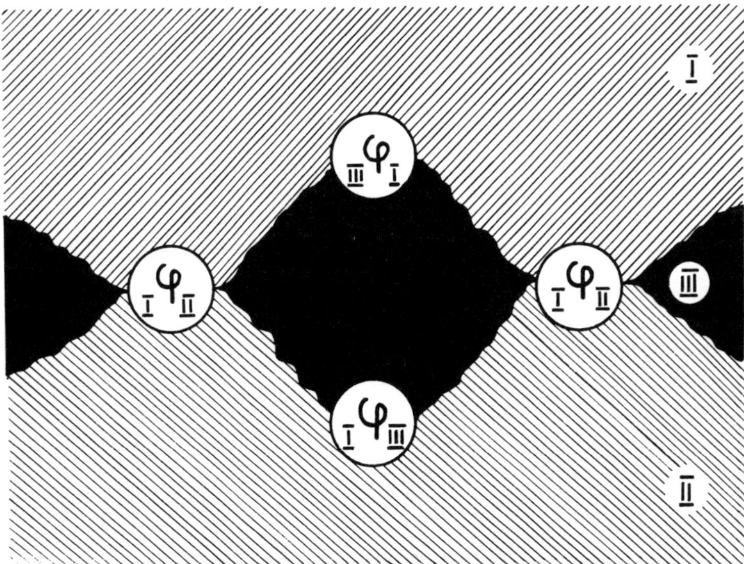

Fig. 2-20. Closed section of the system metal (I)–electrolyte (III)–metal (II).

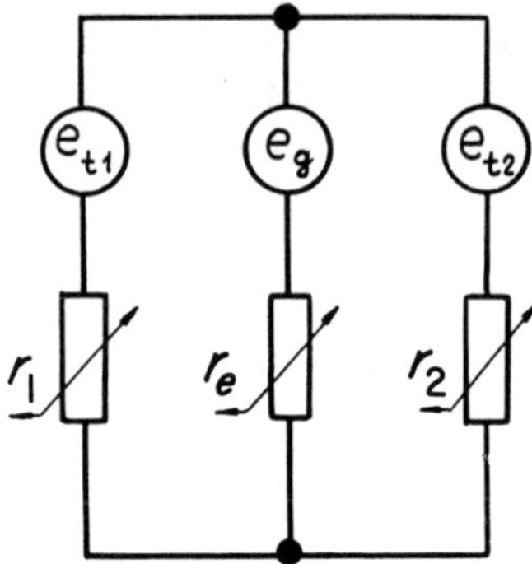

Fig. 2-21. Equivalent electric circuit of a contact cell in relative sliding of metal surfaces in the presence of an electrolyte.

In the relative sliding of surfaces, along with the galvanic emf, thermal emf's will also act (Fig. 2-21). These evidently influence the rates of the electrode reactions, interfacial (surface) tension, and so on.

The author once emphasized that Fig. 2-21 does not indicate the electrical capacitances associated with the interfaces between the components of the frictional couple and the electrolyte solution (double-layer capacitances), their leakage resistances, or the interelectrode capacitance.[15] Melnichenko considered all these points in developing a refined equivalent circuit for a frictional contact applicable to the case of friction with selective transfer (see Section 2.7).[16]

If the electrochemical behaviors of the metals are identical over their entire surfaces and there are no contacting islands at all, then the emf of the galvanic cell is

$$\mathscr{E}_g = \mathscr{E}^0 + \frac{RT}{F} \ln \frac{a_2^{1/n_2}}{a_1^{1/n_1}} \tag{2-26}$$

where $\mathscr{E}^0 = \mathscr{E}_2^0 - \mathscr{E}_1^0$ is the emf of the cell with ionic activities $a_1 = a_2 = 1$; and n_1 and n_2 are the ionic charges.

The mean value of $\mathscr{E}_g$ is determined as the difference between electrode potentials, with allowance for the nonuniformity of the electrode processes on adjacent portions of the surfaces. These values can be judged from the data of Table 2-6.

For the simultaneous action of a galvanic emf, $\mathscr{E}_g$, and a thermal emf, $\mathscr{E}_t$, when the appearance of the thermal cell indicates that the galvanic cell has become short-circuited, the corresponding equivalent circuit for the three-phase frictional system is that of Fig. 2-22. There, R_e is the resistance of the electrolyte; the capacitance C reflects the abilities of the rubbing surfaces to accumulate electrical charges when a third phase is present in the gap of the frictional capacitor. The emf $\mathscr{E}_f$ (potential difference $\varphi_a - \varphi_b$ at $R_0 = \infty$) is a sort of inte-

Table 2-6. Potential differences between specimens placed in different media (in mV).

Medium	Materials of Specimens		Resistance of External Circuit (kΩ)
	Steel-Cu	Steel-Steel	
Pure paraffinic-naphthenic mineral oil (completely free of polar components)	0	0	0.1
Distilled water	1.2	0.3	0.1
3% solution of sodium dodecyl sulfate, $Na[CH_3(CH_2)_{11}]SO_4$	11	5	0.1
3% solution of sodium chloride	238	21	5

gral parameter characterizing the conditions under which the friction (cutting) process takes place. If this emf is recorded with an instrument having no time lag (amplifier A, oscillograph O), then in particular it is possible to obtain information on the state of thin lubricant films on the areas of contact (see Chapters 3 and 4).

Inductive Component

Besides the seats of emf already mentioned, a special role in friction may belong to something resembling a unipolar generator; the frictional couple is transformed into such a generator if at least one of its elements is magnetized. Reasons for the natural magnetization of bodies may include the magnetic fields of nearby ferromagnetics, which have appeared because of saturation of mechanisms and

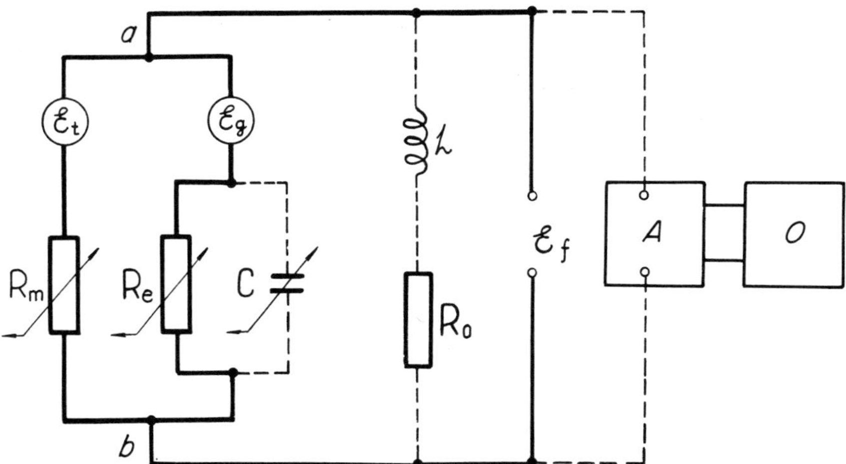

Fig. 2-22. Schematic diagram of the electric circuit of a three-phase frictional system. A = amplifier; O = oscillograph.

devices in the electrical equipment; fields coupled to closed thermoelectric circuits; the elastomagnetic effect (see Section 7.4); the gyromagnetic effect (in very rapid rotation); the geomagnetic field; and so on.

There are two distinguishing features of the emf due to electromagnetic induction $\mathcal{E}_i$ as opposed to $\mathcal{E}_t$, $\mathcal{E}_a$ and $\mathcal{E}_g$. First, the existence of $\mathcal{E}_i$ does not depend on direct contact between the bodies or the presence of a lubricant phase. Second, $\mathcal{E}_i$ is induced in the bulk of the sliding body that moves relative to the magnetized counterface (the source of the inhomogeneous field). Therefore, Fig. 2-22 must not show the seat of $\mathcal{E}_i$ connected in parallel with the seats of $\mathcal{E}_t$ and $\mathcal{E}_g$, as it could show, for example, the seat of $\mathcal{E}_a$.

Without stopping here to refine some of the details (it will be much easier to do this later), we turn to an essential characteristic of a frictional contact, the electrical resistance.

2.4. CONTACT CONDUCTANCE

Metals have finite electrical conductivities because of thermal oscillations and imperfections (e.g., vacancies and impurities) in their lattices. If the interatomic separation becomes smaller, the Bornian repulsive forces between particles rise considerably faster than do the attractive forces. For this reason, external pressure has little effect on the electrical properties of solids, provided, of course, that polymorphous transitions do not occur upon compression. According to Holm, the contact surface that takes up the applied force is only partly conductive.[119] Current flows chiefly through "a-spots," whose appearance is due to the destruction of screening oxide films or to tunneling through the very thin adsorbed and passivating films; these processes provide a quasimetallic contact. Because the a-spots are small, the current lines become curved and hinder the free migration of charge at the interface (Fig. 2-23). This curvature results in the appearance of an additional resistance, called constriction resistance, in the electrical circuit. Destruction of the oxide films requires that the contact

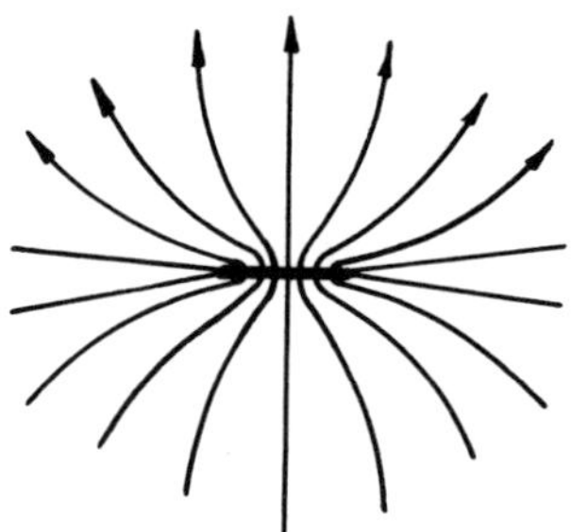

Fig. 2-23. Simplified model of the constriction of current lines in a conductive a-spot (after R. Holm).

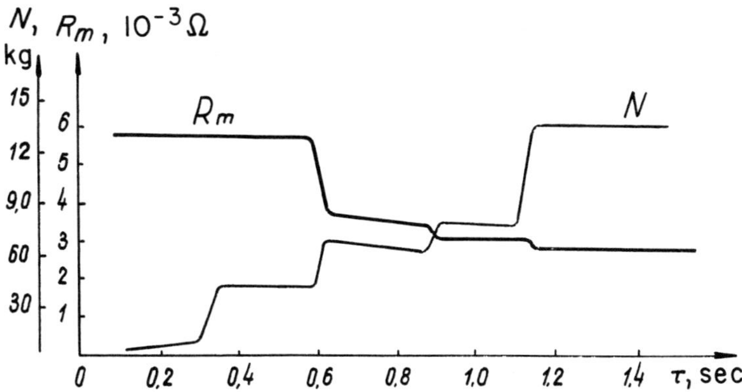

Fig. 2-24. Record of contact resistance and normal load in the compression of aluminum specimens.

pressure at points where the films are damaged reach the yield limit. Therefore, measurements of the electrical conductance of the contact make it possible to gain information about the change in the character of deformation of metals, from elastic to plastic. This is easy to demonstrate for the example of the destruction of the strong oxide film* that covers soft aluminum (Fig. 2-24). Interest in measurements of the interfacial electrical conductance is growing because this quantity, characterizing the state of the contact (which, in general, is physically nonuniform), thus becomes a measure of the real area of contact.[121] Even after careful degassing of a clean metal surface, adsorbed or chemosorbed layers of oxygen atoms remain. Upon the formation of the a-spots, however, current flows through them almost as if it were flowing between different crystallites in a compact metal. This is why recording the resultant constriction resistance makes it possible to trace the establishment of a metallic contact between surfaces and to obtain a numerical estimate of its real area.

The electrical conductance of a contact was investigated on a special friction test machine built to the author's draft design at the Cavendish Laboratory. The following units are mounted on a two-level welded frame (Figs. 2-25 and 2-26): a composite axle with the friction disk fixed to it; a lever-type loading device; a phototachometer; a mercury contact; and an electric drive motor communicating rotary motion to the axle through a variable-speed transmission. An adjustable-voltage system provided a smooth start and speed control for the rotating element of the frictional assembly. Cutting fluid was supplied from a reservoir through tubes in the framework of the loading device. A glassed enclosure with a drain cock prevented the liquid from splashing away when sliding took place.

*Data on oxide hardness and the critical loading, at which the destruction of oxide films on electrolytically polished surfaces begins, are given by Wilson.[120]

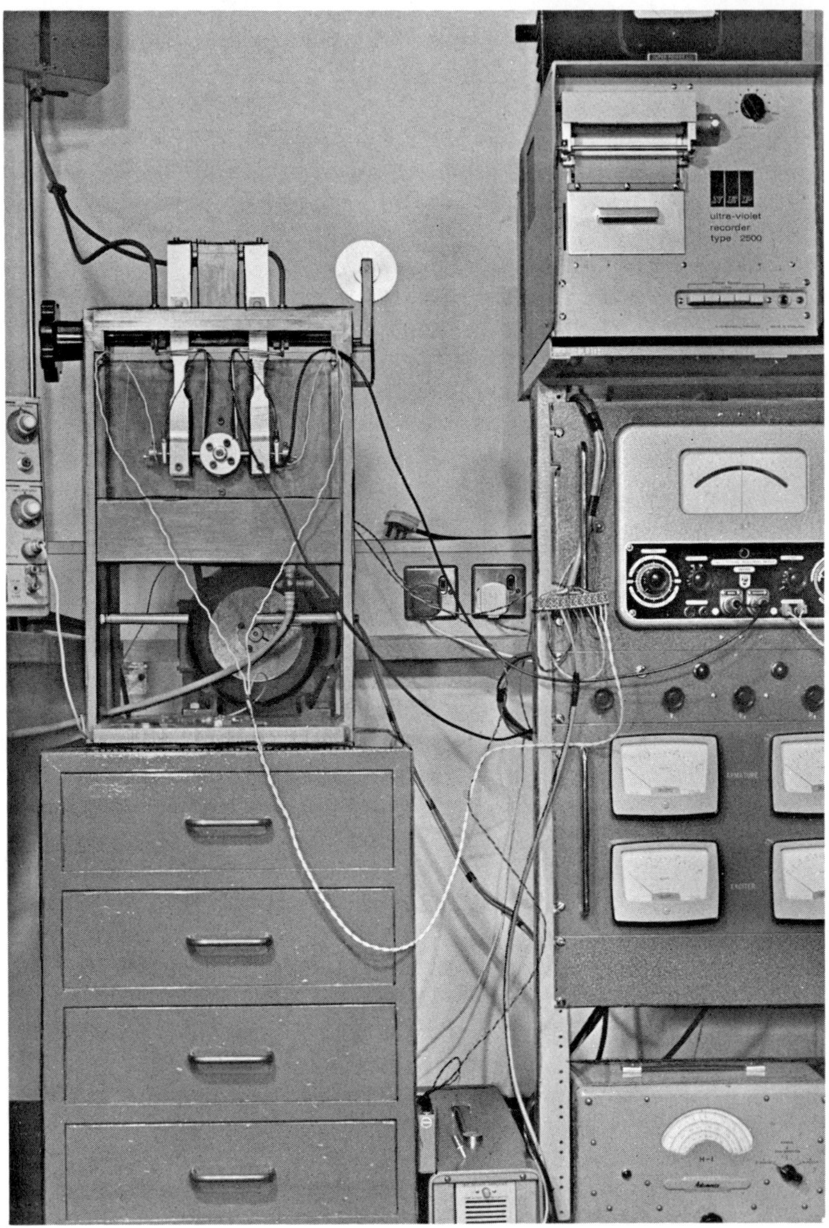

Fig. 2.25. Overall view of the friction test setup.

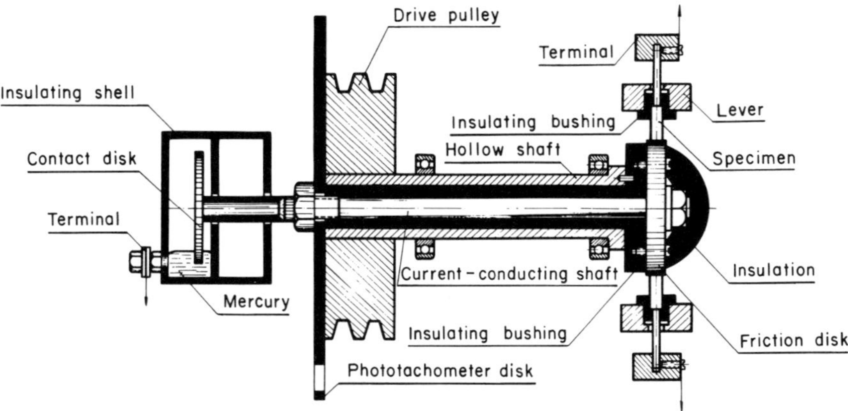

Fig. 2-26. Cutaway drawing of the frictional assembly.

The rubbing specimens, the current-conducting shaft, and the electrical terminals were electrically well insulated from the rest of the apparatus. The normal load N was applied to the specimens by means of a pilot wheel. A screw carried nuts with left-hand and right-hand threads; when the screw was turned, slides rigidly attached to the nuts and bearing vertical levers moved in opposite directions. The specimens, mounted to the lower arms of the levers, made contact with the surface of the friction disk at practically the same time. This happened at the moment when balls sunk in the upper arms of the levers moved right against the ends of a spacing cylinder. The cylinder was free to move horizontally; thus, the nominal pressures acting on the two areas of contact were balanced, provided the disk specimen did not move. When sliding took place, the moments of the frictional force about the axis of rotation of the levers partly upset the symmetry of loading. The degree of asymmetry in the frictional forces and normal loads was determined both by calculation and through check trials. Information on the behavior of the normal load could arrive at the input to the measuring bridge from any one of the strain gages situated in recesses in the sections of reduced cross section in the lower lever arms.

At the outset (in the first series of trials), a bidirectional contact was set up between stationary specimens, one disk-shaped and two cylindrical. (The cylindrical specimen had a radius of 4 mm; the disk had a thickness of 10 mm and a radius of 25 mm.) The surfaces were prepared (the thicker oxide films were removed and the surfaces were degreased) with grade 0000 abrasive paper and acetone.* In order to estimate the nominal dimensions of each contact area, it can be assumed as a guideline that the length equals the diameter of the cylindri-

---

*Mechanical polishing with highly dispersed abrasive and dissolving of the multimolecular portion of the adsorbed layers assured adequate reproducibility of the starting physicochemical properties and the state of the sliding surfaces.

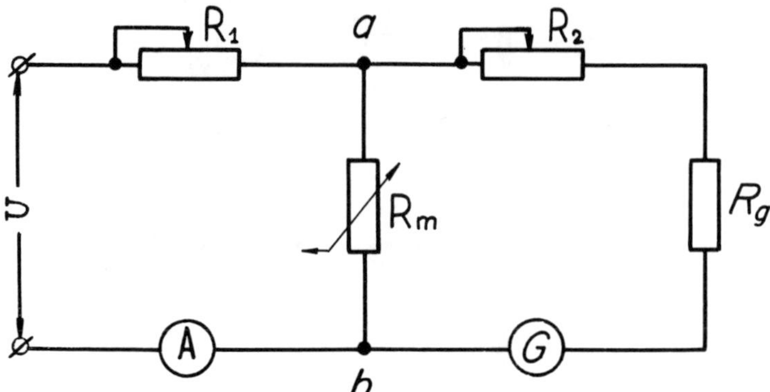

Fig. 2-27. Schematic diagram of the circuit for measuring contact conductance.

cal specimen and the width equals the diameter of a single spot. This diameter is calculated by the formulas of Hertz (elastic contact) and Tresca (plastic contact).

Fig. 2-27 gives a schematic diagram for the trials aimed at determining the electrical conductance of a contact. In that figure, R_1 is a variable input resistor; R_2 is an adjusting resistor; R_g is the resistance of the galvanometer coil; and R_m is the resistance of the contact.* Since the contact shunted the input of the most sensitive galvanometer (1.6 μA/cm; 72 μV/cm) of the six-channel recording galvanometer, the instrument reacted to the tiniest separation between specimens. An elementary calculation of the diagram yields the following expression for the contact conductance:

$$G = U/I_g R_1 (R_2 + R_g),$$

where I_g is the current flowing through the galvanometer and U is the voltage at the terminals of the source.

Curves of $G = f(N)$ were obtained for four couples of metals: copper-copper, molybdenum-copper, steel-copper, and constantan-copper (Figs. 2-28[a], [b], [c], and [d], respectively). We will show that an effective metallic (or quasi-metallic) contact exists on the portions of the surface that take up the applied force. If γ_1 and γ_2 are the electrical conductivities of the metals in contact, then the contact conductivity is

$$\gamma = \gamma_1 \gamma_2 / (\gamma_1 + \gamma_2).$$

*Here and below, the author omits the term *bidirectional*. The cylindrical specimens were in physically equivalent conditions, so that the character of the function $R_m = f(N)$ does not depend on which is being measured: the total resistance R_m or the resistance of one contact region $R_m/2$. Series connection of the two interfaces just multiplies the effect under study.

Table 2-7.

Couple Described by $G'(\gamma')$	Couple Described by $G''(\gamma'')$	G'/G''	γ'/γ''
Cu-Cu	Constantan-Cu	12.8	14.8
Cu-Cu	Steel-Cu	7.3	7.6
Cu-Cu	Mo-Cu	2.1	2.2
Steel-Cu	Constantan-Cu	1.7	1.9
Mo-Cu	Steel-Cu	3.5	3.5
Mo-Cu	Constantan-Cu	6.2	6.2

For two arbitrary couples, if we compare the ratio of conductances G'/G'' obtained by experiment with the ratio of electrical conductivities

$$\frac{\gamma'}{\gamma''} = \frac{\gamma_1'\gamma_2'}{\gamma_1''\gamma_2''} \cdot \frac{\gamma_1'' + \gamma_2''}{\gamma_1' + \gamma_2'},$$

we see in Table 2-7 that the values agree (to within the limits of error). This fact indicates that current flows through a given interface without encountering a marked transfer resistance, as if it flowed between different crystallites in a compact metal.[14]

Rough calculations show that complete destruction of the oxide films (above all, tarnish films on copper) took place at a load of 6 kg.

Reasons for the formation of metallic spots include not only the destruction of films at the contact, but also fretting.* Thus, it is natural that contact conductance improves significantly with a rise in voltage (Fig. 2-28).

It is known that the relation between electrical conductance and load has the form $G = kN^x$, where k depends on the mechanical characteristics and conductivity of the material and the geometry of the asperities on the contact surface.[121-123] If we follow Bowden and Tabor, then we can assess the state of a contact from the value of the exponent x: x is a constant equal to $\frac{1}{3}$ for elastic or $\frac{1}{2}$ for plastic contacts.[121] In Kragelsky's judgment, however, x is a function of the load and decreases, as a rule, with increasing N.[122]

The author and Yashina used the following method to determine x and k:[14]

1. Taking x constant and equal to x', we find its value for two points chosen arbitrarily:

$$G_1 = k'N_1^{x'} \quad \text{and} \quad G_2 = k'N_2^{x'}.$$

2. We assume that the corresponding value k' is a constant over the whole range of loads, and construct the graph of the function $x' = \Psi(N)$.

*According to Holm, true fretting means the destruction of bonds in a dielectric with the formation of a conductive channel inside the film.[119]

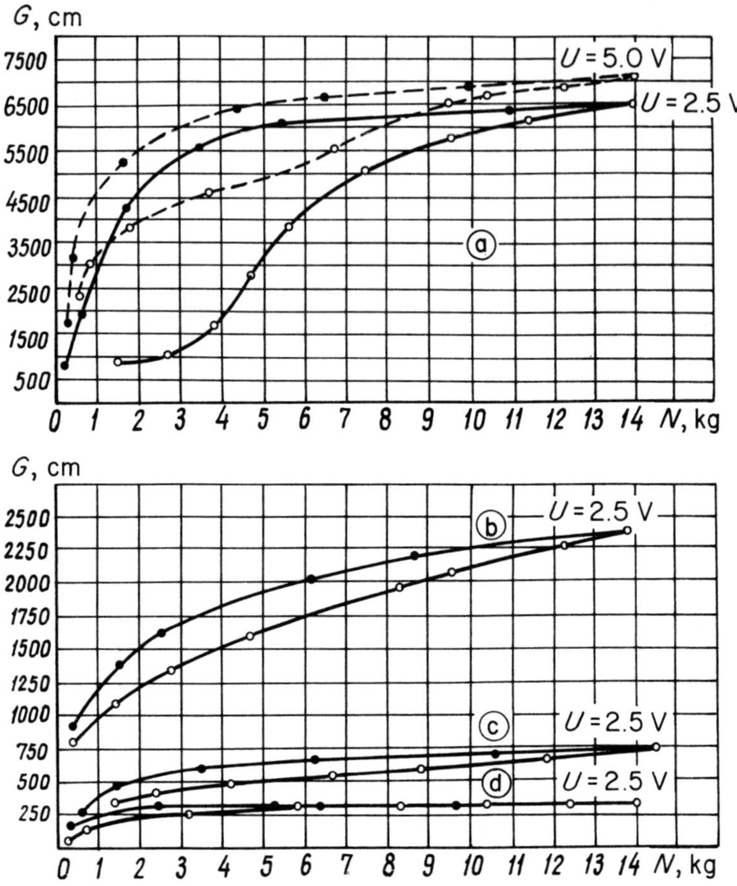

Fig. 2-28. Contact conductance as a function of normal load.

3. We pick out the interval on which x' and thus x are indeed constants; we determine the x value on this interval and the true value of the coefficient k.
4. We find x for all remaining points.

Figure 2-29 presents curves of $x = \Phi(N)$ for the metal couples under consideration. As we see, the exponent in the relation between conductance and pressure remains constant on some interval of loads. The instability in x at small loads is due to the presence of films on the contact areas. The explanation for the decrease in x at large loads is that, as the real area of contact increases, its conductance becomes "saturated" and

$$\frac{dG}{dN} = kN^{x-1}\left(x + N \ln N \cdot \frac{dx}{dN}\right) = 0.$$

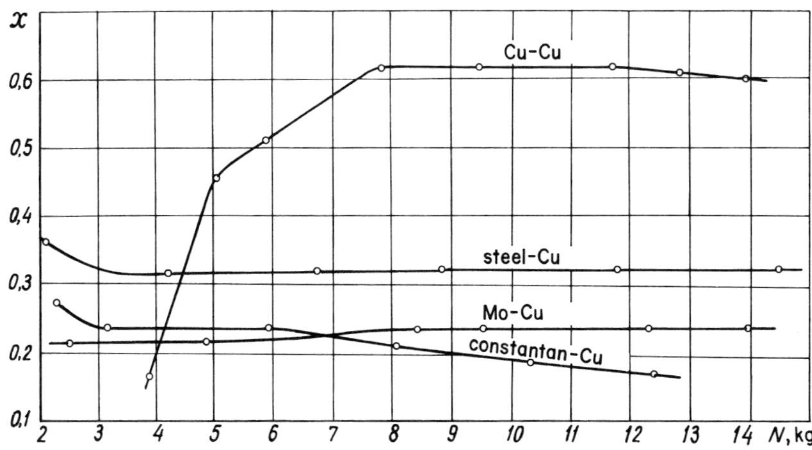

Fig. 2-29. Exponent x in the conductance-load relation as a function of the normal load.

Hence, $x = -N \ln N \dfrac{dx}{dN}$ or, after integration,

$$x = \text{const} \cdot \frac{1}{\ln N}.\tag{2-27}$$

Thus, $\Delta x < 0$ for $\Delta N > 0$ in the region under consideration.

In conclusion, we point out the hysteresis in the conductance curves (Fig. 2-28). If the contact conductance increased comparatively rapidly with rising pressure, then on a smooth release of the load, it decreased significantly more slowly, so that the points on the return curves did not lie on the initial curves. This phenomenon, in all likelihood, is caused by adhesion between the surfaces followed by strengthening of the adhesion joint. The tendency of the metals to cohere can be assessed from the ratio of the area of the loop S_1 to the area S_2 bounded by the initial curve (Table 2-8).

The development of adhesion and the possibility that a cohesion joint will form obviously depend on the plasticities of the materials and the cleanness of the conjugate surfaces.

Another subject for study was the contact of contaminated surfaces bearing

Table 2-8. Area ratios of hysteresis loops S_1/S_2, for N = 6–14 kg.

Couple	S_1/S_2
Cu-Cu	0.154
Steel-Cu	0.073
Mo-Cu	0.051
Constantan-Cu	0

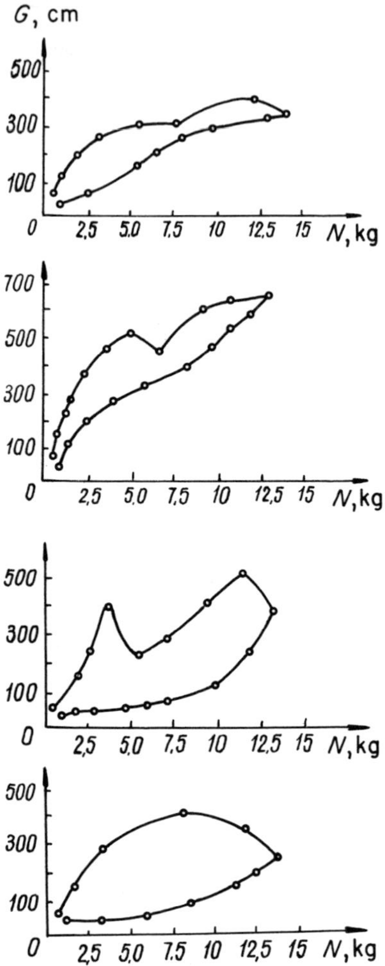

Fig. 2-30. Hysteresis in the contact-conductance curves under cyclic load variation, without preliminary cleaning.

boundary lubricant layers of high-molecular-weight fatty acids. Here, the function $G = f(N)$ was seen to behave far from monotonically upon unloading of the couples (Fig. 2-30). The "distension" of the curves is paradoxical at first glance, and the reason for it is not yet altogether clear, but it is appropriate to suggest that its roots lie in the allotropic rearrangement of the structure of the contaminant layer (see Section 3.2).

In the second series of trials, the variation of contact conductance was recorded in the friction of specimens, where the conditions of sliding were severe, approaching those in the flow of a chip over the contact faces of a tool.* The resistance R_2 (see Fig. 2-27) was selected so that the current flowing through the galvanometer coil would not exceed a permissible value even if the rubbing surfaces separated completely. In practice, this means that the sensitive galvanometer was adjusted so as to detect possible disturbances, and equally the subsequent restorations, of the actual contact between the metals.

The records of R_m showed that the principal reason for significant weakening of the contact, up to the point of breakdown, is vibrations. The cylindrical specimens vibrated both transversely to the vector v (elastic vibrations in the lever system) and in the longitudinal direction (sliding with skips). One phase of each vibration was accompanied by a reduction in the applied compressive force, whereas the deflection of the specimen in the opposite direction resulted in increased pressure on the disk.

The aim of the present work does not include analyzing the reasons for mechanical relaxational vibration in frictional systems. Therefore, we will limit ourselves to specific remarks on the features of the discontinuous motion of rubbing bodies.

The author shares completely the view that changing the mechanical parameters of the system (for example, increasing the stiffness) cannot completely eliminate relaxational vibrations of the elements of frictional couples.[124] One of the main reasons for the occurrence of noticeable vibrations is the decreasing character of the $F(v)$ curve on individual sections.† To argue for this supposition, advanced by Kaidanovsky and Haikin,[125] we point out the following experimental fact. If the pilot wheel used to apply the load remained at some intermediate position, then, as the specimens wore down, the resistance to sliding would decrease considerably, and the disk, freed from the "embrace" of the cylindrical specimens, would speed up accordingly. Similar conditions in friction are extremely conducive to self-excitation and the development of high-frequency vibrations in a light indenter.

Let us examine the records of contact electrical resistance (1) and normal load (2) in Fig. 2-31. First, we should remark that at some values of N there were something like antinodes in the vibrations. These alternated with sections where the amplitude of the vibrations fell off severalfold (Fig. 2-31[a]). These burstwise vibrations are probably connected with repeated breaks in the buildup formed at the surface of the slider. The transfer of copper to steel and constantan, for example, went so rapidly that the usual stage of surface spreading of the metal went over to the formation of a foreign solid (Fig. 2-32). Another reason

---

*Such specific conditions are of course not typical of the operation of rubbing couples in mechanisms and devices, but rather correspond to the beginning of catastrophic wear of these units. For this reason, they are of interest.

†By the elementary theory of friction, the $F(v)$ curve as a whole is complex in character (see Section 1.4).

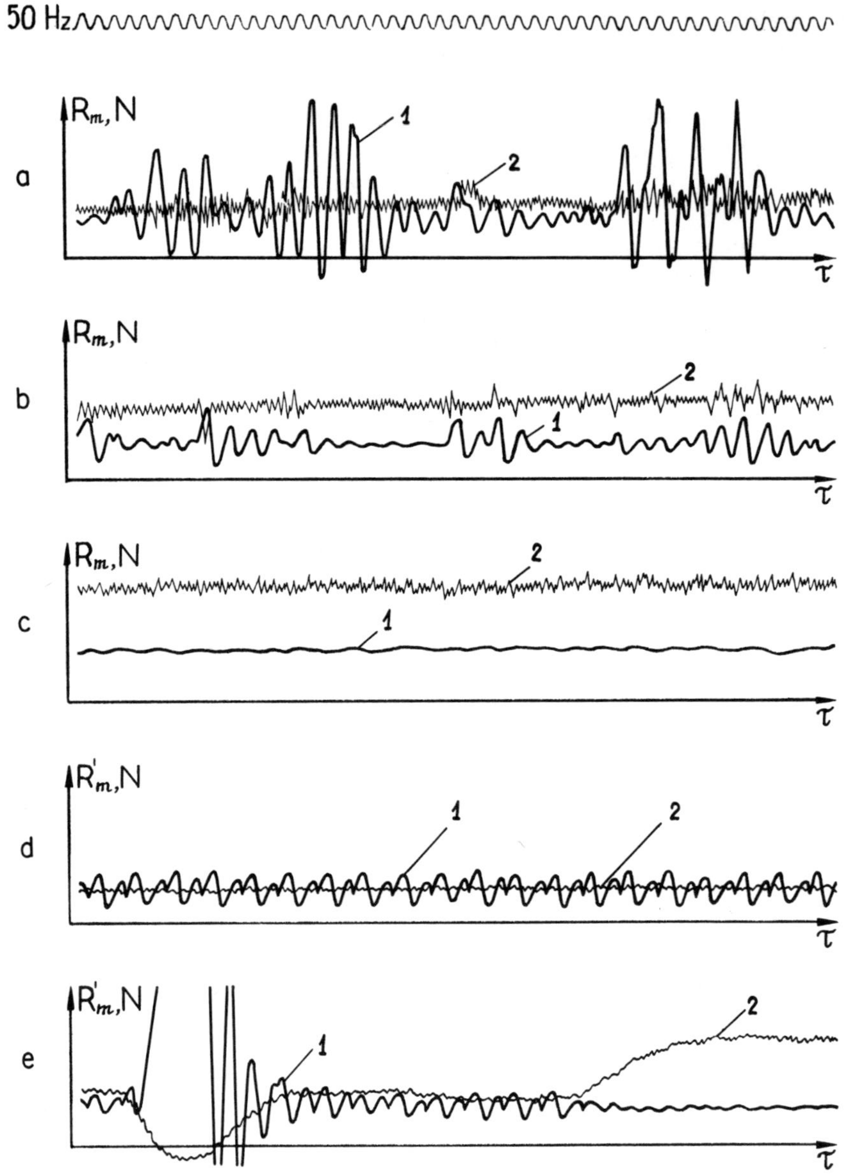

Fig. 2-31. Records of the contact resistance and normal load in friction, reflecting the effect of lubrication on the character of the frictional oscillations.

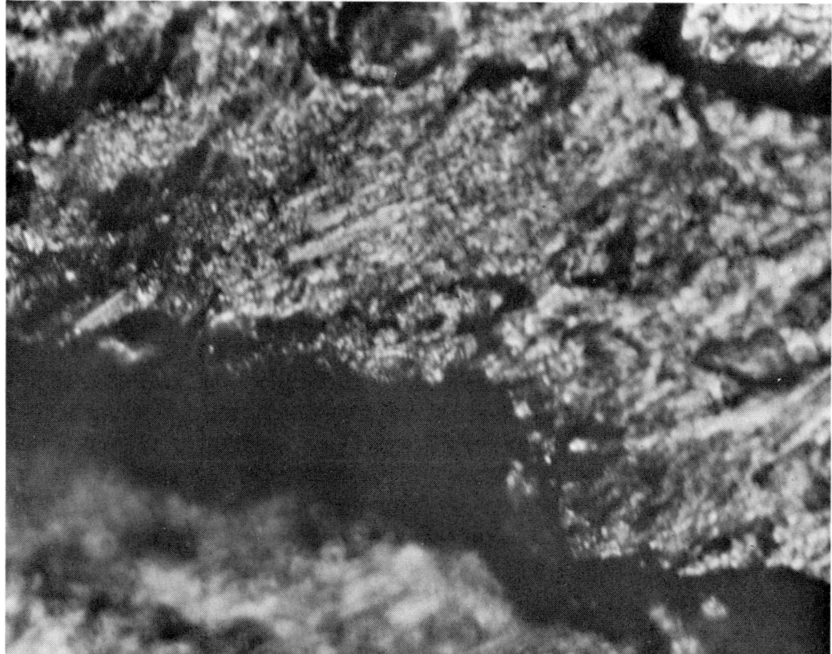

Fig. 2-32. Crustlike buildup of copper on a constantan surface (200×).

for the sharply pronounced relaxational vibrations may be the cutting (or shearing) of prominently sized individual asperities (Fig. 2-33). As the load increases, the vibrations become weaker (Fig. 2-31[b]) and finally almost disappear (Fig. 2-31[c]). Incidentally, records of N obtained with a sensitive input device, such as a strain gage, are themselves an excellent source of information on burstwise transverse vibrations. If the response time of the recording channel is not limited by the time lag of the galvanometer, then the load-variation curves can even be used to estimate the period of the longitudinal relaxational vibrations, which depends on the stick time in the stick-slip process.[26] A stable lubricant film applied in an arbitrary way causes a radical change in the type of vibrations that occur in friction: the film damps the high-frequency vibration of the cylindrical specimens, whereas the low-frequency vibrations become more regular (Fig. 2-31[d]). If the load is artificially removed and then smoothly restored, it is not difficult to find a value of N at which these vibrations are especially distinct (Fig. 2-31[e]). The author has compared records corresponding to the resonance state in a frictional system and to emergence from this state, in the presence and absence of lubricant.[26] The results suggest that the lubricant "shifts" the spectrum, with its principal frequency of vibration, in such a way

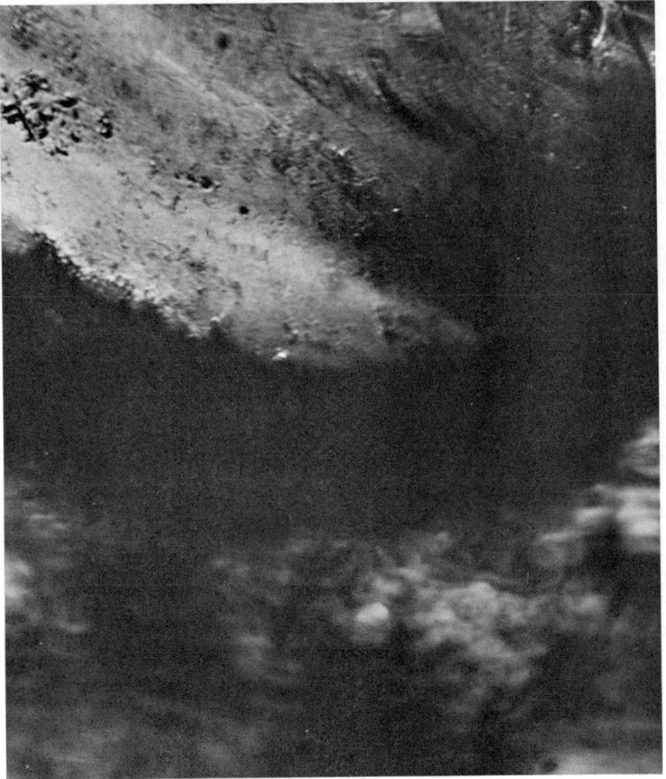

Fig. 2-33. Solidified micro-ingot on a copper disk specimen after friction against constantan with scratches (200×).

that, at relatively high sliding speeds, the mechanical parameters of the system as a whole are determined by the parameters of the subsystem, disk-shaped specimen/axle. The boundary lubricant layer assumes the role of something like a semiharmonic analyzer, selecting from the external influence on the indenter the characteristic periodic oscillations of the massive element.

The author by no means rules out the possibility that at small, and sometimes even moderate, loads, an experimental situation favoring hydrodynamic and mixed friction could arise. This offers even more convincing proof that Tolstoy and Kaplan's conclusion is correct: "In dry or pure boundary sliding friction, as in mixed friction, the freedom of the normal motions of the slider is an obligatory condition both for a decreasing velocity characteristic of the frictional force and for frictional auto-oscillations."[126]

The value of R_m depends on the temperature of the constrictions in the sliding process, and on the crystallographic matching and cleanness of conjugate surfaces. We have already noted that, because of the tunnel effect, current readily passes through thin adsorbed and passivating films. A monomolecular layer of organic lubricant in the boundary state has a finite, but comparatively small, transfer resistance. Well-defined oxide films are semiconductors with high resistivity, so that under ordinary conditions they act as insulators. In friction, however, semiconducting ingredients may have a marked effect on the electrical conductance of the contact: heating of the contact sharply increases the concentration and kinetic energy of free charges in the semiconductors. It is also important that under severe conditions of sliding, the preservation (not to mention the new formation) of a well-defined oxide is often impossible. Despite the rapid regeneration of oxides on denuded contact surfaces, which contributes to the vibrations of the rubbing bodies, the oxide films may vary widely in thickness, structure, chemical composition, and other features. If we take the data of Fig. 2-34 into account and remember that the resistivities of semiconductors vary from 10^{-5} to 10^{10} Ω-cm, while those of dielectrics lie in the range of 10^{10}-10^{15} Ω-cm,[127] then we can consider the range of variation of R_m unbounded $(0, \infty)$. But most often, even if a contact does exist, its resistance does not exceed hundredths or even thousandths of an ohm. The resistance is the less, the larger the dimensions of the conductive a-spots on the surfaces taking up the applied force.

The considerations developed in this section have a direct link with an important rheological feature of the contact: the change in constriction resistance when a comparatively large electric current flows through the interface.

For any given current, there exists a certain critical constriction resistance.[128, 129] As long as the real area of contact gives a constriction resistance not exceeding the critical value R_c, the current will not affect the state of the permanent metallic contact. But as soon as the constriction resistance becomes larger than R_c, the metal in the contact areas begins to soften. Because of the rapid evolution of heat in the immediate vicinity of the interface, physicomechanical collapse of the specimens takes place, and the constriction resistance declines to the value R_c as the conductive spots grow (Fig. 2-35). The critical value of constriction resistance varies in inverse proportion to the maximum current flowing across the contact (Fig. 2-36). The value does not depend on the length or shape of the current pulses when clean metal surfaces are pressed together. Thus,

$$R_c \cdot I_{\max} = C \tag{2-28}$$

where C is a constant equal to the maximum possible voltage drop across the constriction resistance; C depends on the material comprising the elements of the homogeneous couple. The rate of collapse of the specimens is determined by the rate of heating of the contact zones and by the variation of mean yield pressure in the metals. The growth in the area of real contact is particularly

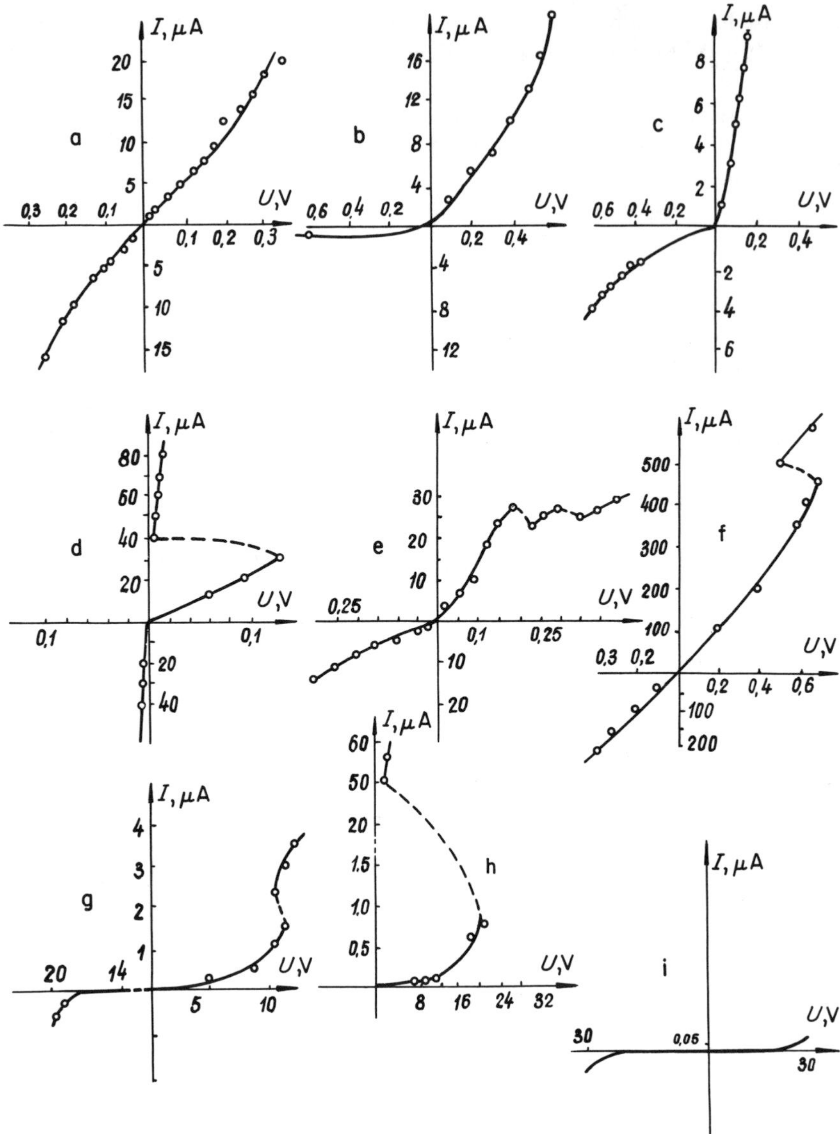

Fig. 2-34. Current-voltage characteristics of oxide films formed on alloy surfaces after 1-hour holding in a muffle furnace at various temperatures (recorded by V. S. Bykova and N. P. Bobrova): (a, b, c, e) type-U8A alloy, 350 °C; (h) type-U8A alloy, 450 °C; (d) type-VNS-5 alloy, 200 °C; (f) type-20 steel, 650 °C; (g) type-OT-4 alloy, 520 °C; (i) type-30KhGS alloy, 600 °C.

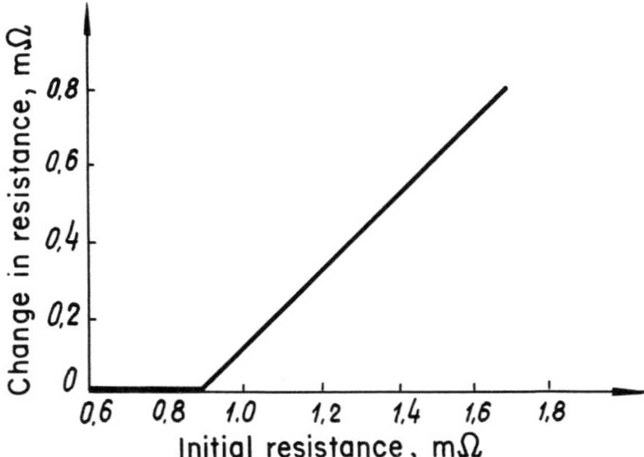

Fig. 2-35. Effect of current pulses (156 A) on contact resistance of pure gold specimens (after F. P. Bowden and J. B. P. Williamson).

marked when a large current flows through the interface, causing practically instantaneous melting of the contact junctions (Fig. 2-37). If the contact is broken, the metal cools and resolidifies, but the fused spots retain the same dimensions.

The device just described naturally enabled the author to trace the effect of an electric current on the conductive contact junctions (Fig. 2-38 [a] and [b]). It is known, however, that effects having to do with the flow of natural electrical microcurrents through the interface between rubbing solids are, as a rule, weakly marked but cumulative. Thus, results obtained when an external source is used to pass the desired current continuously through a contact must not be extrapo-

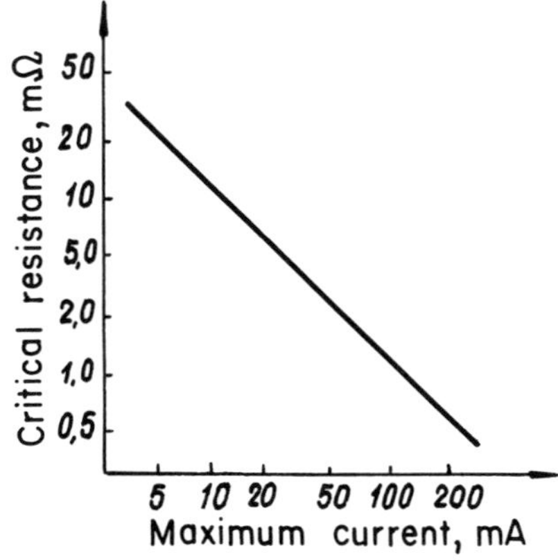

Fig. 2-36. Log-log plot of constriction resistance and maximum electric current through the constriction.[129]

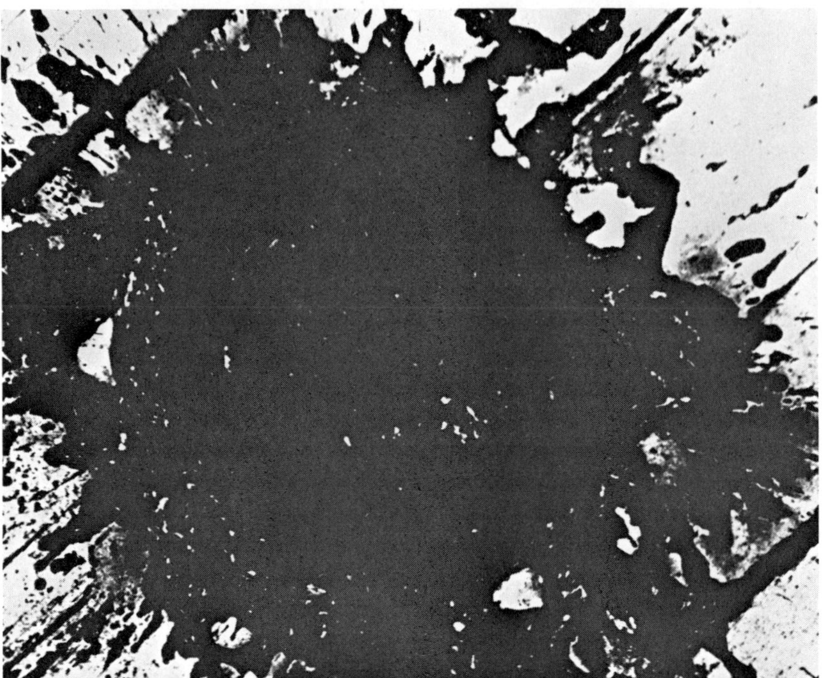

Fig. 2-37. Photomicrograph of a gold surface, showing damage in the contact region due to the passage of a large current under a load of 10 g (120X).[129]

lated to another source (e.g., a natural frictional thermocouple) without additional argument to show that such large thermoelectric currents can flow in friction (cutting). We will take up this question later on.

2.5. FEATURES OF THERMOELECTRIC PHENOMENA IN FRICTION

The measuring unit of the system in Fig. 2-25 included an ultraviolet recording galvanometer. Along with the normal load and sliding speed, this instrument recorded the potential difference between the elements of the frictional couple. When the sensitivity of the galvanometer was not great enough, the signal to be measured was amplified. The metals and alloys used differed in their physico-mechanical properties: aluminum, copper, molybdenum, constantan, and carbon and alloy steels.[26] If the internal voltage drop of the source is neglected, then

Fig. 2-38. Photomicrographs of aluminum surfaces after the passage of direct currents of 5 A (a) and 10 A (b), illustrating the phenomenon of fused contact spots (200X). The load was 2,500 g.

(a)

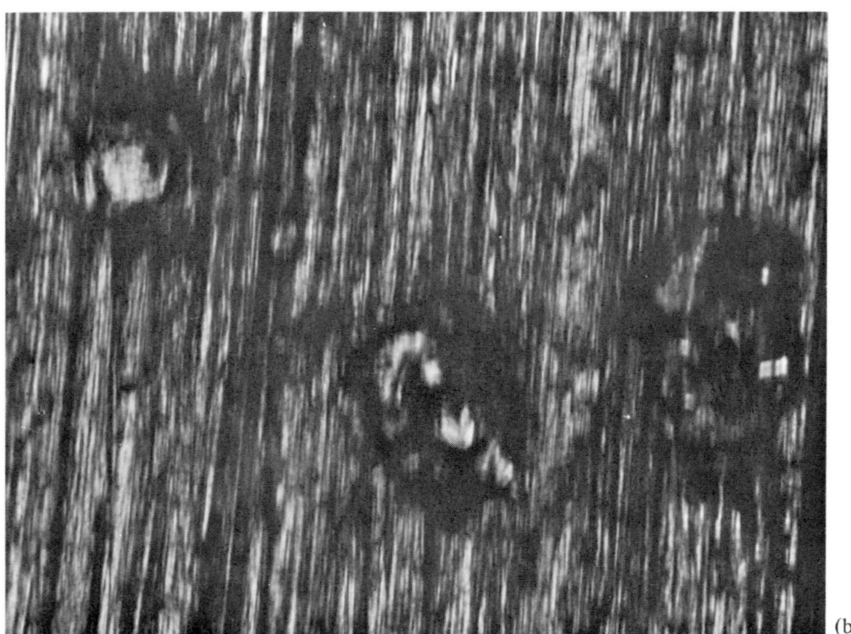

(b)

Fig. 2-38.

the potential difference being measured can be treated as the integral electro-motive force of the frictional system. From what has been said, it is clear that the integral emf, $\mathcal{E}_f$, is due both to thermoelectric and thermionic phenomena at the metal-metal interface, and to the acoustoelectric effect, the Kramer effect, and possibly other factors. However, the entrainment of charge carriers by mechanical stress waves was treated as a second-order effect, as was low-temperature electron emission from the solid surfaces as a result of deformation, structural transformations, and adsorption-oxidation and other physicochemical processes (these were discussed in Section 2.2). Therefore, it was assumed[27] that $\mathcal{E}_f \approx \mathcal{E}_t$ and that, as the records showed, the geomagnetic field, the magnetic field of the motor, and the stray thermal emf's at connections 1, 2, and 3 (Fig. 2-39) had no significant effect on $\mathcal{E}_t$. (The author will later use precisely the same simplification when it cannot affect the results of the qualitative analysis.)

When the conditions of sliding are severe, near those in the flow of a chip over the contact faces of a tool, the average thermal emf, $\mathcal{E}_t$, is 0.15–15 mV. It varies substantially as a function of sliding speed (Fig. 2-40[a]) and displays a tendency to rise in proportion to the normal load (Fig. 2-40[b]). On unloading of the specimens, the $\mathcal{E}_t = f(\tau)$ curve is significantly displaced in time with respect to $N = \varphi(\tau)$ (Fig. 2-40[c]); the author suggests that this behavior is a direct result of adhesional-cohesional hysteresis (see the hatched region of Fig. 2-40[c]).

Figure 2-41(a) shows records of the emf and normal load for friction between specimens of bright-drawn carbon steel (British Standard 970:1955, Specification En8) and a copper disk (steel-copper-steel arrangement). It is easy to see that the maximum value of $\mathcal{E}_f(\mathcal{E}_t)$ decreases on each successive loading of the frictional couple. As the rubbing bodies heat up, $\mathcal{E}_t$ varies widely, up to a change of sign (polarity change of the seat). An inversion of the thermal emf occurs.[130]

Since the sliding conditions selected were very severe, it can be assumed that

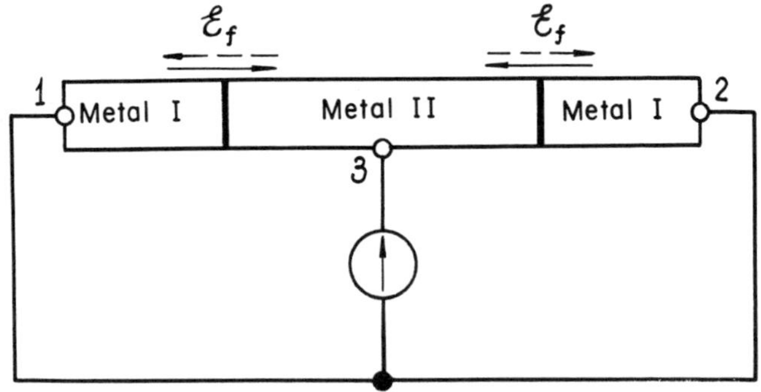

Fig. 2-39. Circuit for measuring emf.

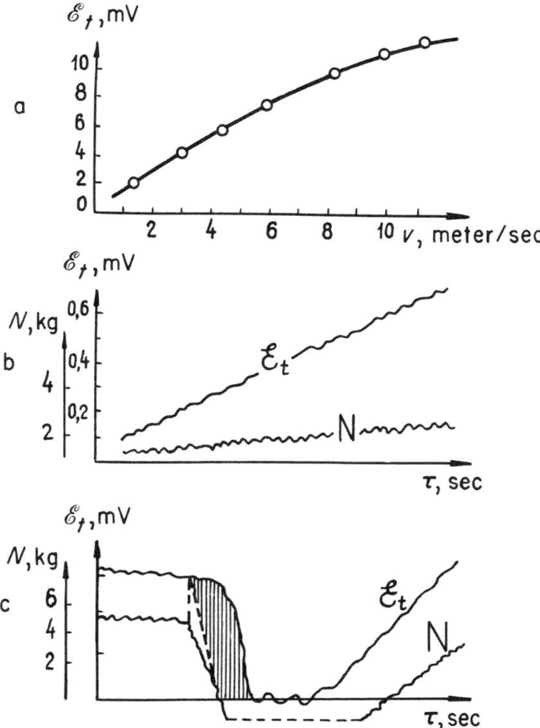

Fig. 2-40. Records of averaged thermal emf in the friction of metals.

the thin oxide films on the contact areas were broken up almost instanta-neously.[120] If we assume that the oxide films are partly preserved on the rubbing surfaces, their effects on the thermoelectric characteristics of the materials are not always significant.[131] The change in polarity of the thermoelectric cell ob-served on the sixth loading is evidently due to the effect of diffusional oxidation on the thermoelectric properties of the metals.

The initial sign of the emf is determined by the positions of copper and steel in the thermoelectric series (steel is positive to copper). Rapid heating of the steel specimens (coefficient of friction $\mu \sim 0.4$-0.8) activates the diffusion process. Asperities on the steel surfaces plow through the softened copper (Fig. 2-42); melting of the copper makes it difficult for oxygen to reach the friction zone. For this reason, the diffusional oxidation of steel goes most rapidly when the specimens break away from the disk as the load is removed. Diffusion of oxygen into the steel brings about a significant change in the Seebeck coefficient (up to a change in sign).

The value of $\mathcal{E}_t$ also depends on the temperature gradients in the disk-shaped

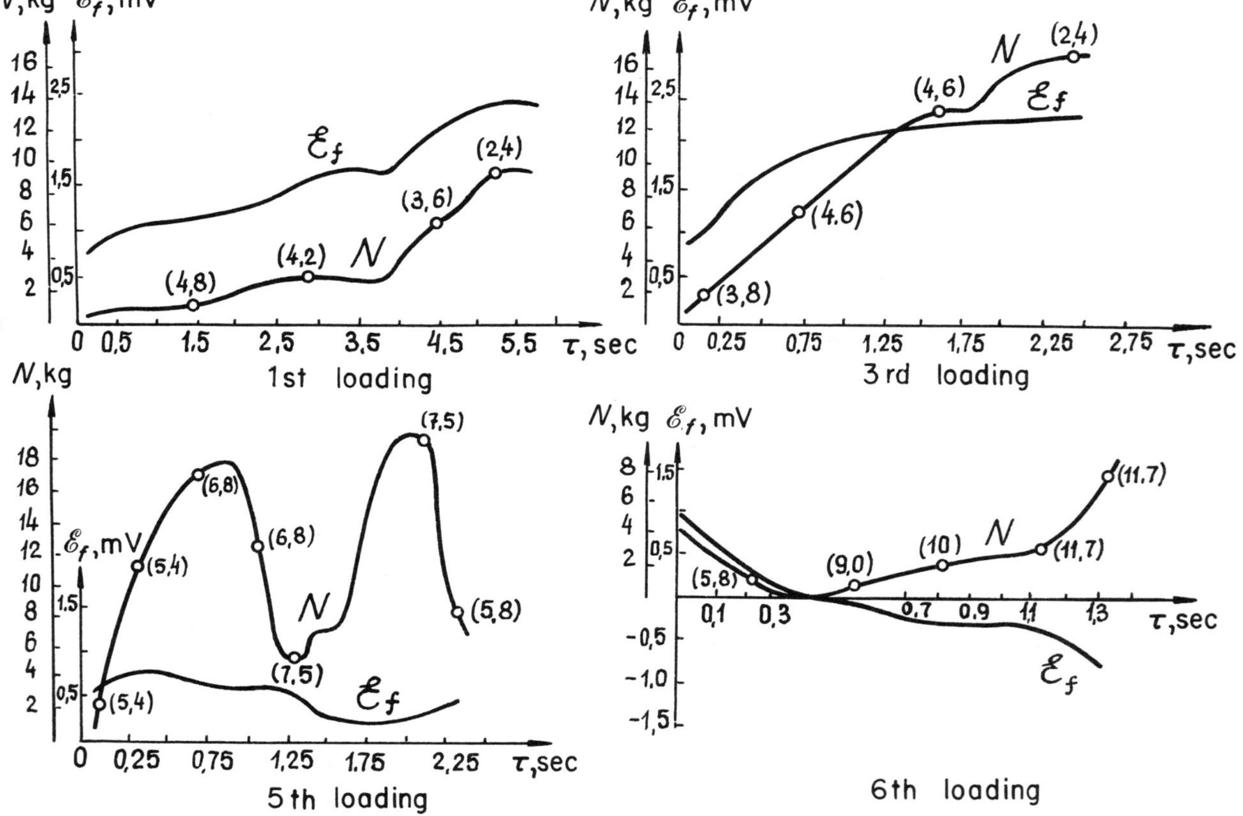

Fig. 2-41. Variation in emf under repeated loadings of a frictional couple. The figures in brackets are the speeds of relative sliding

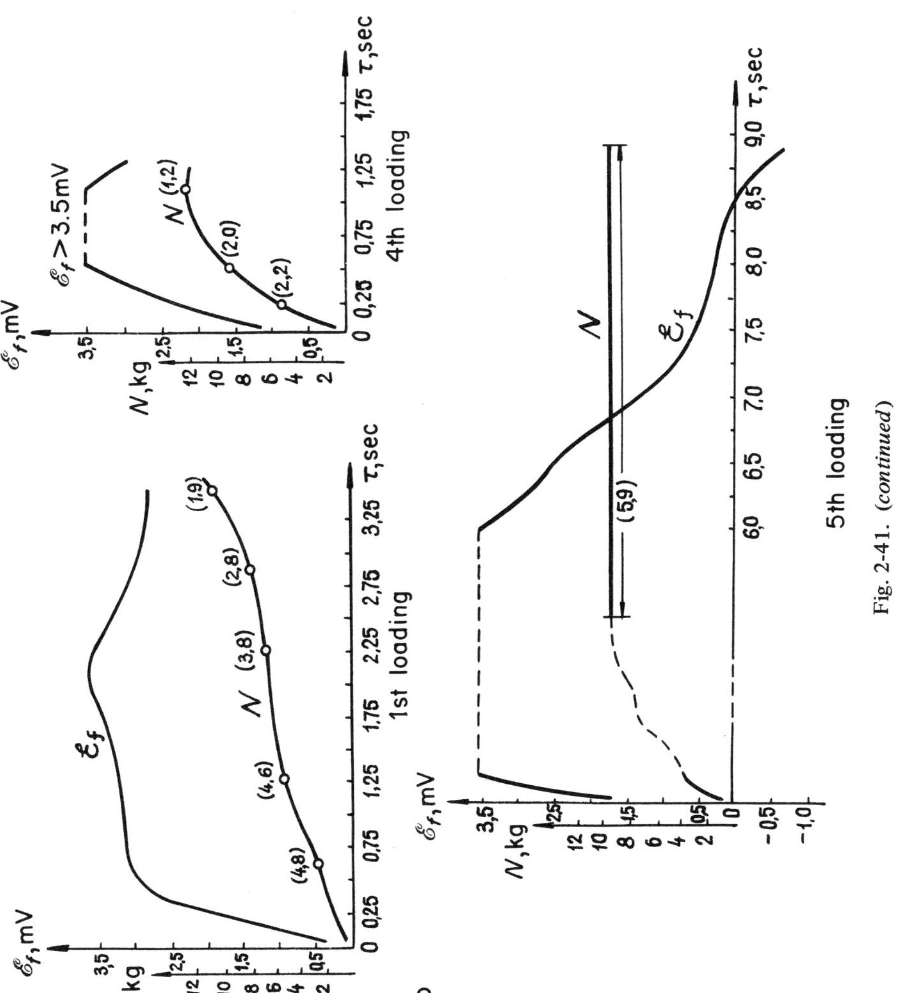

Fig. 2-41. (*continued*)

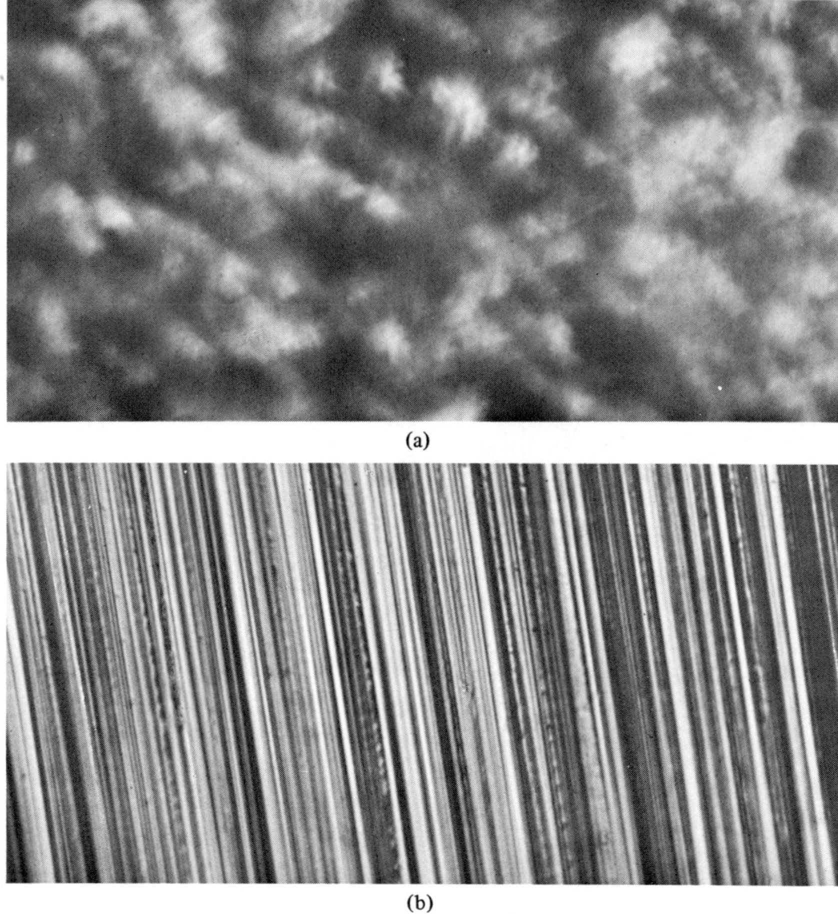

(a)

(b)

Fig. 2-42. Photomicrographs of the surfaces of steel (a) and copper (b) after several loading cycles (200×).

and cylindrical specimens. The difference in temperature gradients results in a nonuniform velocity of charge propagation through the contacting surface of the metals, and in the accumulation of charge on one of the bodies.[132] When sliding takes place, the accumulation of positive charge on the cylindrical specimen and of negative charge on the disk-shaped one reduces $\mathcal{E}_t$ in the steel-copper-steel arrangement and increases it in the reversed arrangement (copper-steel-copper).

Figure 2-41(b) shows records of the emf and normal load in friction between copper specimens and a steel disk.

Free access of oxygen to the steel surface creates favorable conditions for the diffusional oxidation of the metal under large loads. The Seebeck coefficient falls off with rising temperature of the steel surface layers, and if the sliding process is prolonged the thermal emf becomes inverted.

In this connection, the author and Yashina consider the following.[130] Inversion of the thermal emf was seen under sliding conditions nearly as severe as those in cutting. Thus, from the readings of a natural thermocouple calibrated by the conventional method, it is impossible to judge the temperature in the cutting region if an oxidant has access to the communicating parts of the space between tool and workpiece (chip). The thermoelectric characteristics may change as a result not only of surface oxidation[133] but also of diffusional oxidation of the metals, when their temperature approaches the melting point. Incidentally, tantalum, titanium, niobium, and some other materials are excellent getters.

The resistance R_0 (Fig. 2-19) is determined by the state of the external circuit of the frictional couple at a given moment. The resistance may be of the same order of magnitude with the contact resistance R_m; in many cases (in particular in the cutting of metals), it is many times larger.[111] Assuming arbitrarily that $R_0 \approx R_m$, the author thought it possible[26] that the thermoelectric currents of several amperes obtained experimentally by Axer and Opitz[24,134] (in the cutting of metals) were real. Indeed, if $R_0 \approx R_m = 10^{-3}$–$10 \ \Omega$ and $\mathcal{E}_t = 0.2$–20 mV, then $I_t = 10^{-5}$–10 A. However, the power of the thermal seat of emf is still an open question. Thus, the author has invariably emphasized the necessity of reckoning with the dynamics of thermoelectric processes in the cutting zone and with the resulting probability that comparatively large impulse currents will flow.[26,30,135] This has been considered a unique feature of the loading of a frictional seat of thermal emf, distinguishing it from the seats used in model tests of how an electric current affects the seizing of rubbing bodies and the wear rate.[136,137] Such models (with external sources) are, instead, analogs of sliding electrical contacts, whose current loading determines the transfer of material in frictional couples in electrical machinery and devices.[138]

On the basis of the relations obtained by Holm,[119] an attempt was made to evaluate the temperature rise $\Delta\Theta$ of the contact surfaces due to channelization of the thermoelectric current through constrictions.* Taking into account the rise in the temperature t of the specimens as a result of their joint surface deformation in friction, we have:[26]

$$\Delta\Theta \approx \frac{1}{8} \cdot \frac{U^2}{\rho\lambda(1 + \alpha t)} \tag{2-29}$$

where U is the voltage across the contact, equal to $\mathcal{E}_t - I_t R_0$; ρ is the resistivity; λ is the thermal conductivity; and α is the temperature coefficient.

---

*Here, the author has considered only Joule-Lenz heating, ignoring without argument that contact spots are cooled by the Peltier effect as soon as they are formed.

The results from a rough calculation for the cutting of steels follow. Titanium-tungsten-cobalt steel, type T15K6, was used as the component with poor electrical conductivity.

U, mV	5	10	15	20	25
$\Delta\Theta$, °C	0.05	0.2	0.45	0.8	1.25

It would seem that the thermoelectric currents flowing through the interface have no relation to phase-coherence events. At the same time, a number of considerations and facts leave the correctness of this conclusion in doubt.

In the running-in period of surfaces with orthogonal placement of asperities, a brief "overvoltage" of up to 70 mV or more was detected at the terminals of a millivoltmeter connected to the specimens.[139] In order to record temperature flashes in friction, Bowden and Ridler took oscillograms of the mean thermal emf and showed that the instantaneous values for a constantan-steel couple reach 45–50 mV.[140] The emf in individual circuits with a single contact spot might be even higher. If $e_i \approx 60$ mV, then $\Delta\theta_i$ (the local rise in temperature of constantan due to Joule-Lenz heating) could be some 35–40° C when $U \rightarrow \mathcal{E}_t$.

The large input resistance R_d of the measuring device connected to a frictional source sharply lowers the current in the hot junction. Oxidation reactions on the rubbing surfaces may slow down; in the final analysis, this results in weakening of the thermoelectric effects, that is, a drop in $\mathcal{E}_t$.[133] What is more, it should be kept in mind that maximum power transfer from the thermocouple to the load requires the internal and external circuits to have equal resistances. For $R_d \gg R_m$, the efficiency of the thermal converter is far from maximal.

Two matters have not been fully explained. The first is the possible peak in the thermal emf e_i in an impulse. The second is the question of what value the brief current surges i_i, which affect the power of the seat, can take on. However, the most important argument for thinking thermoelectric currents play a part in adhesion phenomena is the tendency, however weakly marked, toward an increase in frictional force upon shunting of a frictional seat. The author could not agree with the judgment of Gordienko and Gordienko, who suggested that the thermoelectric current in the friction of metals, like the current from an external source, has an eroding, disrupting effect on the rubbing surfaces, and that this effect is similar in nature to that seen in the electric-spark machining of metals (see Section 5.1).[22] On the other hand, there were inescapable reasons for considering thermoelectric currents as a factor contributing to the union of crystal lattices in friction. Therefore, with allowance for the process dynamics, since Peltier heating is a linear function and Joule heating a quadratic function of current, it was seen as possible for additional, narrowly localized current heating of the metal to occur upon stretching of the contact junctions. This heating would be due to the reverse transformation of electrical to thermal energy. For example, a portion of the Joule heat may be the "last drop" needed to overcome the threshold energy of seizing,[27] which from the very first we have understood to mean the value of activation energy necessary for the collectivization of valence electrons in the quantum-mechanical interaction of the bodies at the interface between lattices. The disappearance of this boundary leads to diffusional sintering of the materials; transfer of the diffusant is due not only to the temperature gradient but also to the electrical force field. The author has given particular attention to the fact that a frictional couple, as a highly dynamic collection of point microjunctions with a complex system of loops of balanced, eddylike thermoelectric currents, will find conditions most favorable when the conduction mechanism is the same for both conjugate elements. This takes on more and more importance in the cutting of metals. For example, the existence of an internal thermoelectric circuit, tool rake face ($\sim 1,200^\circ$ C)-chip-workpiece-tool flank ($\sim 900^\circ$ C), must not be forgotten. If this circuit comprises electron and hole conductors, then a large current may flow across the interfaces.

Shamshur holds a view that is quite close to this, but that features a very pronounced—one might say a *hypertrophied*—thermal effect of the thermoelectric currents:

By analytical calculations and experimental data it has been established that, when circulating thermoelectric currents flow through a constriction, the constriction may be heated (by liberation of Joule heat) to its melting point. Under dynamic loads, with relative displacement of the contacting surfaces, heating may even reach the boiling point. The considerable rise in temperature at constrictions as a result of Joule heating by circulating (looped) thermoelectric currents leads to the development of seizing and other thermal processes, which contribute to the rapid destruction of the contacting surfaces.[141]

The author's point of view, set out here, has been subjected to correction, in which the role of Joule heating (which has turned out to be a very important factor indeed in regulating temperature fluctuations in the contact zone) was interpreted somewhat differently. The reader can find this interpretation in Section 5.2. The common physical essence of the thermoelectric phenomena in friction and cutting suggests that the study of these phenomena has common theoretical, methodological, and experimental features.

2.6. ROLE OF ELECTROCHEMICAL PROCESSES

It was mentioned above that a three-phase frictional system with an ion-containing lubricant interlayer can be modeled as a short-circuited galvanic cell. Using this model, the author has put forward a number of *a priori* considerations respecting the role of electrochemical processes in the friction and cutting of metals. In particular, it has been shown that the independent (separate) or combined (parallel) action of thermal and galvanic emf's changes the gradient of the fluctuating electric and electromagnetic fields at the interface.[113] Consequently, this action cannot fail to affect specific features of physical adsorption and the kinetics of chemical reactions (including soap-forming reactions), or the mechanism of mobility and adaptation of lubricant molecules in the formation of a boundary layer. It has been seen as all the more justified to consider as parameters of lubricant components such factors as the electric moments of polar molecules,[33] the static charges of liquid and metal ions,[142] the sizes of the ions,[143] the penetrating powers of the media,[18] and so on. With the aim of controlling the mechanisms of triboelectric phenomena, emphasis has been put on the potential of research aimed at developing such methods as the passage of weak currents through the interface from an external source, artificial alteration of the electrode potentials of the contacting metals, and preionization of the lubricant particles as they are introduced into the contact zone.

According to the author and coworkers, an automatic electrochemical mechanism of wear control comes into play in the operation of a frictional couple.[31] * The mechanism depends on such factors as the emf of the frictional galvanic cell, the shift in potentials when the cell is shorted, the polarizability of the electrochemical circuit, the rates of the electrode processes, and so on. It follows that the addition of an oxidant or reductant, an activator or an inhibitor to a lubricant should lead to one of two situations. In the first, the newly formed lubricant, namely the product of the electrode reactions, will behave as a destructive, aggressive medium. In the second, on the other hand, conditions favor the passivation of the juvenile surfaces and the formation of stable protective coatings. Life tests of type-45 steel specimens in corrosive media have shown that the formation of a plastic screening film to lower the surface energy requires an alkaline medium, whereas the films of secondary structures formed in an acidic medium do not hinder corrosive attack on the surface or prevent the emergence of dislocations there.[144]

Bowden and Young observed the effect of polarization on the coefficient of friction and the damage suffered by inert-metal surfaces.[145] They studied the behavior of a platinum-platinum couple immersed in dilute sulfuric acid. Platinum is a good model for a theoretical study of the effect of surface state on the friction process.[80] This is true because both chemosorbed, monomolecular layers with well-defined chemical characters and films of phase oxides can be formed on a platinum surface in a reproducible way by electrochemical means. For example, according to the data of Frumkin and coworkers, the following behaviors are seen on the surface of platinum in contact with an electrolyte (potentials are measured relative to a reversible hydrogen electrode in the same solution):[146] at potentials $\varphi_e \leqslant 0.3$ V, atomic hydrogen is adsorbed; at

---

*The introduction of the cited article repeats the substance of a paper delivered by the author at the Second Scientific Seminar on Electrical Phenomena in Friction and Cutting (Moscow: Institute of Mechanical Engineering, December 17–19, 1969).

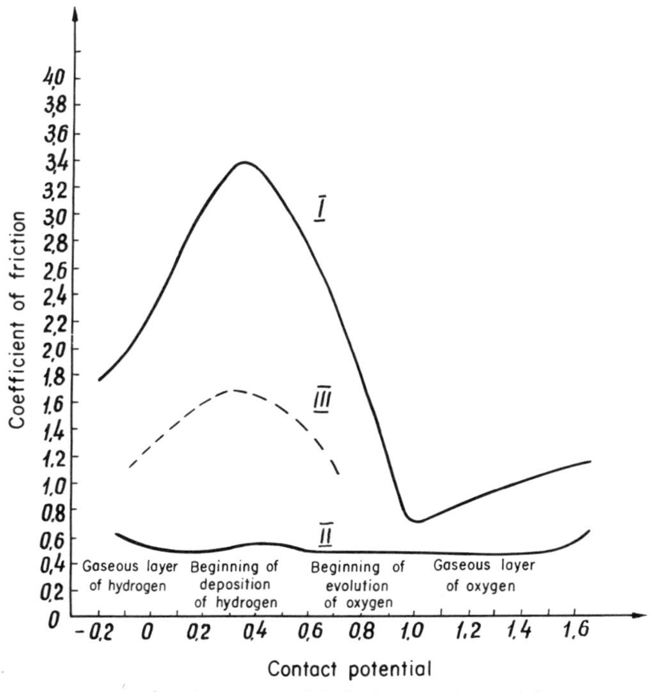

Fig. 2-43. Effect of contact potential (against reversible hydrogen electrode) on the friction of platinum surfaces immersed in a normal sulfuric acid solution.[145]

$\varphi_e \geqslant 0.7$ V, oxygen atoms are adsorbed (at higher potentials they may enter into chemosorbed surface compounds, phase oxides, etc.); for $0.3 \text{ V} \leqslant \varphi_e \leqslant 0.7$ V, the platinum surface is practically free of adsorbed gases (electrical-double-layer range).

Figure 2-43 presents Bowden and Young's results. If the solution contains no incidental impurities (curve 1), the coefficient of friction has its maximum in the double-layer range; the appearance of adsorbed hydrogen and oxygen result in a decline. The dependence of the coefficient of friction on the quantity of oxygen liberated is not, however, monotonic: it passes through a minimum at $\varphi_e \approx 1.0$ V. Here, chemical bonding between atoms and molecules of the primary monolayer and the clean metal surface seems to distinctly predominate over the comparatively weak van der Waals interaction.

The variation in coefficient of friction with contact potential is accompanied by a corresponding change in the character of surface damage. If the potential is gradually lowered from 1.0 V to 0.3 V, marked wear traces can be seen in the form of scratches.

Curve 2 was obtained in friction under the same conditions when the sulfuric acid solution was contaminated with traces of H_2S. Experiment says that the decisive influence on the frictional force is not the contact potential but the nature of the screening films, with their ability to alter the work function of the metal.

Electrocapillary curve 3 was constructed by Bowden and Young from the data of Frumkin and coworkers, who had measured the contact angles between gas bubbles and the platinum surface in a solution of sodium sulfate. It is not difficult to see that the maximum value of surface tension between platinum and electrolyte (arbitrary units) corresponds to the largest value of the coefficient of friction. Bowden and Young explain the relation between friction and surface tension in the following way: if there is a concentration of positive charge (or electrons) at the surface, repulsive electrostatic forces reduce the surface tension at the metal-electrolyte interface; at the same time, the mutual repulsion of the electrical double layers should lower the attraction and adhesion of the metal surfaces, and thus the friction. The greatest surface tension and the corresponding maximum of adhesion (friction) should be expected at the point where the surface charge actually vanishes.

Interpretation of the data obtained by Bowden and Young is complicated by the presence of the electrolyte, whose lubricant function may depend on the surface charge. For this reason, a group in the author's laboratory studied the frictional activity of anodically oxidized platinum in air.[80]

Friction trials were preceded by a study of the laws governing the buildup and interrelations of various types and forms of oxides on the platinum, as a function of oxidation potential, composition of the solution contacting the electrode, and temperature. Measurements made by two electrochemical methods* (fast potentiodynamic I-φ_e curves and alternating-current bridge) support the proposition that the stable oxide formed above ~ 1.5 V is a chemosorbed compound of $Pt(IV)$, with oxygen and the electrolyte components forming ligands in the chemosorbed complex; at the same time, the metastable oxide can be considered a monolayer of chemosorbed oxygen atoms.[148] Therefore, the range of potentials $\leqslant 1.5$ V was regarded as the range of existence of chemosorbed oxygen atoms; the range of potentials $\geqslant 2.2$ V as the range of existence of a monolayer of chemosorbed $Pt(IV)$ complexes; and the transition range of ~ 1.5–2.2 V as the range of coexistence of these forms.

Tyurin and coworkers remark that the appearance of the chemosorbed surface compound begins with the formation of individual molecules or small isolated islands of the compound, localized on active portions of the surface.[149] A rise in potential causes these islands to join up in a monolayer covering the electrode

---

*Tyurin and Volodin[147] describe the methods of measurement.

($\varphi_e \sim 2.2$ V), after which the oxide can compact itself by two-dimensional quasicrystallization.[150]

The formation of polylayers of phase oxides takes place in a limited range of potentials.[151,152] It has become clear that only the lower bound of the range lies at the same potential (~ 2.1 V) independent of the electrode oxidation conditions (galvanostatic or potentiostatic). In other words, it is as if the potential is automatically controlled by the phase-nucleation process alone.

An analysis of the I–φ_e curves, which reflect the low-temperature reduction of the phase oxide formed at a comparatively high temperature,[80] showed that this oxide occupies an extremely small portion of the surface. The centers of its growth are most likely defects in the dense layer of quasicrystalline oxide (or *limiting oxide*).

A special device was used to determine the coefficient of static friction between an oxidized platinum specimen and a glass plate. (A group in the author's laboratory described the device and the method used for measurements on it.)[80] Platinum specimens were prepared from wire 0.5 mm in diameter with one end partly fused. Only those specimens with no defects on the working section of the surface were selected for the trials. After washing in hot acids (nitric and sulfuric) and doubly distilled water, the specimens were dried in a stream of hot air at 60–65° C. The optically polished glass plate was degreased in a mixture of potassium dichromate and concentrated sulfuric acid, then washed in doubly distilled water and dried in air.

Figure 2-44 gives the results of trials conducted to measure the coefficient of static friction. Preliminary oxidation of the platinum specimens at anode potentials (potentiostatic regime) corresponding to the experimental points was conducted in normal sulfuric acid for 2 minutes, at temperatures of 20° C (curve 1) and 80° C (curve 2). The sections of the μ–φ_e curves that vary in the same sense do not coincide. There are probably two reasons for this discrepancy: the

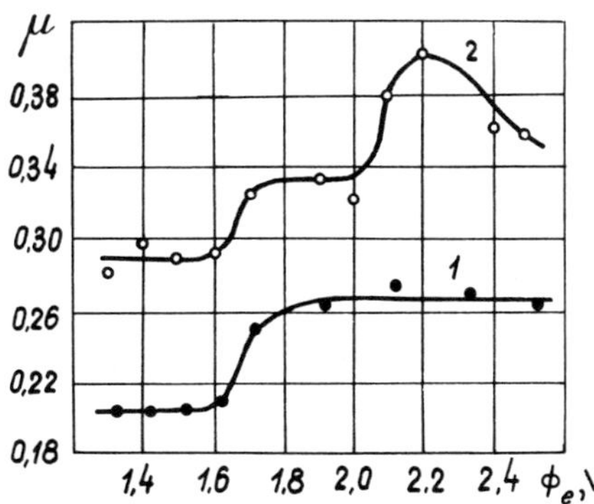

Fig. 2-44. Static coefficient of friction as a function of the oxidation potential of platinum

curves were obtained on different platinum specimens, and the oxidation temperature affects the coverage and packing density of the chemosorbed forms.

The μ-φ_e curve taken on a specimen oxidized at $20°$ C displays two sections of constant μ, given by $\varphi_e \leqslant 1.5$ V and $\varphi_e \geqslant 1.9$ V. It was remarked above that the first of these corresponds in essence to the range of chemosorption of oxygen atoms and the second to the formation of a monolayer of the quasicrystalline Pt(IV) surface compound. The transition section with 1.5 V $\leqslant \varphi_e \leqslant 1.9$ V coincides in potential with the range in which the monolayer of oxygen atoms is replaced by the limiting oxide. The increase in static friction on this section, it might be suggested, is linked with the change in configuration of the molecular field of the platinum surface as it makes its transition to a qualitatively new state.

The μ-φ_e curve for the specimen oxidized at $80°$ C displays, along with the plateaus and transition section indicated previously, a maximum of μ in the range of potentials corresponding to the formation of the phase oxide. It seems that the physical nature of the frictional forces is marked by significantly greater nonuniformity in a limited range of potentials about the maximum:[80] the presence on the platinum surface of microscopic asperities, in the form of polylayers of phase oxides on defects in the limiting oxide, magnifies the role of the mechanical component of frictional force; this is reflected in the value of μ.

Thus, the coefficient of friction is a structure-sensitive characteristic, both in relation to chemosorbed two-dimensional layers of various types and for the three-dimensional phase.

Besides their theoretical value, the results under consideration may also present some practical interest. Platinum is in ever-increasing use in engineering and is found in the most vital frictional assemblies. Much closer to practical problems, however, are studies on the use of polarization to alter the mechanical and physicochemical properties of rubbing surfaces in order to control the wear of frictional couples in corrosive media.[153-155]

For example, Porter and coworkers have demonstrated the possibility of using polarization to vary the surface energy of solids and the kinetics of electrochemical processes, with the aim of obtaining wear-resistant surface structures.[153] They studied the ways in which anodic and cathodic polarization in weakly acidic (pH 6.5) and basic (pH 11) media (and also in a 45 percent aqueous solution of carbohydrates) affect the frictional force, roughness, and wear (measured by weight loss) of rubbing surfaces; the electrical conductivity of the frictional contact; and the way in which the dislocational structure develops in the region of mechanical activation. Table 2-9, in particular, reflects the results of their experiments with a spherical steel indenter sliding on bronze, cast iron, and steel in an alkaline medium. The effect of polarization was most pronounced for the friction of cast iron in a weakly acidic medium: the microhardness of the surface almost doubled.

Polarization effects on the state of the rubbing surfaces and the properties

Table 2-9. Effect of polarization on frictional force and contact-surface microhardness.[153]

Material on Which Indenter Slides	Anodic Polarization		Cathodic Polarization	
	F/F_0	H_m/H_{m0}	F/F_0	H_m/H_{m0}
Bronze, type BrAZhMts	1.2	–	$\simeq 1$	1.26
Cast iron, type SCh12-26	>2	0.83	$\simeq 1$	0.82
Steel, type 45 (HB 240)	0.9	–	$\simeq 1$	0.81

Note: The spherical indenter, of type-ShKh15 steel, slides over the materials in an alkaline medium, at a potential of 1,300 mV. F_0 and H_{m0} are the initial values (before polarization), and F and H_m the final values (after polarization), of the frictional force and the microhardness of the contact surface.

of the films of secondary structures were apparent in both the magnitude and the character of the change in contact resistance. It turned out that fluctuations of the contact resistance depend on the sign of polarization and the magnetic state of the specimen being polarized.

A polarization effect had manifested itself in the variation in the degree of wear of the metals and the roughness of the contacting surfaces. Porter and coworkers observed this effect in concentric-cylinder friction between specimens of type-45 steel and bronze in an alkaline medium.

In order to obtain the clearest picture of how polarization affects the way in which the dislocational structure forms in friction, those authors chose silicon single crystals as their object of study. (Peierls-Nabarro forces are quite significant in these crystals.) In this way, the possibility of excluding random deformation effects was almost assured.

The trials were conducted with an indenter made of type-VK6 hard metal-ceramic alloy. Preis and coworkers summarize the results:

Investigation of the polarization effect on the dislocational structure of the deformed region in silicon single crystals showed that cathodic polarization aids the plasticization of the contacting surface and lowers the nonequilibrium concentration of dislocations and the depth of the deformed region. With anodic polarization, the screening surface films dissolve, and this contributes to increasing the local concentration of deformations and embrittling the friction surface.[154]

Yakunin and Mirbabaev mention the formation of workpiece-tool galvanic couples, but do not touch on the possibility that oxidation-reduction reactions can occur in the cutting zone.[133] They limit themselves to verifying the distorting effect of galvanic emf on the readings of a natural thermocouple. The author emphasized the need for reevaluation of the established views, which ignore the electrochemical nature of the phenomena in cutting with electrically conductive cutting fluids.[26]   He remarked that the effectiveness of these fluids may have a direct link to their electrolytic properties, if the tool and the workpiece in a given medium can be regarded as two electrodes whose surface states and non-equilibrium electrical potentials largely predetermine the character and rate of the processes taking place on them. At the same time, it was pointed out that specific features of the operation of a natural galvanic cell in cutting are related to the continuous variation in the physicochemical state of the rubbing surfaces and of the way in which the boundary films are structured. As it approaches the cutting edge or, more precisely, the sites of intensive heat evolution, the lubricant phase is more and more rapidly converted from the liquid to the liquid-drop and vapor states.*   Under conditions corresponding to those of the flow of a chip over the contact faces of a tool, it is not even out of the question for the lubricant molecules to undergo complete thermal and catalytic decomposition. Therefore, the most stable part of the galvanic cell is the cutter-electrolyte solution-workpiece system at the boundaries of the chip–rake-face and flank-workpiece contact region. At places not far from the cutting edge, oxidation-reduction reactions may occur on one of the electrodes. In essence, this is a corrosion phenomenon.[15, 113]

The majority of the author's considerations relating to electrochemical problems in friction and cutting were of an *a priori* character. Some of them have already received experimental confirmation, especially in the model trials conducted by Korobov and coworkers.[156]

A fundamentally new and, in our view, the most interesting result of these tests is the following conclusion. The cutting process in an electrolyte can be effectively controlled by polarization (with the help of an auxiliary electrode) of one of the elements of the tool-workpiece couple, when the other element is not electrically conductive (in abrasive machining, when powdered-ceramic sheets are used, or in cutting of dielectrics with a metal tool).

We have already dwelt on the question of how the sign alternation of friction is reflected in the thermodynamic state of the metal surface layers subjected to mechanical and thermal activation (Section 2.2). Measurements of the electrode potential have made it possible to show that sign alternation in sliding friction considerably increases the electrochemical activities of steels.[157]

Electrochemical reactions, which form part of the complex of phenomena characteristic of friction and cutting processes, are more often thought of as detrimental than as beneficial, for it is usually the corrosive damage to the surfaces that is evaluated. However, these same reactions make possible a new approach to combating wear, thanks to the phenomenon of selective transfer in friction, observed by Garkunov and Kragelsky.[158] Because work toward a com-

---

*Strictly speaking, the state of aggregation of the dissolved substance may be quite different (melt) from that of the solvent (vapor).

prehensive study of conditions favoring selective transfer is important, we will dwell on some of it in more detail.

2.7. ELECTRICAL PHENOMENA IN FRICTION WITH SELECTIVE TRANSFER

In selective transfer, contacting surfaces form spontaneously and friction takes place between the mobile layers of oxide-free, copper-enriched films. The fluidized state of these films distributes the pressure over quite a large contact area, from the initial stage of the process on. The coefficient of friction declines sharply; the rule that the gradient of the mechanical properties is positive with respect to depth contributes to freedom from wear.

The rate of formation of servo-vital films on the conjugate metals or alloys depends on the reducing ability of the lubricant, the presence of oxidant ions in it, the activity of the end groups of polar molecules (which make surface dispersion easier), and so on. But the decisive factor, which governs both the possibility of creating conditions favoring the initiation of selective transfer and the rate at which those conditions can be established, is the operating conditions of the frictional contact in the initial period. As the author has shown, for example, mobile metal layers can form on the surface (Fig. 2-45) when the sliding speed, specific load, and temperature are high, even when the lubricants responsible for selective transfer are absent.[26] At the same time, for an absolute majority of frictional couples there are no such convenient conditions under which a thin molten layer acting as a natural lubricant can form. The appearance of such a layer is preceded by catastrophic wear of the surfaces (Figs. 2-46 through 2-48). In the friction of metals with a low threshold energy of seizing, the plucking out of particles becomes so rapid that smearing of the liquid phase over the solid surface masks the deep wounds (Fig. 2-49). The use of a lubricant can radically

Fig. 2-45. Sliding marks on a fused copper surface, formed by asperities on a molybdenum indenter (200×).

Fig. 2-46. Copper cladding of a constantan indenter (200X).

alter the character of the frictional interaction. However, even in the case of the tnree-phase system, copper alloy–glycerin–steel, in which the most important physicochemical processes are those peculiar to selective transfer, large specific loads result in pronounced seizing of the metals and spreading of the copper alloy onto the steel surface. Only some time later, when quasiliquid copper films have appeared on the friction surfaces through selective dissolution of the copper alloy, does the rapid wear stop. A normal frictional interaction is actively aided by chemosorption of the products formed in the mechanical degradation and thermochemical conversions of the glycerin. Adhesion of the rubbing bodies weakens markedly at comparatively small loads if the temperature in the contact zone is not high enough to overcome the threshold energy of seizing. Then, in the presence of the glycerin lubricant, whose reducing ability decreases as the concentration of aldehydes falls off, a servo-vital copper film forms directly on the steel as a result of atomic transfer.[79]

Under the action of the oxide-free metal surface and atmospheric oxygen, the decomposition and conversion products of the lubricant molecules polymerize. If the lubricant includes resin-containing fractions, then polycondensation reactions can take place at the contact; the resulting formation of a polymer "cushion" is typical of high temperatures.[159, 160] When pure paraffinic-

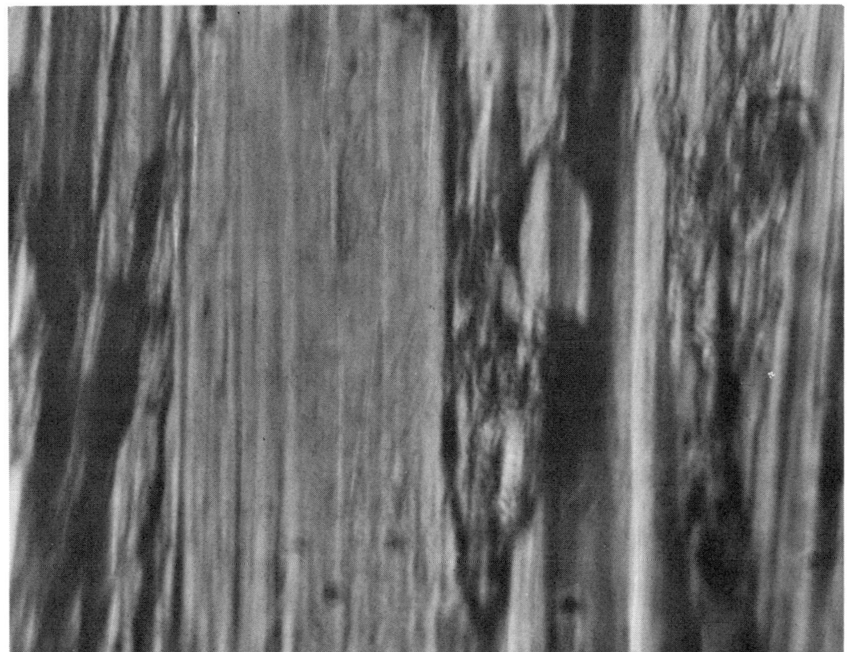

Fig. 2-47. Scratches on an aluminum film coating a copper surface (200X).

naphthenic mineral oil (quite free of polar components) is used in place of a surface-active lubricant (sodium dodecyl sulfate solution), the coefficient of friction decreases by a factor of 10;[161] this effect is probably linked with reactions of the type mentioned. The change of state of the copper surface, expressed above all in the change in degree and character of oxidation of the surface, can be assessed from the photomicrographs of Fig. 2-50.

As a consequence of the automatic electrochemical mechanism for controlling wear between the phase surfaces of metals, a specific boundary complex with a layered structure is formed (Fig. 2-51).[159, 162] This mobile "sandwich" assures (as one would quite logically expect) that the wear and coefficient of friction will be minimal under selective-transfer conditions.

In the running-in of the surfaces, the stage most responsible for the establishment of selective transfer, thermoelectric phenomena may play a highly significant role. Indeed, thermoelectric current is a stimulator of adhesion and also affects the kinetics of oxidation processes.[1, 2]

Melnichenko, furthering the author's idea of using equivalent electrical circuits to analyze the operation of a frictional system, began studying the specific features of friction with selective transfer by using electrical modeling of the

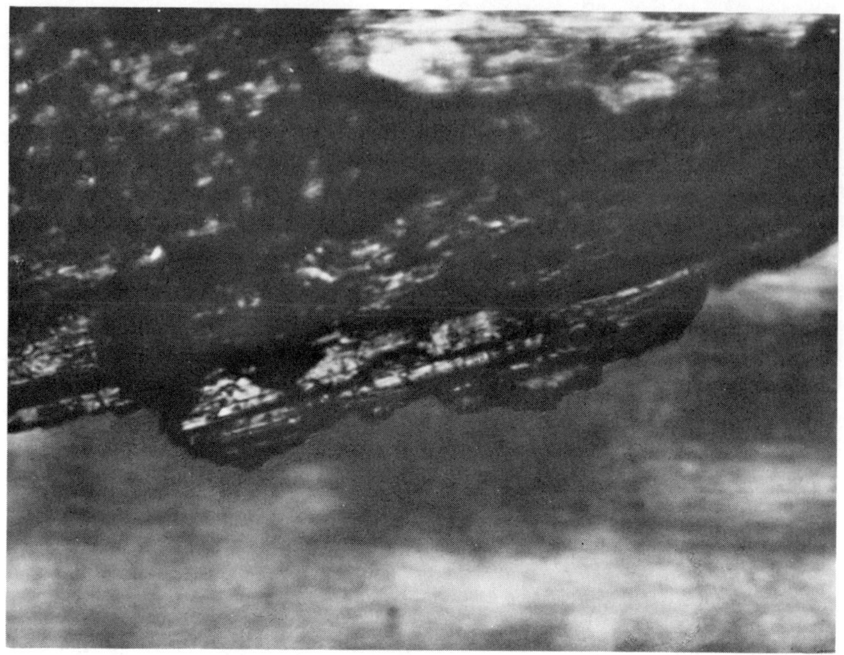

Fig. 2-48. Photomicrograph of a steel surface, with particles of a softer steel that have stuck to it (200×).

frictional contact.[16] His starting points were, first, the adsorption theory of the electrical double layer (according to Stern, the structure of the layer should correspond to some combination of the models proposed by Helmholtz and Gouy) and, second, the structure shown in Fig. 2-51 for the boundary layer in selective transfer.

The equivalent circuit of Fig. 2-52 corresponds to the simplest case. Here, unlike metals make contact over the entire surface, and the frictional couple can be regarded as a seat of thermal emf with an internal resistance equal to the resistance of the real contact.

Considerably more complex is the system model consisting of two electrodes separated by a lubricant layer that displays something similar to electrolytic properties (Fig. 2-53). We mention, incidentally, that even distilled water can take on the properties of a weak electrolyte, because it dissolves gases from the air, in particular carbon dioxide. This gas, forming carbonic acid, markedly increases the electrical conductivity of water:

$$H_2O + CO_2 \rightleftharpoons H_2CO_3$$

$$H_2CO_3 \rightleftharpoons H^+ + HCO_3^-$$

$$HCO_3^- \rightleftharpoons H^+ + CO_3^{2-}$$

Fig. 2-49. Blister remaining on a copper specimen under a layer of natural lubricant (200×).

In devising this model, Melnichenko took into account the capacitances of the double layers. These capacitances consist of diffuse and Helmholtz components* connected in series; adsorptive capacitance due to the concentration (retention) of ions at the metal surfaces because of specific adsorption forces (i.e., non-Coulombic forces); and the interelectrode capacitance, reflecting the abilities of the rubbing surfaces to accumulate electric charges. Here, R_1', R_2', R_3', and R_4' are the corresponding leakage resistances.

The complete equivalent circuit corresponding to the actual state of the contact is represented as a combination of the circuits that have been examined, as characterizing the limiting cases (Fig. 2-54). Seats of thermal and galvanic emf's connected in parallel, with their internal resistances, are shown as shunted by the transfer resistance of the external circuit (R_0); the inductance of the system (L), although quite small,[27] is also indicated.

As Melnichenko remarks, stabilization of the friction parameters under selective-transfer conditions makes it possible to neglect the reactive contact resistance and simplify the equivalent circuit (Fig. 2-55). Then it is far easier to analyze and calculate the circuit. We now proceed to do this, following Melnichenko.[16]

---

*They correspond approximately to the diffusion and barrier capacitances of a *p-n* structure.[16]

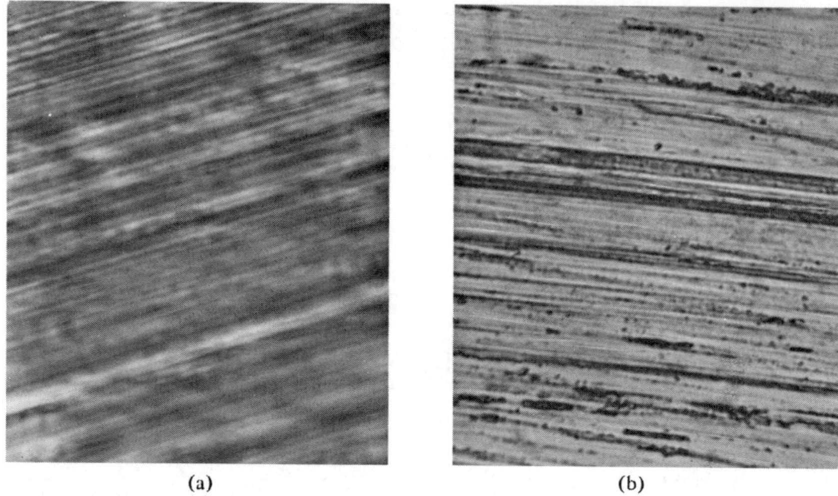

(a)   (b)

Fig. 2-50. Photomicrographs of the surfaces of copper specimens after operation in couple with steel in (a) a 3% solution of $Na[CH_3(CH_2)_{11}]SO_4$ and (b) liquid paraffin of moderate viscosity (200×).

The internal resistances R_1 and R_2 are unknown, while the resistance of the real contact is much smaller than that of the electrolyte (glycerin). This is why the contribution of thermal emf to the integral emf of the system cannot be neglected. Using the principle of superposition and assuming $g_1 + g_2 = g$, Melnichenko obtained the following expressions for the conductances g_1 and g_2 in terms of the integral emf $\mathcal{E}_f$ and the overall conductance g, both of which are accessible to direct measurement:

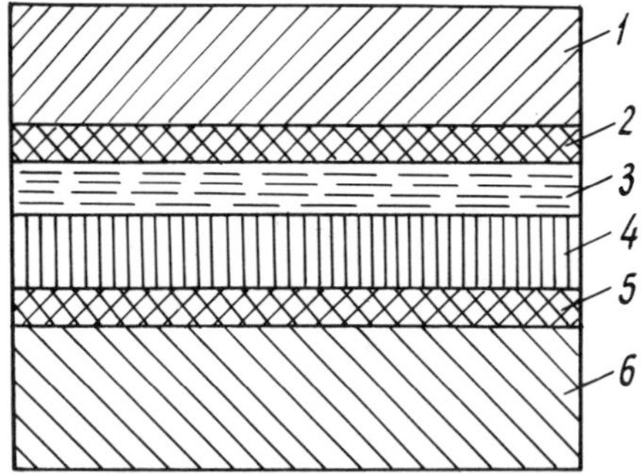

Fig. 2-51. Structure of the boundary layer in selective transfer (after D. N. Garkunov): (1) steel; (2) mobile, polymicellar copper layer; (3) polymerized molecules; (4) layer of adsorbed glycerin molecules; (5) plastic, mobile copper layer; (6) copper alloy.

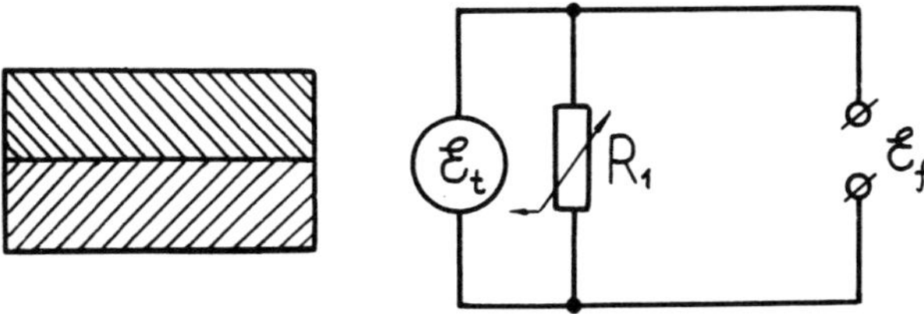

Fig. 2-52. Model of a frictional contact with direct contact of the rubbing surface.[16]

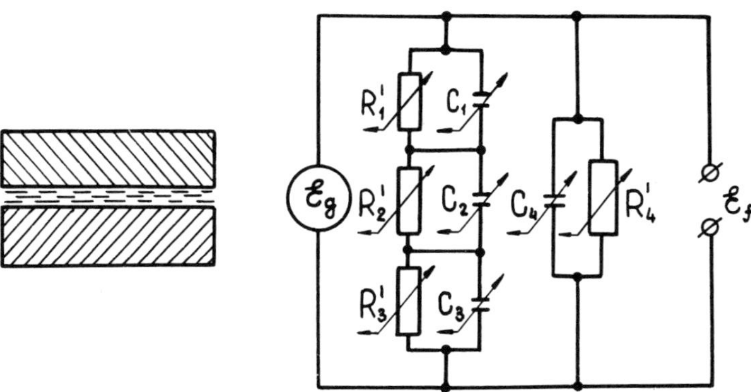

Fig. 2-53. Model of a frictional contact when an intermediate phase is present between the conjugate elements of the frictional couple.[16]

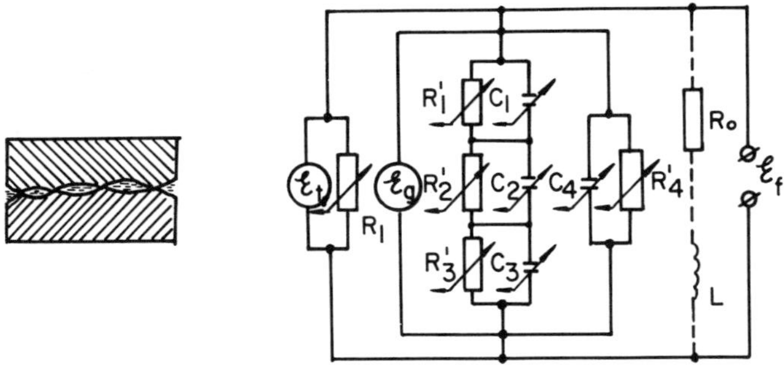

Fig. 2-54. Equivalent electric circuit of a frictional contact.[16]

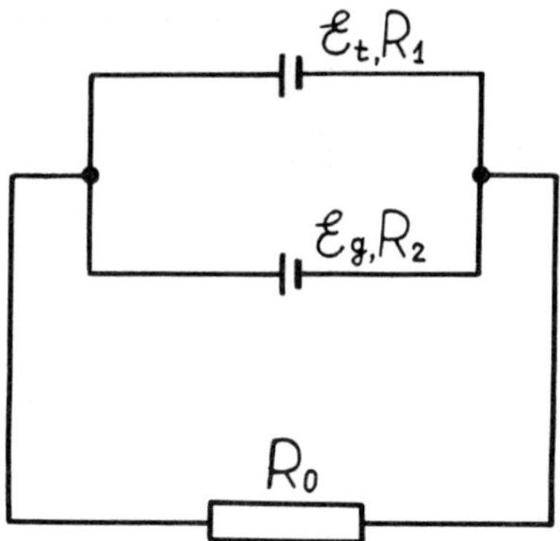

Fig. 2-55. Simplified equivalent electric circuit of a frictional contact.[16]

$$g_1 = \frac{\mathcal{E}_f - \mathcal{E}_g}{\mathcal{E}_t - \mathcal{E}_g}\, g; \qquad (2\text{-}30)$$

$$g_2 = \frac{\mathcal{E}_f - \mathcal{E}_t}{\mathcal{E}_g - \mathcal{E}_t}\, g. \qquad (2\text{-}31)$$

If estimates of $\mathcal{E}_t$ and $\mathcal{E}_g$ are obtained experimentally or by calculation, then through measurements of $\mathcal{E}_f$ and g it is possible to trace (by computation) the variation in the resistances of the metallic contact and the frictional galvanic cell under concrete conditions of sliding.

From Equations 2-30 and 2-31 we have

$$\mathcal{E}_f = \alpha \mathcal{E}_t + \beta \mathcal{E}_g \qquad (2\text{-}32)$$

where

$$\alpha = R/R_1 = 1 - R/R_2 \text{ and } \beta = R/R_2 = 1 - R/R_1.$$

The quantities α and β characterize the relative contributions made by thermo-electric and electrochemical seats to the integral emf. It is not difficult to verify this by determining the values of α and β for the limiting cases. Indeed, if $R_1 = R$ and $R_2 = 0$, then $\alpha = 1$ and $\beta = 0$. This is equivalent to the absence of any seat of galvanic emf in the region of contact interaction. On the other hand, if $R_1 = 0$ and $R_2 = R$, then $\alpha = 0$ and $\beta = 1$, so that $\mathcal{E}_f = \mathcal{E}_g$.

Melnichenko emphasizes that the sign of the $\mathcal{E}_f$ value recorded depends on the sign of the algebraic sum $\alpha \mathcal{E}_t + \beta \mathcal{E}_g$. Passage of this sum through zero is the condition for inversion of the integral emf; this fact can also be used in calculating R_1 and R_2.

Now a new possibility opens up, that of using the equivalent electrical circuit of a three-phase frictional system for monitoring the behavior of the real area of contact (S) and the average thickness of the lubricant layer (H) between the rubbing surfaces. Of first importance here is that the method proposed by Melnichenko and explained below makes it possible to evaluate quantitatively the variations in S and H during the friction of opaque solids (real machine parts). This is an advantage of the method over prior ones.[121, 124, 163–166] To achieve the goal set forth, it is sufficient to measure $\mathcal{E}_f$ and R in the presence of a lubricant that prevents the formation of oxide films on the friction surfaces (prevention of oxide film formation is precisely what happens in selective transfer), then calculate g_1 and g_2 by Equations 2-30 and 2-31.

The relations of interest to us are those between conductance g_1 and the area of real contact and between conductance g_2 and the thickness of the lubricant layer. These relations can be seen, for example, in the following formulas:

$$S = \rho_1^2/4\pi R_1^2 \tag{2-33}$$

and

$$H = R_2(S_n - S)/\rho_2 \tag{2-34}$$

where ρ_1 and ρ_2 are the resistivities.

Equation 2-33 was written for the case of contact between a sphere and a plane.[119] In other types of contact, the formula will be different, but this does not affect the final result if $S \sim 1/R_1^2$.

According to Equations 2-33 and 2-34, evaluating the variations in the real area of contact between opaque rubbing bodies and in the thickness of the lubricant layer between them requires the use of the obvious relations

$$S/S_0 = g_1^2/g_{10}^2; \tag{2-35}$$

$$H/H_0 = g_{20}/g_2. \tag{2-36}$$

No proof is needed of the importance of having information on the behavior of these parameters as functions of external factors (sliding time, load, etc.), especially if this information is obtained during the operation of the frictional system.

Melnichenko notes that, in the investigation of the variations in S and H under selective-transfer conditions, the stabilization of the values of these parameters (which are different for different frictional couples, types of lubricant, and process conditions) begins during running-in of the surfaces and actually comes to a stop in steady-state selective transfer.[16] He concludes,

By monitoring the way the parameters S and H vary, it is possible to forecast the performance of the sliding contacts in various lubricants. A small S indicates boundary friction; if S increases and H decreases to some value, constant for given conditions, the friction will go over to steady-state selective transfer. This behavior confirms the possibility of making a more valid choice

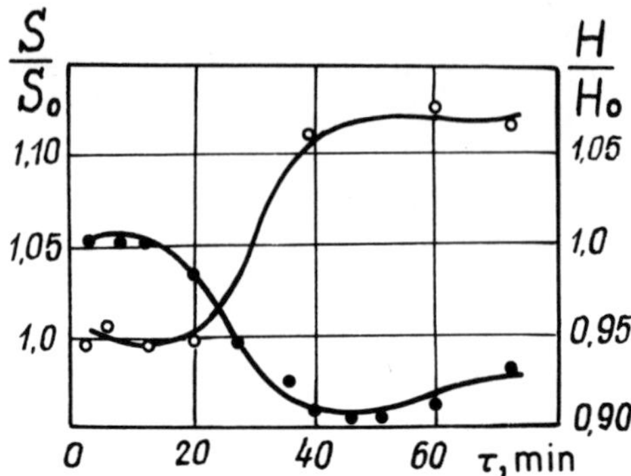

Fig. 2-56. Area of real contact (○) and thickness of lubricant interlayer (●) in relative units, as functions of the process time, for the frictional couple of type-L63 brass and type-U8 steel in glycerin.[16]

of materials, lubricants and friction conditions, which will help to establish conditions favoring selective transfer.

A good illustration of the application of the new method appears in Fig. 2-56, which shows the results of tests on the frictional system brass-glycerin-steel at comparatively high specific loads.

The results of Melnichenko's experiments (discussed in personal correspondence with the author) have shown that the $\mathcal{E}_f = f(\tau)$ and $R = \varphi(\tau)$ curves themselves can be used to monitor the transition processes in the establishment of selective transfer, and thus to predict the wear resistance of the materials making up the frictional couple. That is, S and H need not be determined.

Values of the integral emf and the contact resistance were found with the help of the device diagramed schematically in Fig. 2-57. The value of $\mathcal{E}_f$ was taken from readings on a type N700 oscillograph with a type F18 galvanometric photoamplifier; that of R was found by the ammeter-voltmeter method. Before testing, the specimens were degreased and run in until the rubbing surfaces fit

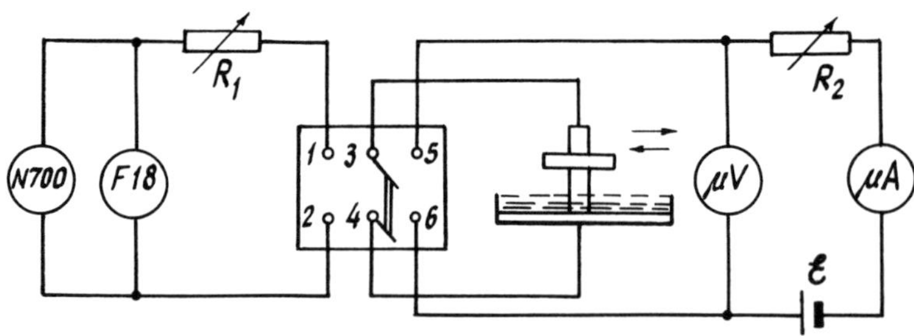

Fig. 2-57. Schematic circuit diagram of a device for studying emf and contact resistance in friction (after I. M. Melnichenko).

completely. In order to minimize the incidental distorting effect of external thermal emf's, the connecting wires chosen for the measuring system were as nearly as possible thermoelectrically identical, and relatively stable thermal conditions were maintained around connections.

The establishment of selective transfer proceeds in different ways, depending on the combination of materials making contact and on the properties and thickness of the coatings on them. However, the most distinctive way in which selective transfer develops under large loads is the spreading of copper alloy on steel, followed by the formation of a quasiliquid copper film on the friction surfaces. When lubricant is plentiful and thermal conditions are favorable, such a film is able to form because of the reducing ability of the glycerin (which keeps the spread layer from oxidizing) and its electrochemical activity in relation to the copper alloy. Frictional brass-plating and the resulting appearance of a clearly visible servo-vital copper film were precisely the principal criteria for differentiating the initial and final stages in the development of selective transfer. In addition, as the establishment of selective transfer proceeded, the frictional moment was seen to decrease monotonically toward a certain fixed value corresponding to the sliding conditions.

Figures 2-58 and 2-59 show typical curves of emf and contact resistance versus time for copper-alloy–steel couples: Fig. 2-58 for low sliding speeds and high specific loads and Fig. 2-59 for comparable speeds and low loads.

A comparison of those curves show that both in the running-in stage and at the time of the final transition to selective-transfer conditions, the emf and contact resistance of rubbing systems are one to two orders of magnitude lower at large

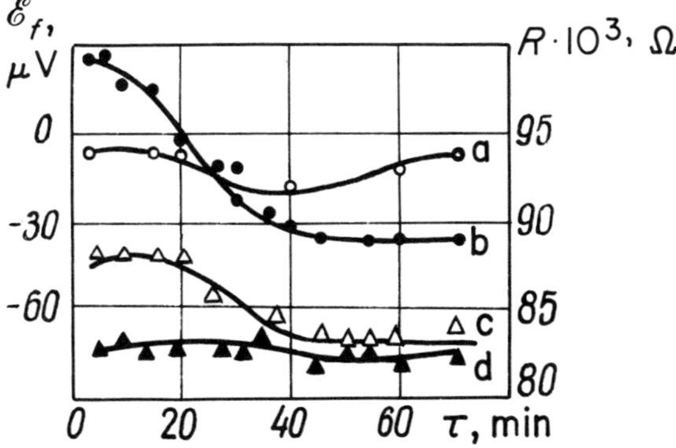

Fig. 2-58. Emf and contact resistance as functions of the time of friction (after I. M. Melnichenko): (a) emf and (c) contact resistance for type-L63 brass sliding on type-U8 steel, specific load $p = 6$ mN/meter^2, sliding speed $v = 0.16$ meter/sec; (b) emf and (d) contact resistance for type-LMTsS 58-2-2 alloy sliding on U8, $p = 7$ mN/meter^2, $v = 0.13$ meter/sec.

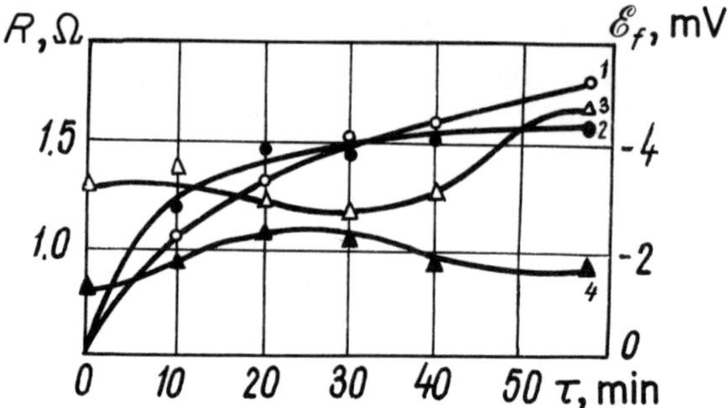

Fig. 2-59. Emf and contact resistance as functions of the time of friction, for the specimens of Fig. 2-58 in a glycerin medium (after I. M. Melnichenko): (1) emf and (3) contact resistance for L63 on U8, $p = 0.2$ mN/meter^2, $v = 0.3$ meter/sec; (2) emf and (4) contact resistance for LMTsS 58-2-2 on U8, $p = 0.1$ mN/meter^2, $v = 0.15$ meter/sec.

loads than at small loads. This difference should not be thought unexpected; it could have been foreseen on the basis of existing ideas on the friction zone as a seat of thermoelectric and electrochemical emf's acting at the same time.[113] As for the effect of other sources on the integral emf, any such effect was assumed negligibly small, on the basis of a comparison of the currents due to exoelectron emission ($< 10^{-10}$ A), triboelectron emission ($\sim 10^{-9}$-10^{-8} A),[167] and the combined action of the sources that have been treated ($\sim 10^{-4}$-10^{-3} A).*

In steady-state selective transfer, the integral emf is on the same order of magnitude as its thermoelectric component. If a change in the conditions of friction, and thus in the state of the interface, results in the internal resistances R_1 and R_2 becoming equal ($\alpha = \beta$), and if $\mathscr{E}_t$ and $\mathscr{E}_g$ have opposite polarities, the integral emf will equal zero. Of course, the reason for the inversion of $\mathscr{E}_f$ is overcompensation of one of its components. In our judgment, however, this is not the sole reason for the inversion: the polarity of a frictional seat of emf may change as a result of asymmetric sliding of the specimens, a change in the thermoelectric characteristics of the materials upon surface and diffusional oxidation (see Section 2.5), physicochemical modification of the surfaces with the lubricant components taking part,[168] and so on.

In a recent unpublished manuscript entitled *Effect of friction on the protective properties of surface structures*, Melnichenko studied the corrosion behavior of copper-zinc alloys under static conditions and the variation in protective properties of servo-vital films in friction. He did this by measuring the electrode potential, a quantity that, as we already know, predetermines the kinetics of the electrode processes and the properties of the surface structures in the region of the frictional interaction.

*Experiments conducted in the author's laboratory have shown that this matter still requires special and very careful work (see Section 4.1).

Taking into account the principal factors causing the potential of an alloy to become more or less noble, Melnichenko first analyzes the relation between the potential-composition curve and the phase diagram.   Then he compares the relation with the data that Polyakov obtained on the correspondence between the phase diagram and the wear resistance of an alloy.[169] This comparison gave a clear correlation between alloy wear resistance and potential for brasses; Melnichenko suggests that this result is due to the dependence of these values on the composition, fine structure, and protective properties of surface layers.

Figure 2-60 is a schematic representation of the device with which electrode potentials were measured in friction.   In order to reduce the ohmic losses, the silver chloride reference electrode, type EVL-1M, was brought as close as possible to the test specimen.   An inverted-U-shaped salt bridge filled with saturated KCl solution and agar-agar was used for this purpose.

Curve 1 in Fig. 2-61 is a typical graph of the time variation of electrode potential as the load is applied and relieved.   The frictional couple comprised type-L62 brass and type-40 steel in a glycerin medium.   The zero reading was taken as the steady-state potential of the specimen rotating in the cell without the application of load.   The corresponding variation in the coefficient of friction was found simultaneously (curve 2).   It is not difficult to see that the largest shifts in electrode potential and the pronounced maxima of friction practically coincide in time.

From his analysis of these results, Melnichenko concludes that an unambiguous relation exists between the electrode potential and the surface state of the alloy. He touches on the question of under what conditions there can be a mobile equilibrium between the destruction and restoration of surface structures.   In particular, he emphasizes that the optimal conditions for selective transfer come into being when the servo-vital film has the smallest thickness still sufficient to

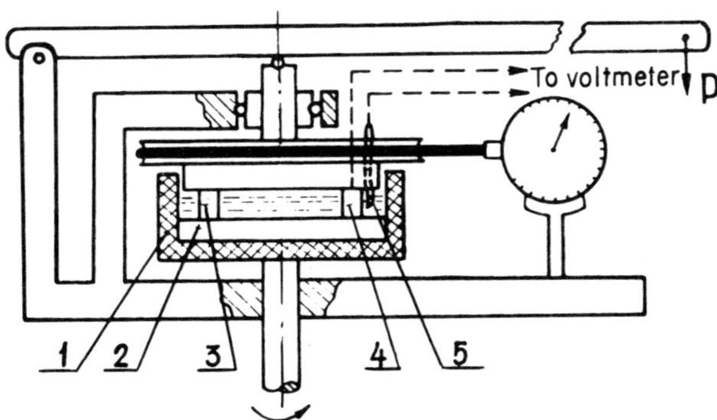

Fig. 2-60. Frictional assembly in the device for measuring electrode potentials in friction (After I. M. Melnichenko): (1) insulated cup; (2) lower rotating specimen; (3, 4) upper stationary specimens; (5) reference electrode.

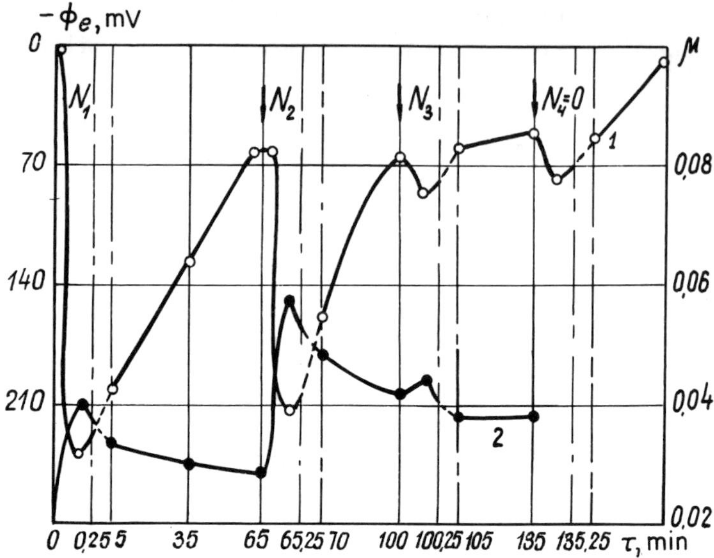

Fig. 2-61. Time variations of electrode potential (curve 1) and coefficient of friction (curve 2) with loading and unloading (after I. M. Melnichenko).

passivate the alloy and provide plasticity. To characterize more precisely the protective properties of films, Melnichenko uses a method widely applied to corrosion research, the cleaning of the electrode surface in an electrolyte medium.

In conclusion, it should be noted that, despite the limited range of occurrence of selective transfer in actual sliding systems (the limitation applies to sliding speeds, temperatures, specific loads, and materials and lubricants),[159] this phenomenon still has much potential for use in the solution of practical problems. Besides the investigations mentioned previously, there are other publications supporting this statement. For example, these publications demonstrate the ability of surface-active additives to cause selective transfer in weakly-active or nonreactive oils and water,[170, 171] increased load-carrying capacity of a lubricant layer containing organometallic compounds (due to frictional metal-cladding);[172] the possibility of determining both selective and overall wear of copper alloys in glycerin lubricant by polarographic methods;[173] and so on.

3

Electrical Properties and Load-Carrying Capacity of Boundary Lubricant Layers

3.1. INTRODUCTION

Without a clear idea of what importance the electrical properties of hydrocarbon chain molecules have for the mechanism of boundary-layer formation, it is impossible to understand the electrical properties of the layers themselves as two-dimensional collectives of molecules. At the same time, unless we know the details of the state of a thin lubricant layer separating solid surfaces and lying within the range of their influence on the lubricant, it is pointless to discuss the load-carrying capacity of the layer. We use the customary definition of *load-carrying capacity*: the ability of a lubricant film to prevent direct contact between rubbing bodies (in the present case, metals).

The situation is eased, though, by the existence of Akhmatov's superb monograph,[33] which in its author's judgment could have been called *An introduction to the study of the boundary state of matter and to the molecular physics of boundary layers formed on a metal surface by long-chain organic molecules.* (This monograph is available in English translation.) Still, it makes sense to give here a very brief, abstractlike exposition of some fragments from it and to apply this exposition to answering some questions that have been raised about one of the extremal regimes of boundary friction (according to Akhmatov's classification of the types of friction).[33] A significant feature of this regime is that oxidized, but still physically and chemically clean, surfaces may make either direct metallic contact or contact "through" a layer of the lubricant phase (single adsorbed organic molecules or isolated packets of such molecules).

Among the factors defining the essence of a frictional interaction between metals in the presence of multicomponent industrial lubricants, the following are currently of most interest:

1. The structure and properties of the lubricant molecules.
2. The way in which the molecules adapt to a solid surface.
3. Special conditions for the existence of a boundary layer enclosed in a crevice between phases.

For these reasons, we will examine classical notions and remark, in particular, on the following:

- Organic compounds are divided into three classes by the structure of the basic molecular skeleton: aliphatic, carbocyclic, and heterocyclic. Compounds of the first group, aliphatic hydrocarbons whose molecules in the normal, *trans*-isomeric state have a filamentary (nematic) shape, are in widest use as lubricants.
- The individual atomic groupings and radicals making up the chain molecules of fatty acids, alcohols, ethers, and some other derivatives of the homologous series of hydrocarbons have electric moments. In electrically neutral chains, a weak induced moment may appear under the influence of a dipolar "side" group.[174] In a methylene chain, induced polarization may also result from mechanical deformation in friction under load.[175]
- Upon the formation of a gas or liquid phase, chain molecules tend to be both displaced as rigid systems and structurally deformed in lateral force fields (the relative positions of the elements cease to be constant, and units in the chain rotate). The ratio of the deformation (degree of departure of the system from its equilibrium position of minimum potential energy) to the forces causing it depends on the type of interaction between the particles.
- Forces acting between large, net electrically neutral hydrocarbon molecules include van der Waals forces and a special category of forces called hydrogen bonding.
- An influence on the way in which organic molecules associate is the nature of the medium in which the major components of the lubricant dissolve.
- The physical properties of a substance nematic in structure undergo basic changes when the interaction between the side groups of neighboring chains (transverse cohesion bonds) is amplified or attenuated.
- The repulsion between the electron shells of atoms when they draw close together substantially limits the freedom of intramolecular rotation of the methylene chains in hydrocarbons (up to complete immobilization of the chains).
- In hydrocarbon crystal lattices, the methylene chains retain only one degree of freedom, that of axial rotation. Under these conditions, the molecules can be

regarded as extremely stiff atomic complexes if the exceedingly large modulus of axial elasticity of the methylene chains is taken into account.

● If sufficient activation energy is present (e.g., if the temperature rises), rotational isomers of the methylene chains are formed, despite the braking of intramolecular rotation. For this reason, chain hydrocarbons in the liquid or dissolved state contain not only straight chains but also entangled coils of chains. The effect of temperature is just to displace the temporary statistical equilibrium, either toward the predominance of linear forms (freezing-out of rotational isomers in crystallization) or toward the appearance of new, complex configurations.

● The components of lubricants and additives are highly diverse; the same can be said of the state of aggregation of a lubricant composition. Lubricants are used as true or colloidal solutions, or as emulsions or suspensions varying in degree of dispersion. Contamination of lubricants with water or solutions of acids, salts, and the like imparts weak electrolytic properties to the lubricants, despite a lack of electrochemically active additives that can be added at will.

● A juvenile metal surface bears a considerable reserve of free energy, especially at crystallite boundaries and at points on the boundaries of vicinal faces. Such a surface is characterized by all-round physicochemical activity. The surface has a large adsorbing capacity.

● The bonding between atoms and molecules of the primary monolayer on a solid surface may be either chemical (covalent) or van der Waals (orientational, inductional, or dispersional).

● These two types of bonding may be present at the same time on the same metal surface, where different portions of the surface (crystallites) have different chemical activities.

● A film of metal soap (product of the chemical reaction between a fatty acid and a metal) can withstand considerable deformations without breaking. This behavior results in a drastic reduction in metallic contact between surfaces. As the temperature rises, soaps soften and lose their protective properties.

● The kinetics of physical adsorption is determined by the conditions under which the force field of the condensed phase interacts with the fields of atoms and molecules in the lubricant. The degree of orientation and the integrity of the molecules on a solid surface depend on the nature of the surface.

● The most valuable components of lubricants are polar chain molecules of organic substances. These molecules become oriented in the field of a solid, move toward the source of the field, and become firmly fixed to the surface of a metal grain by their centers of electrostatic attraction. The energy fixing them is that of adsorptive bonding. These molecules are capable of stable vertical orientation.

● The electric moment of nonpolar hydrocarbons in mineral lubricating oils and greases equals zero. The molecules of nonpolar hydrocarbons go into horizontal orientation, as the state of minimum potential energy of the system. The molecules are not bound to the solid surface by special forces of interaction. For this reason, the physical properties of a nonpolar substance under boundary conditions do not differ in important ways from the bulk properties.

• In the case of multicomponent systems, such as the majority of industrial oils, surfactant molecules are adsorbed concurrently (selectively), with a preference for linear forms.

• As the adsorbed layer becomes compacted, the layer as a collective of molecules acquires special properties under the action of the strong local field of the metal surface. This field is capable of disrupting the charge distribution, and thus the structure, of the molecules.

• Thermal encounters between adsorbed molecules have two effects. First, there is a two-dimensional migration of the molecules along the equipotential levels of the metal surface. Second, molecules transfer from one energy level of the crystal-lattice field to an adjacent level.

• Any given state of the layer should be regarded from the standpoint of the layer's seeking a dynamic equilibrium, in which there is molecular exchange with the external medium and the rates of adsorption and desorption are equal.

• The character of the interaction between polar organic molecules and a metal surface is profoundly influenced by the presence of films of water (a stronger adsorbent) and by the acidity of the water.

• In the field of a metal surface, organic substances built up from dimers (fatty acids) acquire a previously-lacking elasticity of shape and take on the properties of crystalline solids. Under boundary conditions, all the principal structural forms are available to a lubricant layer: solid, crystalline, liquid, liquid-drop, gas, and other states.

• When the adhesion of a liquid exceeds its intrinsic cohesion, the liquid begins to spread over the metal surface. For example, nonpolar oils can spread without limit over a steel surface,[46] while polar substances have no such property.

• When two metal surfaces, each bearing a boundary layer, are placed one against the other, special processes take place to form and stabilize a combined structure in the intense, fluctuating electromagnetic field of the crevice-shaped capillary space. The physical properties of quasicrystalline molecular collectives under the combined influences of condensed phases (even at a distance of hundreds of thousands of angstroms from them) may differ significantly from the properties of boundary layers formed on the surface of a single phase.

• The boundary layer between the peaks of metal surfaces brought close together has the thickness of a monolayer. Films of organic compounds have exceptionally high compressive strength because of an atomic-elasticity mechanism in carbon chains oriented vertically to individual areas of solids under relatively uniform load.

It is becoming clearer and clearer what a complex set of phenomena is involved in the friction of rough polycrystalline surfaces in the presence of incomplete boundary lubrication, where the microgeometric profile is not leveled out by the lubricant. Under these conditions, the density of adsorbed molecules does not, as a rule, attain values corresponding to a condensed state in the layer. The active lubricant is, at best, in the liquid-crystal state, and there is no reason to expect boundary structures with very high mechanical strength to form on the metal surface. The possible forms in which a substance can exist in the crevice-shaped capillary space include, along with single islands of oriented molecules and liquid films of the nonpolar solvent medium, communicating portions of a gaseous interphase layer. The time required for adsorptive coverage of denuded metal areas after local breaks in the coating (latent period), the rate of surface migration of the molecules and the character of their contact accommodation, and the energy of bonding between the particles and the surface—all these are of great technical importance from the standpoint of the recovery (multiple regeneration) of lubricant films in the friction of metals. If a boundary system in the extremal regime can easily be brought out of a dynamically stable state by a slight thermal, mechanical, electromagnetic, or other stimulus, then it is necessary to look at the instantaneous states and nonequilibrium-hysteresis processes in the molecular structure of a gas-liquid organic film, on the phase surfaces of a solid, and in the surface layers of the solid.[26]

Of course, in view of the possible destruction of the lubricant film (up to thermal breakdown of the boundary layer), it is desirable to add surfactants to the nonreactive hydrocarbon medium. These substances are better able to go from an adsorptive (van der Waals) to a chemical interaction with metals; the chemical interaction involves a considerably larger bond energy. Molecules of a metal soap, for example, are rather strongly bound to the crystal lattice of a metal or its refractory oxide, even at high temperatures. At the same time, these molecules prevent the corrosion process from propagating into the bulk of the solid phase. We remark, incidentally, that for this reason it is indisputably of interest to investigate topochemical reactions in which fatty-acid boundary layers are formed on a metal surface, and to develop methods for evaluating the maximum lubricating ability of industrial oils.[176-178]

For the example of the antiwear properties of phosphonates and chlorophosphonates, which are used as additives for oils, it has been shown that to explain some features of the behavior of the additives, it is sufficient to study their adsorbability by measuring the surface potential of the metallic adsorbent.[179] Shor and Lapin are systematically studying the functional properties of lubricant materials (detergent, anticorrosion and antiwear materials, film strength additives), using measurements of the contact potentials arising at the metal-oil interface; they are also investigating the changes in service properties of lubricants due to polarization of the lubricated surface by an external source.[180-182] They have shown, in particular, that in nonaqueous electrolytes, such as doped oils, the dependence of metal wear rate on the magnitude and sign of the potential resembles an electrocapillary curve whose zero-charge potential is linked with the work function. This result, like the conclusions Shor and Lapin draw from it, has a direct relation to the subject of the preceding chapter. One practical consequence of these studies has been the ability to control the mechanism of the electrochemical interaction between industrial oils and the components of sliding systems. There can be no doubt of the versatility of this procedure if it is taken into account that applying an electric field makes it possible to shift the dynamic equilibrium between the rates of formation and destruction of a servo-vital film, destroy or create an adhesional contact, and alter the intrinsic viscosity of collectives of polar molecules, disperse systems, and the like.

In one of the publications cited, Shor and Lapin tried to find the reason for the electrical charging of a liquid hydrocarbon.[180] This effect was due to friction against a rotating metal electrode, and the degree of charging depended on the presence of polar components, micelle-forming impurities, and so on in the liquid. They found that the magnitudes and signs of the electrostatic charges of liquids with unequal polar characters correspond to the positions of the metals in order of increasing work function. Then the investigators proposed two principal charging mechanisms, consisting in the accumulation of ions of opposite charges: negative charges accumulate through the capture by liquid molecules of exoelectrons emitted by the metal; positive ones accumulate through the ionization of molecules by friction where the motion of the near-surface lubricant layer is turbulent. The competition between these mechanisms determines the degree of charging in any concrete case.

This point of view, even if it seems plausible, requires additional explanation of the nature of electron emission from the electrode surface.[167, 13]

According to Salomon, electric charging depends on the presence of a disperse phase in the system.[183] This phase may consist of bubbles dissolved in the oil under intense compression and released upon a sudden drop in pressure. Charges resulting from the breakup of a gas-saturated oil film formed under an earlier load now discharge, as soon as the potential difference across the layer exceeds the dielectric strength of the layer. The discharges, if they occur in spark form, cause a narrowly localized increase in temperature and pointwise melting of the metal.

According to Lazarenko and Lazarenko, the breakdown of a liquid dielectric consists in the evaporation and ionization of the dielectric, the formation of a constricted conduction channel, and the practically instantaneous transmission through this channel of the energy stored in the system.[184] The brisance of this "electrical explosion" is increased by incidental phenomena of the water-hammer type, which occur because of the exceedingly slight compressibility of the liquid in comparison with the gaseous medium.

Experimental data obtained by Raiko and Pavlov indicate that electrical discharges in thin lubricant layers develop in gas bubbles, while the oil serves only to stabilize the discharge, taking no direct part in it.[185]

To prevent the buildup of electric charges in oils (which, incidentally, leads to rapid oxidation of the oils), Salomon proposed lowering the ohmic resistance of the lubricant by adding to it an electrically conductive material. As Ventsel and coworkers found,

> At low resistances of the lubricant film, "silent" discharges occur. . . . As the viscosity of the oil rises, the resistance of the lubricant film increases, causing the accumulation of charges, with subsequent breakdown. The breakdown is accompanied by mass transfer, elevated local temperatures on asperities of the friction surfaces, and increased wear.[186]

A very important result that appeared in the same publication is the reduction in wear rate of internal-combustion-engine parts (pistons, piston rings, cylinder sleeves, crankshaft bushings, etc.) seen in plant bench tests when the electrical circuit between the conjugate elements of the rubbing couple was artificially shorted in order to prevent spark discharges.

In the bounds of the present chapter, the author cannot dwell on research into the electrical charging of plastic disperse systems subject to electrokinetic phenomena,[187–188] which is related to the subject at hand, or on discussing new methods of following molecular transformations in liquid hydrocarbons.*

Leaving aside, in essence, many questions on the electrical properties of boundary lubricant layers, we will report only on those experimental results that are most interesting in our judgment, and on some results of the author's own research.

3.2. ELECTRICAL PROPERTIES OF BOUNDARY LAYERS FORMED BY CHAIN MOLECULES

It is natural to relate the electrical properties of boundary lubricant layers to the polar character of the chain molecules. Oriented adsorption of chain molecules with pronounced active centers alters the physical properties of hydrocarbons under boundary conditions. When a monolayer of an insoluble aliphatic compound (e.g., a fatty acid) is formed, all the dipole moments tend to be oriented the same way relative to the solid surface. The lubricant film takes on the properties of an electrical double layer; this is indicated by the asymmetry in its electrical

---

*Ventsel and coworkers have reported on the results of applying one such method, nuclear magnetic resonance.[189]

conductivity (Vieweg rectifier effect)[190] and by the change in interfacial potential (on the order of millivolts or tens of millivolts).[191,192]

Of fundamental value in the development of research in this area are the results obtained by the application of the "blowing-off" method[193] to boundary films of liquid hydrocarbons on metals.[194,195]

It has turned out that, with approach to the metal substrate (right down to the surface), the viscosity of the liquid does not change if the liquid is free of polar components. But when such a liquid (Deryagin and Pichugin used carefully purified vaseline oil) contains even very small quantities of polar molecules (e.g., fatty acids or ethers), then at a layer thickness of about 10^{-5} cm, the viscosity changes discontinuously, then remains constant at the new value. In other words, a boundary layer (or layers) with a different viscosity appears on the metal surface. Since the interface with the bulk phase is rather clearly marked for such layers, they have come to be called boundary phases.

Although the observed viscosity changes were not large (the value was still on the same order of magnitude as for the bulk phase), the very fact of a discontinuous change in one of the properties of a liquid in the boundary state is quite important. It serves

> ... to indicate a peculiar, oriented molecular structure in boundary phases, differing sharply from the disordered arrangement of molecules in the bulk of the liquid. Without some such assumption it is impossible to explain simultaneously the constant viscosity in boundary phases and the discontinuous transition to the bulk-phase viscosity.[5]

Using this interpretation of boundary-layer formation as a phase transition occurring under the influence of a solid surface, Snitkovsky arrived at a very interesting conclusion:

> There is reason to suppose that a boundary phase should have a structure consisting of individual volumes within which the intrinsic or induced dipole moments are ordered. The reason is the short-range order in liquids: their tendency, near a phase-transition point, either to undergo macroscopic density fluctuations or to form regions of like-oriented spins. This kind of structure falls into the category of "domain structure" and is characteristic of ferroelectrics and antiferroelectrics with their large permittivities. . . . The energetically most favorable phase transition is one accompanied by an increase in permittivity. This follows from an inspection of the problem of finding the total free energy when a dielectric is inserted into a field, less the free energy of the field in all space in the absence of the body. This problem is considered by Landau and Lifshitz.[196]

An experimental investigation of the charge-voltage characteristics of boundary layers precipitated from liquid hydrocarbons on metal substrates showed that these curves have the form of double or single hysteresis loops.[196]   This fact is considered confirmation of the domain structure of boundary phases. Permittivity values calculated from these characteristics turned out rather large ($\sim$230 for oleic acid in a layer thickness of 2 μm). As Snitkovsky remarks, "The increase in permittivity upon the phase transition indicates that this transition is initiated by the electric field."

The notion of a nonspontaneous phase transition under the influence of the near-surface field of the substrate does not offer any grounds for thinking about a domain structure in boundary phases just because of their unusual boundary state. The author considers, for example, that in the conditions under consideration there might be assumed regions of induced polarization, "induced" domains (if it is desirable to call them domains at all), instead of the regions of spontaneous polarization, the "intrinsic" domains. Induced domains appear in the inhomogeneous electric field of a rough polycrystalline surface, because the orientation of the axes of polar molecules adsorbed on the crystallites is just as disordered as is the arrangement of crystallographic axes of the crystallites themselves (see reference 33, p. 220). In other words, the electric moment of a packet of dipole systems oriented alike is a distinctive mark of the packet, as if it characterized the appearance of a domain induced in the field. A collection of these packets, which resemble liquid crystallites, determines the polycrystalline structure of the boundary phases. It is still an open question whether the angles between axes of the molecular crystallites gradually decrease as adsorptive coverage of the boundary space advances[33] or the orientation of the molecular structure falls off at some distance from the metal surface.[5]

It is clear that when the external field varies cyclically, the rearrangement of the molecular structure of the boundary layer will continue to show traces of a lag (especially at high frequencies). This behavior has become a subject for study with the help of hysteresis loops.

As for changes in the permittivities of substances (paraffinic hydrocarbons, fatty acids, alcohols, ethers, etc.), the value of ϵ for fatty-acid boundary layers, for example, may turn out unusually low.[33]   This indicates precisely that the boundary existence conditions are energetically unfavorable for the liquid hydrocarbon.

All this does not, however, mean that the transition of the near-surface liquid layer to a mesomorphic state is not accompanied by acquisition of ferroelectric or antiferroelectric properties by the boundary phase. (An antiferroelectric may, incidentally, have a negligibly small dielectric susceptibility; Snitkovsky did not take this into account).[196]   The author would only wish to emphasize once again the difference between a ferroelectric boundary phase and a conventional ferroelectric (a crystalline substance with no center of symmetry). The difference is that not only the appearance but also the continued existence of such a phase requires, in our view, proximity to a polycrystalline solid surface.

The nonlinearity of the current-voltage characteristics of boundary phases has been verified under conditions that excluded the effects of chemosorption products and oxide films of significant thickness.[197] The trials were conducted with electrodes galvanically coated with gold. The lubricants used were monobasic carboxylic acids of the saturated series (propionic, butyric, caproic, enanthic, and caprylic) and a higher unsaturated acid with one double bond (oleic). It was shown that the semiconducting properties of boundary layers are directly linked with their detailed state; both the Schottky and Frenkel conduction mechanisms are at work in them.

The presence of a saturation region on nonlinear current-voltage characteristics taken with an alternating-current voltage indicated that S-shaped static characteristics could be obtained; that is, it demonstrated the probability of reversals in liquid-hydrocarbon boundary layers under the action of external fields.[198]

Yuriev observed such reversals, from the low-resistance to the high-resistance state and back, where the electrical resistance of the layer varied by four to seven orders of magnitude.[199] The reversals were caused by ac, dc, and impulse fields and took place at voltage values representing thresholds for given layer thicknesses. In trials with vaseline oil and oleic acid, for example, reversals in layers 4 μm thick were seen at voltages of 2.5 V and 1.9 V, respectively. After studying the character of the reversals in friction (which, depending on the thickness of the boundary layer, made it either a barrier layer or a conductive layer), Yuriev pointed out the special role played in these processes by contact electric fields and fields due to the accumulation of static charges or the connection of external sources. He also considered how the parameters of the external circuit of the frictional couple affect the electrical properties of the lubricant layer; the performance of the frictional system depends to a great extent on this question.

In connection with what has been said, we must not fail to consider one very important fact. This is that the reaction of boundary-layer electrical resistance to even an extremely slight change in the field is typical of liquid crystals, whose properties may change drastically upon insignificant variations in the external conditions. For example, it is widely known that the color of some liquid crystals depends in a structure-sensitive way on the temperature or the presence of negligibly small concentrations of chemical substances in the atmosphere.

Let us turn to other interesting facts, which show that deformation—more precisely, the internal-stress field—can "switch" the electrical state of a lubricant film.

In control trials checking the continuity of a stack of layers, the electrical conductivity of boundary layers was observed, in a determination of their elastic constants,* to depend on the pressure.[33] In Akhmatov's judgment, this behavior

---

*L. V. Panova (Koshlakova) carried out the measurements. The stack-of-layers method, which in the literature outside the Soviet Union has received the name "Akhmatov sandwich," is set forth in detail in Akhmatov's monograph.

is explained by a rearrangement of the molecular structure of the substance under the compressive load. Migal and Pavlov observed a significant change in the electrical resistance of a lubricated contact upon a deformation of the lubricant film between burnished roller surfaces.[200]

Now let us examine a curve showing how the integral emf $\mathcal{E}_f$ varies in the friction of carbon steels in liquid paraffin (Fig. 3-1).[27] Because of complete internal compensation of the electric moments of atomic bonds in methylene and methyl groups, the paraffin molecules are nondipolar. It follows that $\mathcal{E}_f = 0$ for complete boundary lubrication, and $\mathcal{E}_f \rightarrow \mathcal{E}_t \neq 0$ when direct metallic contact occurs on the interface. It would seem that, even in the case of nonpolar hydrocarbons, recording the integral emf would make it possible to monitor the transition from pure boundary friction to mixed sliding conditions. However, the records suggest that as the load N on the epilamen increases, an additional discontinuity in the potential appears. This potential discontinuity can be linked with induced polarization of neutral paraffin molecules and with their orientation in the interfacial boundary field. As Akhmatov remarks, when a methylene chain is compressed, a resultant electric moment of deformational origin appears in it.[33]

All this taken together is in agreement with previous findings on analogies in the variation in boundary-layer molecular structure under the influence of mechanical deformations and electric fields (after Frenkel). Thus, it is necessary to take up the question whether mechanical hysteresis must be studied along with electrical hysteresis.

Studies of the conditions favoring the formation of ordered structures with minimal surface energy on rubbing surfaces are exceptionally important. This

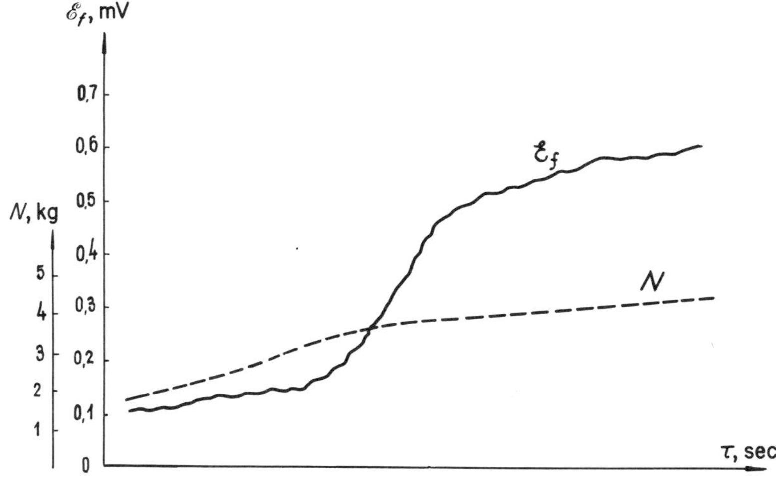

Fig. 3-1. Oscillogram of the potential difference between carbon-steel specimens sliding in paraffinic-naphthenic mineral oil ($v \approx 3.5$ meter/sec).

became particularly clear when, subsequent to the phenomenon of super-easy sliding, discovered by Hardy[201] and observed by Akhmatov and Panova,[33] the phenomenon of ultralow friction upon intense irradiation of a solid surface with a stream of accelerated particles was discovered.[54]

3.3. EFFECT OF SIZE OF CHAIN MOLECULES ON THE LUBRICANT PROPERTIES OF LIQUIDS

Using conductivity measurements on a lubricated frictional contact, Lunn investigated the load-carrying capacity of oil boundary layers.[202] To evaluate the ability of a lubricant film to hold rubbing surfaces apart, he used the frequency and the spacing of states corresponding to either the presence or the absence of point metallic contacts. A steel ball executed reciprocating motion on a flat surface (copper-, aluminum-, tin-, and lead-based alloys and cast iron). The patterns in Fig. 3-2 show what kind of information Lunn got from the oscilloscope screen; from left to right, they correspond to the transition from "complete metallic contact" to interaction through a lubricant layer ruptured in a few places.

He thus concluded that the nature of these solid substrates had a substantial effect on the load-carrying capacity of films formed on them. He says that the "consistency of the near-surface layer" is quite different from that of the bulk oil; the layers have thicknesses of $\sim$1,000 Å; and, for a well-chosen oil-metal combination, the layers can hold the rubbing surfaces completely apart, even if the sliding speed is extremely low and the specific load is correspondingly high (e.g., up to 1,500 kg/cm^2 for soft babbitt metals).[202]

Using a method similar to that just described to record the number and repetition rate of the electrical signals (pulses appearing at microruptures of the boundary lubricant layer), Hanmamedov and coworkers evaluated the frictional force and the wear values for the elements of the frictional couple.[203]

Unfortunately, both the first and the second of these studies used lubricants of complex composition, whereas the study of phenomena at the molecular level requires the use of chemically well-defined substances and mixtures of two or perhaps three components for which we know well the "biography" of the molecules.

What is more, the studies cited have a common (and, given the existing hardware for physical experiments, probably an unavoidable) shortcoming: they were

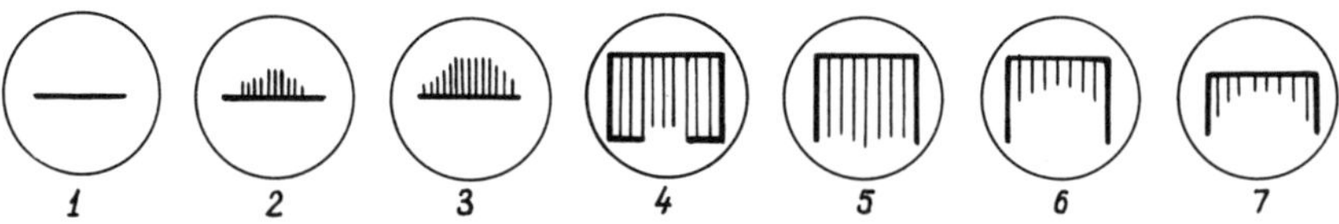

Fig. 3-2. Patterns seen on the oscillograph screen in the evaluation of load-carrying capacity of boundary layers of oils (after B. Lunn).

conducted under such conditions that, in determining the mechanical strength of lubricant layers, the investigators had to reckon with the dielectric strength of the layers and with possible threshold switching in them. These phenomena were mentioned previously.

As we will see later on, this shortcoming automatically disappears when a method proposed by the author is used to determine the load-carrying capacity of boundary phases precipitated from a layer of electrolytic lubricant. A few preliminary remarks will make the matter even clearer.

Besides the load-carrying capacity of a lubricant film, the following are known: the screening ability of the film, characterizing the degree of distortion and extinction of the residual field of a solid already coated with a screening oxide film;[33] the lubricating ability of the liquid (one recent definition of this term is "adsorbability and chemosorbability on solid surfaces");[204] and the penetrating power of the medium (Section 4.4). It is perfectly clear that each of the properties just named (for example, the oiliness [Schlüpfrigkeit, **маслянистость**, onctuosité] of a liquid), all of which act together to give an antifriction effect, is a complex characteristic. Thus, each depends not only on the structure and properties of the lubricant molecules but also on other factors. These factors include the adsorption potential and chemical activity of the rubbing surfaces, the temperature gradient and electric field strength in the crevice-shaped space between phases, special properties that the lubricant layer takes on in the boundary state, and so on. For these reasons, it is customary in comparing the lubricant functions of liquids to use a parameter, such as the coefficient of friction, whose value is affected by any change in the three-phase boundary system.

As the molecular weight of the lubricant (length of polar chain molecules) increases, friction declines; Hardy was the first to obtain this result.[205] Within a single homologous series, the coefficient of friction μ is a linear function of the molecular weight M:

$$\mu = a - bM. \tag{3-1}$$

The ordinate intercept a is a measure of the frictional interaction of clean surfaces. The slope b indicates the change in friction associated with each methylene group in the given chemical series (normal paraffins, fatty acids, alcohols, etc.). The results of Hardy's experiments were confirmed by the investigations of Bowden and Tabor[176] and, somewhat later, by the work of Zisman.[206]

It should be noted that Equation 3-1 is valid only up to some definite member of the homologous series; after that the value of μ does not depend on molecular weight. It turned out not to be so simple to identify the homolog beyond which all members of the series have the same antifriction activity. For example, in the series of monobasic carboxylic acids the "critical" homolog is the twelfth according to Hardy, the fifth according to Bowden, and the fourteenth according to Zisman.

As Fuks showed, the phenomenon of predictable variation in boundary friction upon a change in molecular structure can also be seen in solutions of fatty acids in various solvents.[207] Fuks concluded that the increase in lubricating ability in the systems he studied was due to the effect of carbon-chain length in the molecules on the disjoining pressure of thin liquid films (after Deryagin).

In Akhmatov's judgment, the results obtained by Hardy, Bowden, Zisman, and Fuks "can be explained on the hypothesis of a residual field of the solid phases."[33]

While hydrocarbons, alcohols, and amines exhibit a tendency to be physically adsorbed on a metal surface, fatty acids may react chemically with the surface to form either a monolayer of metal soap or a viscous layer of considerable thickness that behaves like an elastohydrodynamic film.[208] In the chemical theory of boundary friction (developed by Bowden and coworkers),[209, 176] it is remarked that saponification of fatty acids takes place with the participation of oxides; however, not enough attention is accorded the fact that the oxides themselves have a powerful screening effect and thus reduce the coefficient of friction.[33]

The chemosorbability of hydrocarbons on a juvenile metal surface was established comparatively recently; it turned out that an alkene (that is, a hydrocarbon containing a double $C{=}C$ bond) was a far more effective lubricant than the corresponding alkane.[210]

It is not nearly always possible to predict ahead of time what will be the joint product of the processes taking place in a three-phase system (thermal degradation of molecules, catalysis, direct chemical reactions, etc.). However, several investigators have already observed the formation of this product in the form of a black polymerlike residue.[211, 212]

Now it becomes clear that the field of boundary phenomena, important in the friction of metals when organic lubricants are present, has not been studied much.* Meanwhile, a method described by the author and coworkers makes it possible to obtain information about the kinetics of formation of thin lubricant films under the bilateral influence of metallic phases and to evaluate the load-carrying capacity of these films.[17–20]

The idea of the method is very simple: recording and calculating statistically the variations of emf $\mathcal{E}_f$ in a three-phase frictional system (Fig. 3-3). In the presence of an ion-containing medium (R_e is the resistance of the electrolyte), such a system is a galvanic cell; when it is short-circuited, the galvanic emf $\mathcal{E}_g$ acts as if it were cut out of the external circuit. The system remains a seat of emf (a natural thermocouple); however, as a rule, the absolute value of the thermal emf is small: $|\mathcal{E}_t| \ll |\mathcal{E}_g|$. Jumps in emf corresponding to the transitions $\mathcal{E}_g \rightleftharpoons 0$ (making and breaking of contact) or $\mathcal{E}_g \rightleftharpoons \mathcal{E}_t$ (the same with sliding) are like commands; thus, the source of information actually controls the communication channel between the pulse generator and the pulse counter. Suppose, for exam-

---

*Readers who wish a closer acquaintance with the state of research in this field can consult the review by D. Tabor, delivered at the Conference on Lubrication and Wear, London, September 1967.[213]

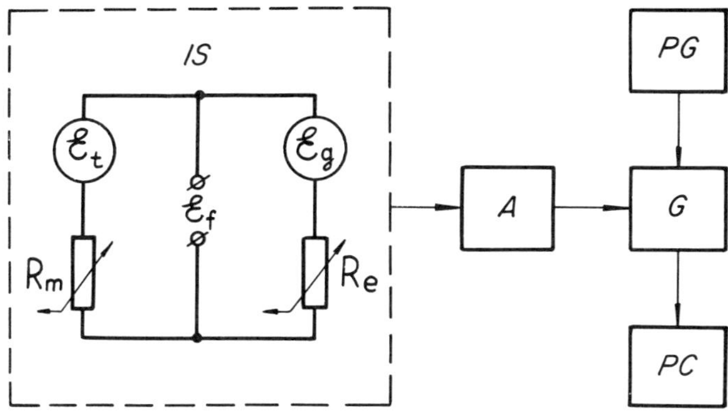

Fig. 3-3. Block diagram of device for statistical evaluation of the lubricant properties of fluids. *IS* = information source; *A* = amplifier; *G* = threshold gate; *PG* = pulse generator; *PC* = pulse counter.

ple, that in the case of direct contact between the metals, when the total resistance of a discrete contact $R_m \rightarrow 0$ and $\mathcal{E}_f \approx \mathcal{E}_t$, the threshold gate is closed. The situation will not change as long as there is a single contacting island in the layer of electrolytic lubricant. But no sooner will all such islands disappear than an amplified signal corresponding to a value $\mathcal{E}_f = \mathcal{E}_g$ will open the threshold gate; after that, the pulses emitted by the generator will begin to reach the counter. From the number of counts recorded in identical times (τ_s), the overall time (τ_e) during which $\mathcal{E}_g$ acts is determined for each of the compositions under study. It is self-evident that the control signal-commands arriving at the threshold gate can easily exchange roles. Then, the number of counts recorded will correspond to the time during which $\mathcal{E}_t$ acted $(\tau_m = \tau_s - \tau_e)$. The τ_e and τ_m values obtained by automatic computation of the discontinuous changes in $\mathcal{E}_f$ are used for a comparison of different lubricants with respect to protective function; in this case, the initial conditions of the trial are kept completely identical. According to Fig. 3-4, for example, of the three test lubricants ($L1$, $L2$, and $L3$), only the last allows the process to be conducted under conditions that completely rule out metallic contact throughout time τ_s. Lubricant $L2$ performs the protective function far less strongly; disruption of the continuity of the film formed by lubricant $L1$ takes place with no "time lag" at all (as in air).

The load-carrying capacity of a lubricant layer is evaluated in the following way:

$$\delta = \frac{\tau_l}{\tau_f} \tag{3-2}$$

where

$$\tau_l = \sum_{i=1}^{N} \Delta\tau_{li}; \; \tau_f = N \, \Delta\tau_f;$$

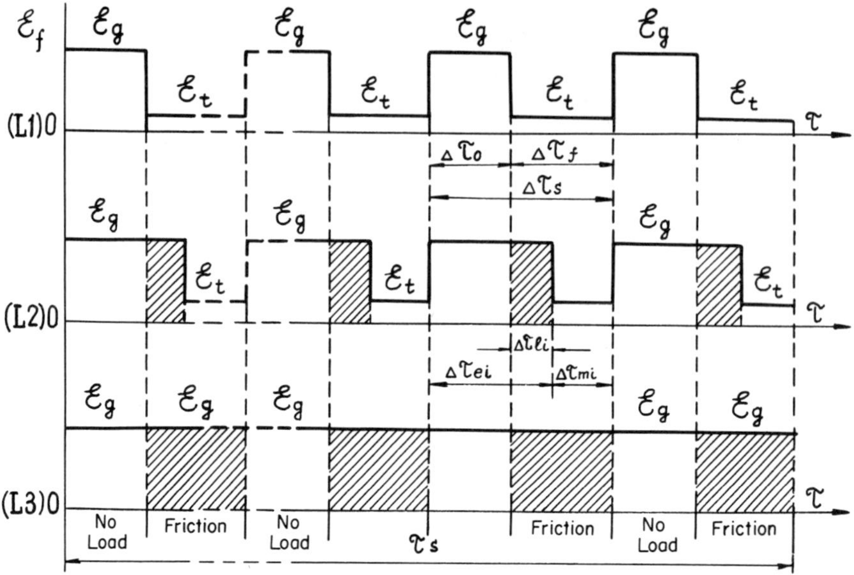

Fig. 3-4. Time variations of integral emf in cyclic contacting of rubbing metals in three electrically conductive media ($L1$, $L2$, and $L3$). τ_s = total duration of the experiment for each lubricant. Time intervals for the i-th cycle of contacting: $\Delta\tau_s$ = total duration of cycle; $\Delta\tau_o$ = time in which the contact is open in air (no-load); $\Delta\tau_f$ = time in which the contact is closed in air (friction); $\Delta\tau_{ei}$ = time in which surfaces are separated by a layer of electrolyte; $\Delta\tau_{mi}$ = time of direct metallic contact; $\Delta\tau_{li}$ = time in which electrolyte acts as lubricant.

and N is the number of contacting cycles.

If an interval $\tau_s = N\,\Delta\tau_s$ is selected and the counter readings are used to find

$$\tau_e = \sum_{i=1}^{N} \Delta\tau_{ei}$$

and $\tau_0 = N\,\Delta\tau_0$ (the constant term), then

$$\delta = \frac{\tau_e - \tau_0}{\tau_s - \tau_0};\tag{3-3}$$

and for known $\tau_m = \sum_{i=1}^{N} \Delta\tau_{mi}$ and τ_f we have

$$\delta = 1 - \frac{\tau_m}{\tau_f}.\tag{3-4}$$

The diagrams of Figs. 3-5(a) and (b) illustrate the determination of τ_e (or τ_m) and τ_0 (or τ_f), respectively.

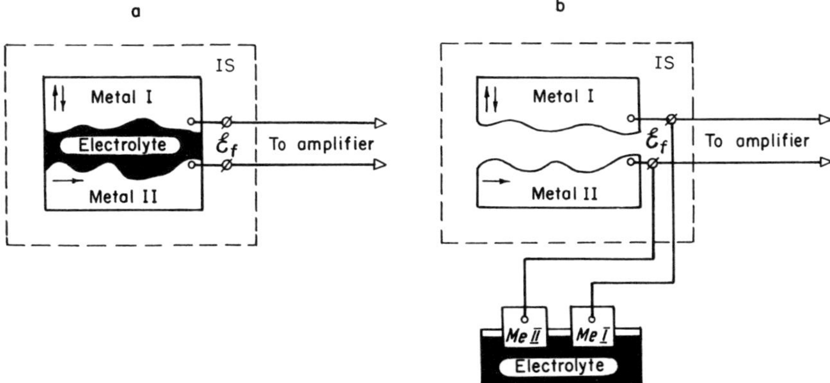

Fig. 3-5. Information source in principal (a) and auxiliary (b) states. The arrows indicate directions in which the specimens may move.

Molar aqueous solutions of the sodium salts of dibasic carboxylic acids were chosen for investigation. According to Table 3-1, galvanic cells with iron, nickel, and zinc cathodes differ in relative stability of the emf values. The relative stability is greater for the Cu-Fe couple, which was used in the first series of trials. The work of adhesion was computed by the Young-Dupré equation

$$A = \gamma_{LA}(1 + \cos \theta) \tag{3-5}$$

where γ_{LA} is the surface tension of the lubricant and θ is the contact angle.

It should be remarked, however, that this calculation is somewhat arbitrary with regard to the systems selected.

Figure 3-6 gives a schematic overall view of the sliding system. Intermittent sliding contact between electrodes 1 and 2 was provided by means of eccentric 3, which periodically pushed against follower screw 4, causing specimen 2 to move out of contact with specimen 1, then allowing it to return, all on a fixed cycle. The eccentric rotated at the same rate (60 rpm) as specimen 1. Thus, the total cycle length $\Delta\tau_s$ equaled the time required for specimen 1 to make one turn, that is, 1 sec. The eccentricity was selected so that the times $\Delta\tau_0$ and $\Delta\tau_f$ were about the same. The maximum clearance between the specimens was controlled by the follower screw and did not exceed 0.5 mm. The force pressing the specimens together during sliding (~9 g) was provided by lever 5 and a string passing over pulley 6 and carrying a weight. Both electrodes were electrically well insulated from the rest of the device. The rotating specimen was connected with the external electric circuit through mercury contact 7. The liquid lubricant entered the friction zone by way of a rubber hose with glass tip 8.

The surfaces were prepared (phase oxides, oils, and greases were removed) with fine abrasive paper and acetone.

Table 3-1. Data on molar solutions of salts belonging to the homologous series $(CH_2)_n(COONa)_2$.

Solution of Electrolyte	Molecular Weight M	Viscosity (Poise) (18–19 °C)	Work of Adhesion A (erg/cm^2)				Galvanic emf $\mathscr{E}_g$ (mV[a])		
			$_{26}$Fe	$_{28}$Ni	$_{29}$Cu	$_{30}$Zn	Cu(+) Fe(−)	Cu(+) Ni(−)	Cu(+) Zn(−)
Sodium oxalate NaOOC–COONa	134	90.8	143	124	146	142	172→	35↓	470→
Sodium malonate NaOOCCH$_2$COONa	148	96.4	160	140	171	165	475	56↓	924→
Sodium succinate NaOOC(CH$_2$)$_2$COONa	162	101.9	98	79	122	104	508	108→	990
Sodium glutarate NaOOC(CH$_2$)$_3$COONa	176	107.5	79	79	96	81	505	83↓	988→
Sodium adipate NaOOC(CH$_2$)$_4$COONa	190	113.1	141	118	152	150	498→	97↓	994↓
Sodium azelate NaOOC(CH$_2$)$_7$COONa	232	123.6	77	81	85	73	195↓	80↓	819↓
Sodium sebacate NaOOC(CH$_2$)$_8$COONa	246	125.3	108	95	110	106	186↓	86↓	905↓

[a]Arrows indicate moderate (→) and rapid (↓) decrease in $\mathscr{E}_g$.

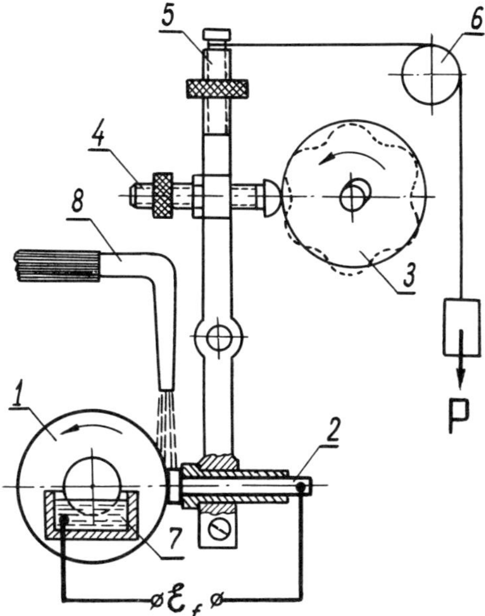

Fig. 3-6. Schematic diagram of the frictional assembly.

Along with trials in which chemically reactive metal surfaces were screened by
the formation of soap films on them, an evaluation was conducted on the protective properties of adsorbed layers between materials with good corrosion resistance.   The materials selected were aluminum bronze and chromium-nickel-
titanium steel, which has especially good resistance to oxidizing aggressive media.
It turned out that one important circumstance helping to explain the differences
in the abilities of van der Waals films to prevent the restoration of contact is the
way in which specimen 2 oscillates when it is free for short times to execute its
normal motions. This is why, before the second series of trials, the eccentric disk
in the friction device was replaced with the stellate cam shown by the dashed
line in Fig. 3-6.

Figure 3-7 presents curves of load-carrying capacity versus time for the case of
distinct chemical reactions between the lubricant and a copper surface. These
curves, especially 6 and 7, are approximate, since the rapid polarization of the
electrodes upon short-circuiting of the galvanic cell results in a decline in emf;
this behavior is reflected to some extent in the relation between τ_l and τ_m. These
curves can nonetheless be used to follow the kinetics of the processes taking place
in a contact interaction of metals in the presence of ion-containing media. According to Fig. 3-7, for example, when solutions of different salts belonging to
the same homologous series $(CH_2)_n(COONa)_2$ are used as liquid lubricants, the
kinetics of the chemical reactions in which the solutions participate are substan-

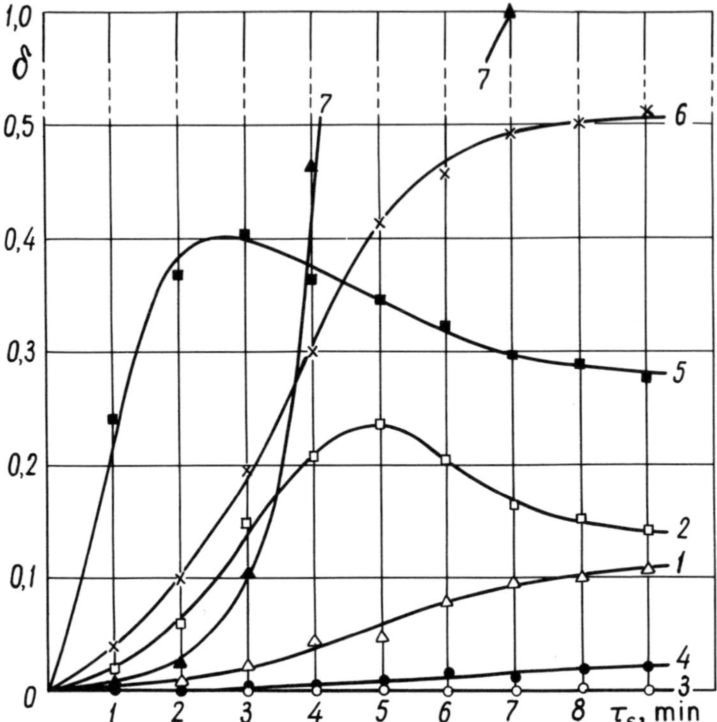

Fig. 3-7. Criterion of load-carrying capacity δ as a function of running time τ for the couple of Cu (disk-shaped specimen) and Fe (cylindrical specimen) in contact interaction in $1M$ solutions of the sodium salts of dicarboxylic acids: (1) oxalic; (2) malonic; (3) succinic; (4) glutaric; (5) adipic; (6) azelaic; (7) sebacic.

tially different. The larger the organic molecules, the greater ability the resulting films have to prevent metallic contact between the surfaces. With a solution of sodium sebacate as lubricant (curve 7), microareas of contact with low resistance were seen to disappear entirely after the friction device had been in operation for 7 minutes. This behavior could have resulted directly from hydrolysis of the sodium soap

$$C_8H_{16}(COONa)_2 + 2HOH \rightleftharpoons C_8H_{16}(COOH)_2 + 2NaOH$$

followed by saponification of sebacic acid in chemical reactions with the metals comprising the frictional couple. Under conditions like these, the lubricant effect is influenced by all types of metal soaps that form, on dissolving in water, a smectic variety of the liquid-crystal state. We note, incidentally, that an aqueous solution of soaps of higher fatty acids, which is alkaline, was also a good medium for realizing the selective-transfer effect.[214]

The law mentioned above is not, however, general for all salts of dicarboxylic acids. It does not apply to solutions of sodium succinate or glutarate, for which $\delta = 0$ (curve 3) and $\delta < 0.05$ (curve 4). These anomalously low δ values are linked, it is natural to suppose, with the conversion of succinic and glutaric acids into their cyclic (internal) anhydrides by splitting off of water. This decomposition goes particularly easily upon heating of succinic acid, whose molecules have four carbon atoms:

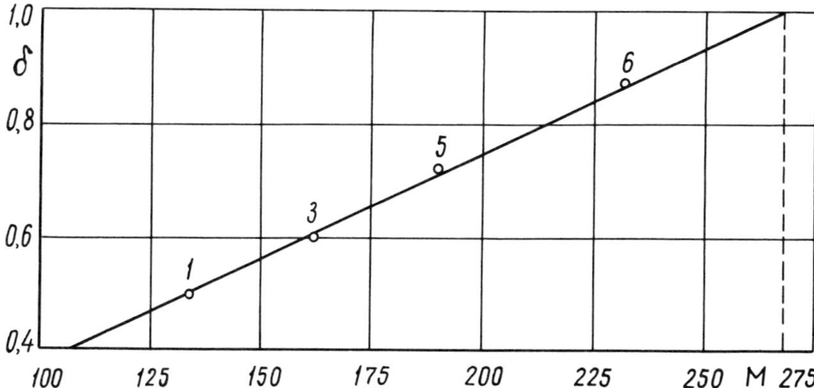

In trials with succinic acid, the author observed only slight tarnishing of the copper surface, whereas the use of malonic and adipic acids was accompanied by obvious decomposition of the chain molecules with the formation of sootlike deposits. The deposits could be freely removed from the solid surface as they built up; this behavior probably does explain the decrease in δ (maxima on curves 2 and 5) after blackening of the friction track on a copper specimen.

In a series of experiments with chemically inactive alloys and a lesser collection of the same acids, a result (Fig. 3-8) was obtained that is in direct agreement with the results of Hardy's trials. It turned out that, under conditions where van der Waals adsorption predominates, the load-carrying capacity of a boundary layer is

Fig. 3-8. Effect of molecular weight of lubricant on the load-carrying capacity of a boundary layer, in the contact interaction of the couple of Cr-Ni-T steel (type Kh18N9T) and aluminum bronze (type BrAZh9-4). The figures identify the electrolyte solutions (see also Table 3-1): (1) sodium oxalate; (3) sodium succinate; (5) sodium adipate; (6) sodium azelate.

an increasing linear function of the molecular weight M and thus of the size of the chain molecules:

$$\delta = kM. \tag{3-6}$$

Here, k is a constant equal to the reciprocal of the limiting value of M, that is, the value of M beginning at which $\delta = 1$ (in our case $k = \frac{1}{268}$).

The most important feature of this result, it seems to us, is that it reflects the mechanical strength and the lubricating function of preferentially reticular structures formed by polar molecules with two active centers located at the ends. The difference between the solutions with respect to dynamic viscosity (Table 3-1) is not as great as that with respect to the value of δ. Hence, it can be concluded that, when the indenter has some freedom to carry out its normal motions, the effect of the extraviscous (boundary) state of a layer on the load-carrying capacity makes itself quite apparent.

With a similar friction device, but with significantly greater loads and larger specimens, a statistical pattern was also obtained for the stability of lubricant films when there is spontaneous vibration of the elements of the frictional couple. According to the experimental data, the load-carrying capacity of boundary layers depends on the nature of the metallic bodies on which they are formed.[20]

In conclusion, let us note that the method described above can be used at will for a rapid search for optimal lubricant compositions to comply with given combinations of materials in the contacting couples.[19]

4

Experimental Investigation of Electrical Phenomena in the Cutting of Metals

4.1. PARAMETERS OF THE TOOL-WORKPIECE-MACHINE THERMOELECTRIC CIRCUIT

Figure 4-1 shows a diagram of the tool-workpiece-machine thermoelectric circuit. Opening the external portion of this circuit is one way to improve the wear resistance of a metal-cutting tool. The principal parameters of the external circuit include:*

(a) The thermal emf acting in it, with a resultant value of

$$\mathcal{E}_t = \sum_{i=1}^{n} e_i g_i / \sum_{i=1}^{n} g_i \qquad (4\text{-}1)$$

where e_i is the emf in the loop g_i–R_0, and g_i is the conductance of the i-th constriction at the interface between the tool (1) and the chip (3) or the workpiece (2).

(b) The resistance of the cutting zone

$$R_m = 1/ \sum_{i=1}^{n} g_i; \qquad (4\text{-}2)$$

*The analogy between the formulas presented here and some of the ones in Section 2.3 is quite natural and underlines that both sets have the same physical meaning.

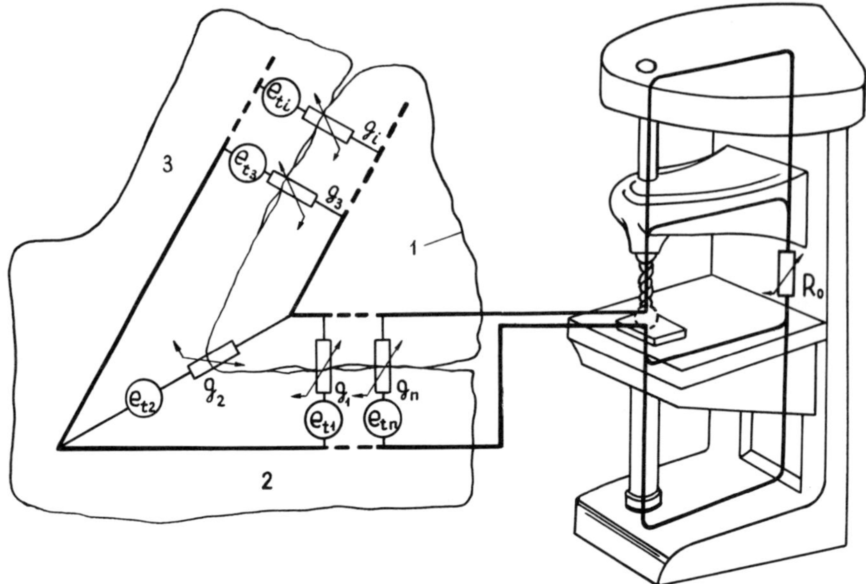

Fig. 4-1. Schematic diagram of the tool-workpiece-machine thermoelectric circuit.

(c) The resistance R_0, which includes those of the machine and all the transfer contacts at points of attachment of the tool, holding device, and workpiece. In that which follows, we will call this just the resistance of the machine, since the predominant influences on R_0 are the type of machine, its degree of wear, the state of the kinematic chain, and similar factors.

The values of $\mathscr{E}_t$, R_m, and R_0 determine the thermoelectric current in the external circuit of interest to us,

$$I_t = \mathscr{E}_t/(R_m + R_0). \tag{4-3}$$

According to the data of Avakov and coworkers, the value of $\mathscr{E}_t$ in the machining of steels with carbide-tipped tools may reach 15–20 mV.[215] The value is influenced by "parasitic" thermal emf's, whose seats are the rubbing contacts in the kinematic chain of the machine, points of contact between the cutting element and the holder, those between holder and machine and between workpiece and machine, and so on.[111, 216]

The present author and coworkers found that in the drilling of difficultly machinable materials with a tool made of high-speed steel, the value of $\mathscr{E}_t$ found by the method of compensation is on the order of a few millivolts (Fig. 4-2).[30] The value depends on the conditions of cutting.

For what follows, it is rather important to note that Semko demonstrates the strong influence of tool heat treatment on the thermal emf.[217] The same publi-

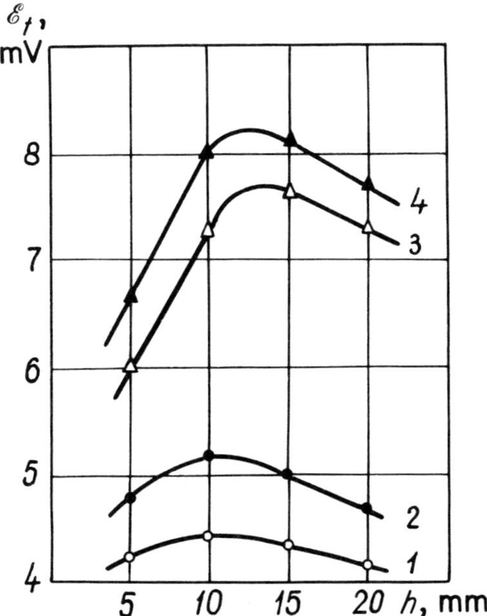

Fig. 4-2. Thermal emf of the external circuit versus depth of drilling of specimens of type-VT5 titanium alloy (curves 3, 4) and type-EI654 stainless steel (curves 1, 2). The tool material is type-R18 high-speed steel, workpiece thickness is 25 mm, and drill diameter is 10 mm. Cutting conditions: (1) v = 4.4 meter/min, s = 0.11 mm/revolution; (2) v = 4.4 meter/min, s = 0.20 mm/revolution; (3) v = 4.4 meter/min, s = 0.15 mm/revolution; (4) v = 8.6 meter/min, s = 0.15 mm/revolution.

cation also indicates that there is an inversion point of the natural tool-chip thermocouple in some range of cutting speeds.

Using sensitive measuring apparatus and special screening arrangements to eliminate external noise, Yakovlev measured the ohmic resistance of the region of the sliding contact in cutting.[218] He found that in the milling and turning of type-30 steel with a carbide-tipped tool made of type-T15K6 alloy, the value of R_m varies over the range of $\sim$ 0.001–0.0025 Ω for a wide range of cutting speeds reaching up to and beyond 300 meters/min. Yakovlev thinks that the increase in contact resistance seen at certain speed values is caused by the formation of an oxide film.

Ryzhkin and coworkers found roughly the same value of R_m ($\sim$ 0.0011–0.0021 Ω) by taking load characteristics of the spontaneous seat of thermal emf formed by the cutting zone.[219, 220] After studying the variation of R_m as a function of operating conditions in the turning of type-45 steel using a cutter with disposable blades of type-T15K6 alloy, they looked into the circumstance that, when the feed increases, the rise in the area of real contact may exceed severalfold the rate at which the cutting-zone resistance decreases. In their judgment, again, this behavior is due to the changes in thickness, structure, and phase makeup of the oxide films as a result of the change in temperature.

The effect of oxidation on R_m is especially great when contact of the tool with the chip and workpiece is intermittent. In milling, for example, with a cutter made of high-speed steel, the resistance of the cutting zone reaches several ohms when a cutter tooth cuts into the workpiece, and falls to 0.1–0.5 Ω when the tooth withdraws; if the tool is carbide-tipped, the respective values can be several tens of ohms and 0.7–1.8 Ω.[221]

According to the external characteristics given in Fig. 4-3 for cutter-workpiece thermocouples, the value of R_m determined from the maximum power dissipated in the load equals 0.03–0.045 Ω.[222] According to Pisarev's data, the resistance of the region of contact between drill and workpiece is about 0.1 Ω when type-U8 tool steel is machined with a tool made of type-R18 high-speed steel; however, an investigation of the couple of type-VK15 alloy and type-18Kh2N4VA steel in turning and friction gives contact-resistance values in the range of 0.035–0.07 Ω.[223]

In order to obtain objective data on the magnitude of the thermoelectric current in cutting, Bobrovsky measured the resistances of various types of milling machines by the ammeter-voltmeter method.[111, 224] It became clear that R_0 can, although it rarely does, reach values of tenths of an ohm when the machine is shut down, whereas R_0 "fluctuates over a wide range, sometimes reaching several hundred ohms"[224] when the machine is running under load. As a rule, the minimum resistance of the machine lies in the range of 2–4 Ω when the spindle is turning at a low speed.

Barrow and Spenser measured steady-state resistances of tens and hundreds of ohms in tests of lathe-type milling machines; however, one of these machines had a resistance of $\sim$ 0.8 Ω under load.[225]

In the light of considerations set forth above on the disruptive function of thermoelectric current,[1] the question of the parameters of the tool-workpiece-machine thermoelectric circuit impelled the author and coworkers to undertake a simple experiment.[30]

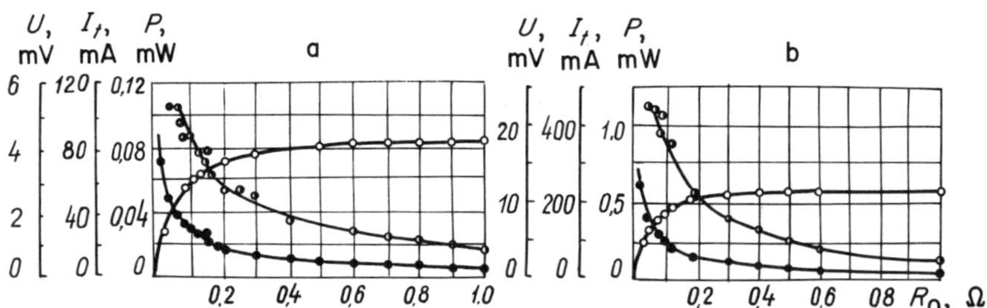

Fig. 4-3. Load characteristics of natural thermocouples formed in the turning of type-18Kh2N4VA steel at a cutting speed of $v = 50$ meter/min.[222] The cutters are one-piece, made of type-R18 steel (*a*) and type-VK15 alloy (*b*).

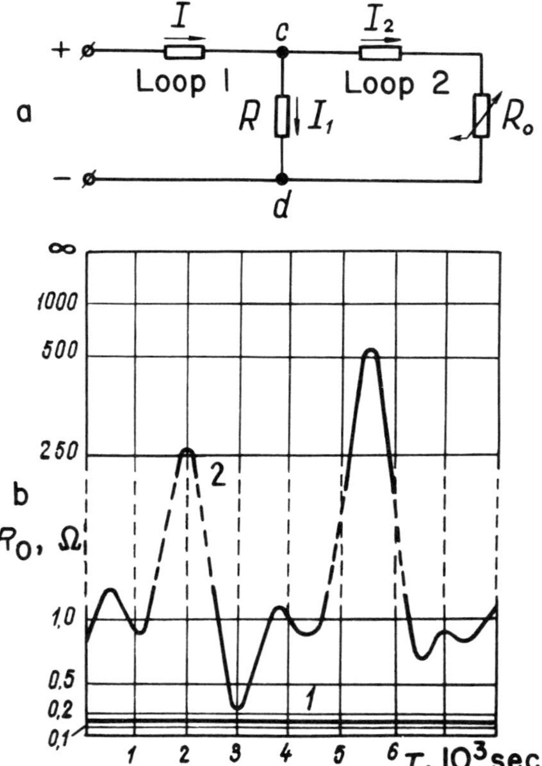

Fig. 4-4. (a) Diagram of circuit for recording machine resistance (I = 1 mA = const, R = 10 Ω = const, U_{cd} = 5–10 mV); (b) oscillograms showing the magnitude and character of the variation in R_0: (1) with machine not running; (2) in drilling of holes (R18-VT5, v = 4.4 meter/min, s = 0.15 mm/revolution, h ≈ 15 mm).

A special circuit—Fig. 4-4(a) shows a fragment of it—was selected for measuring the resistance of the machine. Stability of the circuit was assured through automatic stabilization of current I, which can be monitored by the first loop of a type-N102 oscillograph. The value of I was selected so that, as the resistance of the machine varied between 0 and ∞, the voltage across constant resistance R corresponded to the thermal emf in the machining of a titanium alloy.

The second loop was used to record R_0; a resistance box was inserted into this circuit for preliminary calibration. Trials were conducted when the machine was not operating and during the drilling of holes. The tool and workpiece were electrically insulated from the machine. A mercury cup was used to connect the current bus with the rotating spindle.

From the two oscillograms obtained, it was possible to follow the variation of R_0 in high and low ranges at the same time (Fig. 4-4[b]). When the machine was not running, its resistance was 0.1–0.2 Ω, while R_0 varied from tenths of an ohm to hundreds of ohms during cutting. However, if allowance is made for the time lag of the loop oscillograph and for the fact that the resistance of the contact depends on the voltage applied to it,[119] then it is reasonable to doubt the reliability of the lower limit of R_0. One of the shortcomings of the circuit described above is that the rise in conductance of the machine is retarded because of its own effect, namely, the decrease toward zero of the voltage across R_0.

It is curious that the drop in resistance of the machine to very small values, which could result from switching in thin lubricant films when they are penetrated by pressure, was seen comparatively seldom. On the other hand, as we have already ascertained, the resistance of the cutting zone is ordinarily on the order of thousandths or hundredths of an ohm. For this reason, it might be thought that the thermoelectric current has an especially large effect on tool wear just at the moments of the rare drops in R_0, that is, during brief surges of thermoelectric current. And while there are no grounds for speaking of a comparatively large thermoelectric current flowing in the external circuit for prolonged times, still any such conclusion with regard to instantaneous impulse values of thermoelectric current would be in error *a priori.*

Let us now turn to data on the magnitude of thermoelectric currents in cutting. The first attempt to evaluate these currents experimentally led, as we already know, to I_t values that seemed implausibly high (5.2–5.5 A).[134]

Indeed, from an inspection of the load characteristics (Fig. 4-3) and of the current-voltage characteristics (Fig. 4-5)—a variety of load characteristics—

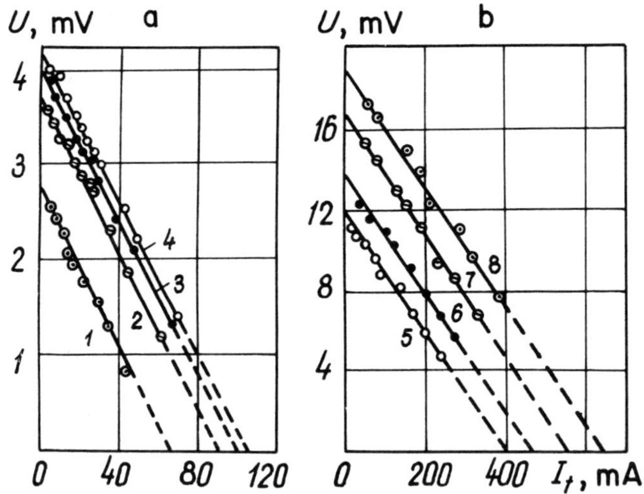

Fig. 4-5. Current-voltage characteristics of cutter-workpiece thermocouples formed in the turning of type-18Kh2N4VA steel at various cutting speeds.[222]   (a) One-piece cutter of type-R18 steel, cutting speeds (1) 21.5, (2) 30, (3) 40, (4) 50 meter/min. (b) One-piece cutter of type-VK15 alloy, cutting speeds (1) 50, (2) 75, (3) 153, (4) 304 meter/min.

obtained by Afanasiev and Bobrovsky on the assumption that I_t is a linear function of $\mathscr{E}_t$,[222] it follows that thermoelectric currents measurable in amperes do not flow in the external circuit. Even in the case of the more powerful thermocouple investigated, the short-circuit current found by extrapolation of the $I = f(U)$ curve did not exceed 700 mA; this means in practice that a thermoelectric generator whose load R_0 is always greater than zero can operate only at smaller currents.

The data of Bobrovsky, according to which the thermoelectric current in the external circuit often lies in the range of 5–15 mA,[224] were confirmed recently by Markosyan, who investigated the following natural thermocouples in various speed ranges: type-R18 high-speed steel and type-45 steel; type-VK8 alloy and type-1Kh18N9T stainless steel; type-VK8 alloy and type-EI835 stainless steel; type-VK8 alloy and type-45 steel; and type-T15K6 alloy and type-45 steel.[226] Using the compensation method for the measurement of weak currents (this method is widely used in galvanometric compensators),[227] she showed that the thermoelectric current flowing through the tool–workpiece–current-collector–tool circuit varies in the range of 8–95 mA. But under real conditions of cutting, the current may be many times less than this, because of the electrical resistance of the machine, which increases markedly on start-up.

However, as what follows will make clear, another result of Markosyan's observations is especially interesting to us. She writes, "The instantaneous thermoelectric-current values go through unaccountable pulsations, characterized by sharp jumps in the oscillogram curves."[228] The maximum value of thermoelectric current may, she remarks, approach the short-circuit value.

Although Bobrovsky's and Markosyan's conclusions on the magnitudes of thermoelectric currents during cutting have been proven, it would be extremely risky to depend on them in experimental situations that can be directly linked to the dynamics of thermoelectric processes. In our view, it is more correct in such cases to work on the basis of information supporting the results obtained by Axer (concerning the order of magnitude of I_t).

By measuring the voltage drop across a very small resistance used to simulate the workpiece-machine contact resistance, Dubrov and coworkers recorded currents in the range of 0.6–0.8 A during milling.[229] Lebedev found roughly the same values of thermoelectric current (0.4–0.75 A) in the planing of metals, and calculated a current density in the cutting zone equal to tens of amperes per square centimeter of tool-workpiece contact area.[114] (He found a similar-sized acoustoelectric current density in sodium when the mechanical-stress waves had an intensity of ~ 1 W/cm^2.)[114]

For the case where the resistance of the external circuit is small, Yakunin and coworkers have measured the thermoelectric currents flowing there in the longitudinal turning of type-9KhS steel with a one-piece high-speed cutter made of type-R18M steel.[230] Figure 4-6 gives a diagram of the trials. The four standard

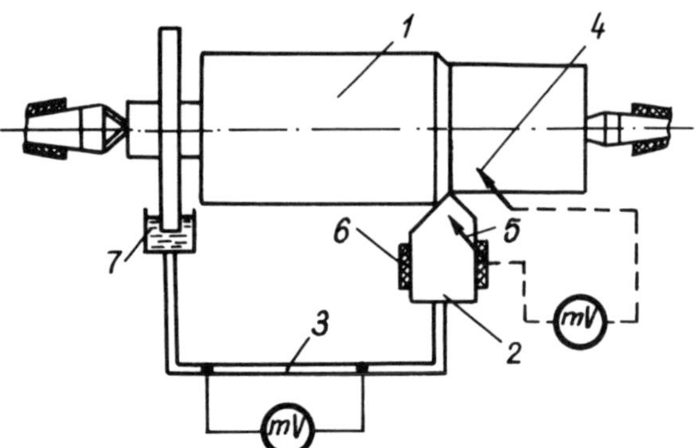

Fig. 4-6. System for determining thermoelectric parameters:[230] (1) workpiece; (2) cutter; (3) standard resistance; (4) probe made of type-9KhS steel; (5) probe made of type-R18M steel; (6) insulating liners; (7) mercury current collector.

resistances were pieces of copper cable selected with $R_{S1} = 8.7 \cdot 10^{-3}$ Ω, $R_{S2} = 5.2 \cdot 10^{-3}$ Ω, $R_{S3} = 8.2 \cdot 10^{-4}$ Ω, and $R_{S4} = 2.0 \cdot 10^{-4}$ Ω. According to the data obtained, I_t reached 4.5 A for the smallest of these resistances (Fig. 4-7).

Yakunin and coworkers also investigated how thermoelectric current flowing through the cutting zone affects the thermal emf of the natural thermocouple. By calculations, they found (for $R_m \sim 0.001$ Ω) that $\mathcal{E}_t$ declines markedly when the current load on the thermoelectric generator increases (Fig. 4-8). The explanation of this result by the Peltier effect is unexceptionable from the thermodynamic standpoint (see also Section 5.1).

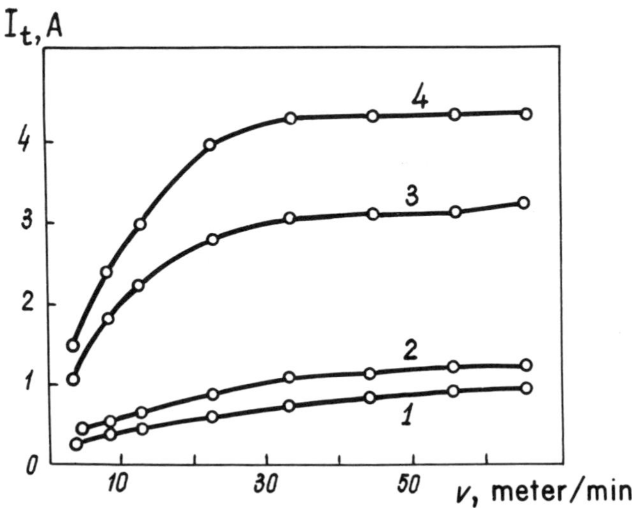

Fig. 4-7. Thermoelectric current as a function of cutting speed:[230] (1) with resistance R_{e1} inserted in circuit; (2) with R_{e2}; (3) with R_{e3}; (4) with R_{e4}.

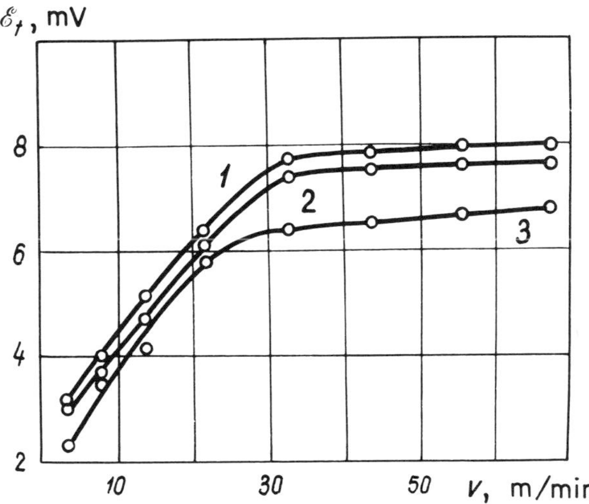

Fig. 4-8. Thermal emf as a function of cutting speed:[230] (1) measured with a millivoltmeter; (2) calculated with R_{e1} and R_{e2} inserted in circuit; (3) the same, with R_{e3} and R_{e4}.

Ryzhkin and coworkers have also reported steady-state thermoelectric currents of several amperes flowing under conditions approaching the short-circuiting of the cutter-workpiece thermocouple.[219,220]

There is no reason to doubt these results, for even in trials on a special friction tester (Fig. 4-9) the author found it easy to obtain a thermoelectric current of

Fig. 4-9. Overall view of the special device for investigation of thermoelectric phenomena in friction.

up to 1.5-1.75 A. For the determination of I_t, the voltage was measured across a type-R310 resistance coil inserted into a thermoelectric circuit comprising two stationary and two rotating specimens arranged in the following way:

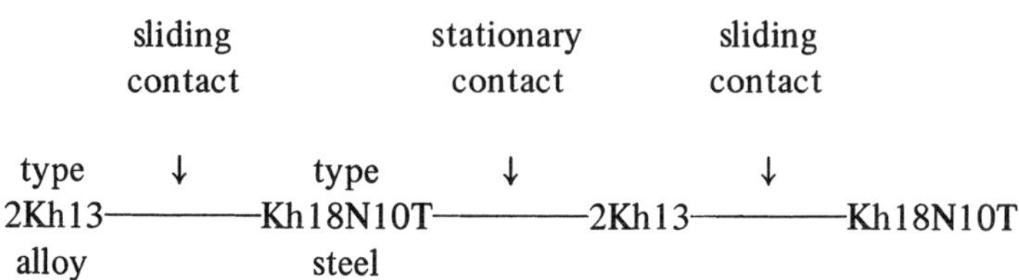

Pairwise sliding was accompanied by intensive wear of the specimens. The sliding speed was ~1.3 meter/sec and the loading, applied symmetrically by levers, was up to 18 kg.

All of the preceding makes it even clearer what role thermoelectric processes can play in the internal, low-resistance circuits of the cutting zone. It is precisely here that the different levels of development of temperature-velocity and temperature-deformation effects on the contact areas create, automatically as it were, the conditions under which quite large local thermoelectric currents can flow. Unfortunately, it is in fact impossible to measure these eddylike currents directly. And as for the parameters of the internal thermoelectric circuits (methods of determining these include drilling workpieces with thin metal foil pressed between them but at the same time electrically insulated from them;[231] separate drilling of a composite of tubular specimens of different diameters;[232] and simultaneous drilling of two cylindrical cups insulated from each other[233]), numerical values of the thermal emf's and currents obtained by the devisers of these special procedures are highly relative in value. They correspond to heat-balance conditions already known to differ from actual conditions, and may be greatly distorted by the effect of static charging due to the cutting of insulated interlayers.[111]

What has been said does not, however, stand in the way of making practical recommendations for controlling the thermoelectric processes in the internal circuits. One such recommendation is the author's suggestion of considering the conduction mechanisms of the tool and workpiece materials in selecting the optimal combination of materials.[26] (The same recommendation applies to the selection of materials in making up frictional couples.) From the standpoint of the consequences of limiting the thermoelectric currents, other conditions remaining equal, the combination of materials should correspond to the smallest value of the relative Seebeck coefficient (see also Section 5.3).[234]

4.2. EFFECT OF THE STATE OF THE THERMOELECTRIC CIRCUIT ON CUTTING-TOOL LIFE AND THE ROUGHNESS OF THE MACHINED SURFACE

When the state of the thermoelectric circuit of the frictional system (tool-workpiece-machine) is altered by opening of the external circuit, compensation of the thermal emf in the external circuit, or passage of an electric current from an outside source, the lifetimes of the elements of the rubbing couple (cutting tool) change. This effect is sometimes called the GAO (Gordienko-Axer-Opitz) effect in the literature. Since in the present case we are concerned not with the discovery of new laws but with the way in which known ones manifest themselves, this name does not seem altogether suitable. Probably the name "GAO method" would fit the actual operation on the external circuit as a way of extending the lifetime. For the results of this and related effects, in what follows we will use the term "effects of changing the electrical state (CES effects)."

The first practically useful CES effect of interest to us was the effect of opening the external circuit (electrically insulating the tool). This effect is characterized by an appropriate criterion of efficiency of the GAO method:

$$K_e = T_o/T_c,$$

where T_o is the tool life when the tool-workpiece-machine thermoelectric circuit is open, and T_c is the tool life under ordinary cutting conditions, that is, when the thermoelectric circuit is closed. But the author's task was not so much to determine quantitative values of K_e as to explain the physical reasons behind these values. The results obtained when the CES effect was pronounced can be analyzed qualitatively, even without the more precise methods of mathematical statistics. These methods must be used, though, for a quantitative estimate of K_e at the stage where recommendations are being worked out for the industrial utilization of the effect. However, the lives of high-speed drills with roughly equal hardnesses and resistances to high temperatures are scattered by more than a factor of three;[235] for this reason, it is doubtful whether carrying the values of K_e to two-decimal-place precision, as Bobrovsky often does,[29, 111] makes any sense.

The method described in Appendix 1 is recommended for a quick and relatively accurate determination of K_e with reduced consumption of materials and time.

This section includes results from experimental studies performed by the author and coworkers.[30, 234, 236, 237] Presentation of these results is preceded by that of some information drawn from the literature on the use of the GAO method.*

• When the thermoelectric circuit was opened or an external seat of emf was inserted in it (of course, we mean the external circuit), the lives of high-speed drills

---

*Markosyan systematizes data of this kind with application to various cutting operations (turning, drilling, milling, etc.).[238]

increased by a factor of 2–4.5 in the drilling of titanium alloys (types VT3-1, OT4, VT22, etc.) and by a factor of 1.5–3.5 in the drilling of steels containing titanium (types Kh18N9T, Kh18N10T, etc.).

• Opening the thermoelectric circuit changes the character of drill wear: wear on the flank faces becomes more uniform, and at the same time the wear on the margin is reduced.

• Opening the circuit has an important effect on tool life when, other conditions remaining equal, machining is done on a machine with a lower resistance or (equivalently from the standpoint of the spontaneous variation of R_0) at a lower speed.[111] The effect of electrically insulating the tool varies with speed (Fig. 4-10); this behavior is also linked with the change in the predominant type of wear, from abrasive-chemical (oxidative) to adhesive to diffusional.[239]

• The results of opening the tool-workpiece-machine circuit include a reduction in the microhardness of the built-up edge and the near-tool part of the chip and a reduction in the degree of cold work of the cut surface.[239,240] This change in the state of the thin surface layers, which has been seen in studies of microsections of chip roots, was confirmed by photostimulation of exoelectron emission. The emission level increased in the following order: annealed specimen, specimen machined with electrical insulation, specimen after conventional cutting.[241] The CES effect is therefore expressed in a change in the ratio of tool hardness to hardness of the contact layers of the chip. In turn, this change in ratio should reduce the transfer of tool material and lengthen the distance traversed by the tool and thus the tool life.[242]

• If current from a chromel-alumel thermocouple ($\mathscr{E} \sim 14$–16 mV) flows through the cutting zone in the same direction as the thermoelectric current, the lives of high-speed cutters made of type-R18 steel in the turning of type-St3 steel in an oxygen medium are increased significantly. At the same time, if the thermocouple is connected in the sense opposite to the tool-workpiece thermocouple, the life is markedly shortened.[243]

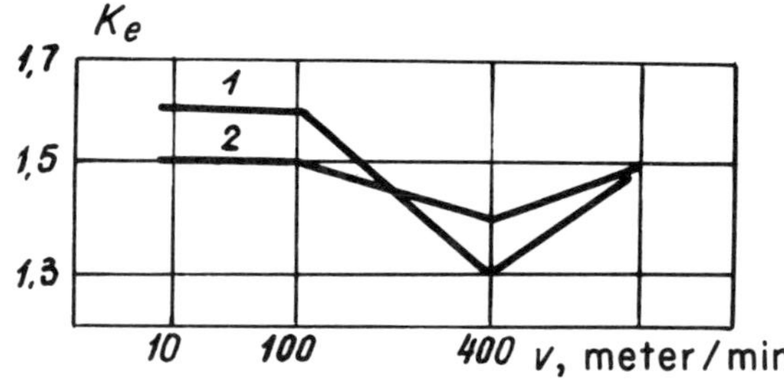

Fig. 4-10. Criterion of efficiency as a function of cutting speed, in the turning of type-30KhGSA steel with a carbide-tipped tool (T15K6):[239] (1) $s = 0.3$ mm/revolution, $t = 0.5$ mm; (2) $s = 0.1$ mm/revolution, $t = 0.5$ mm.

• The GAO method makes it possible to improve the cleanness of the machined surface by an average of one grade. The reasons are, first, that the thin oxide films formed on contact areas when oxidation is less rapid (that is, when there is no thermoelectric current) are better lubricants;[244] second, that adhesion between the metals, which is stimulated by the thermoelectric current, is weakened[239] (this view is in agreement with what the author said earlier); and third, that the harmful effects of impulse (spark) electrical discharges (which are thought[23] to contribute to "adhesional seizing") are limited.[23, 245, 246]

• If the thermoelectric circuit is opened and direct currents of 0.06, 0.12, 0.6, and 1.2 A are passed through the cutting zones in different directions, the cutting forces either change slightly or are not affected at all.[229, 244]

• The use of cutting fluids, as a rule, perceptibly lowers the criterion of efficiency of the GAO method.

• Negative assessments of the GAO method found in a number of publications are based on turning experiments with a carbide-tipped tool.[225, 228, 247–250] As far as the author knows, they do not apply to drilling with a drill made of high-speed steel.

In the very first series of trials conducted to verify the effect of electrically insulating drills made of type-R18 high-speed steel, we used two materials with different conduction mechanisms: type-VNS-5 stainless steel and type-VT5 titanium alloy. The measurements showed that, at a given temperature difference between the hot and cold ends of the thermocouple, the value of $\mathcal{E}_t'$ in the circuit of the R18-VT5 couple (p-n transition) is considerably higher than the corresponding value $\mathcal{E}_t''$ in the circuit of the R18–VNS-5 couple (p-p transition). It follows that electrical insulation of a tool made of high-speed steel has a greater effect when the workpiece is type-VT5 titanium alloy than when it is type-VNS-5 steel.

Tool-life tests were conducted on drills 10 mm in diameter with standard geometry satisfying specification RTM (Engineering Guideline) 5-65 (normal grinding). The trials took place on a model-6M12P vertical milling machine, without cooling. Table 4-1 describes the experimental conditions. The criterion of dullness was optimal wear on the flank face at the drill periphery δ_o. A Brinell magnifier was used to measure δ_o. A Textolit (resin-dipped fabric laminate) bushing fitted to

Table 4-1.

| Type of Material | Ultimate Strength σ_B (kg/mm^2) | Thickness of Workpiece h (mm) | Cutting Conditions | | Optimal Criterion of Dullness δ_o (mm) |
			v (meter/min)	s (mm/rev)	
VT5	95	25	8	0.10	0.2
VNS-5	120	25	6	0.09	0.4

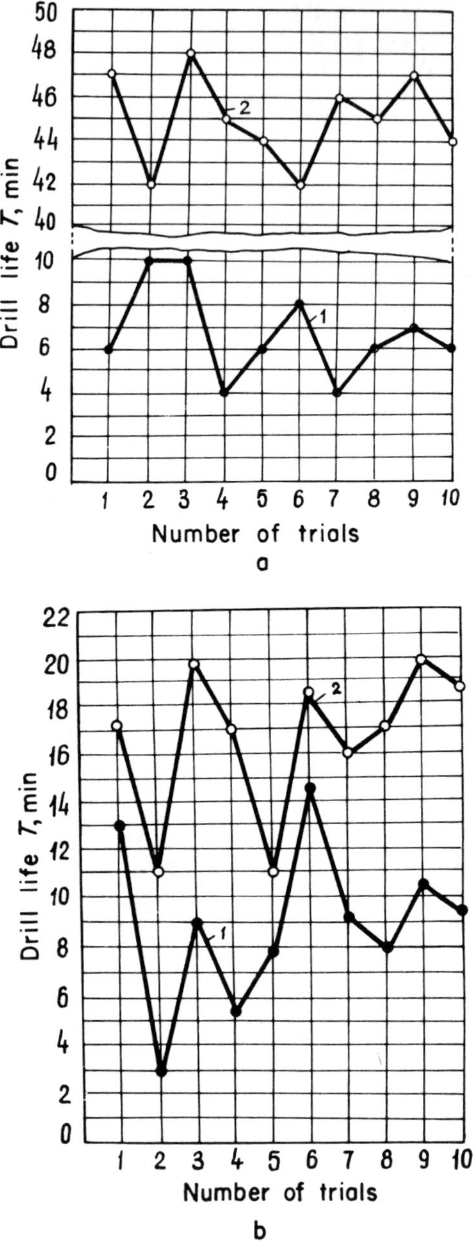

Fig. 4-11. Increase in drill life through electrical insulation of drills—(1) circuit closed, (2) circuit open—when the tool material was type-R18 high-speed tungsten steel, a p-type conductor, and materials with different conduction mechanisms were machined: (a) type-VT5 titanium alloy, n-type; (b) type-VNS-5 stainless steel, p-type.

the cylindrical drill shank was used to open the tool-workpiece-machine electrical circuit.

Figure 4-11 presents the test results. An approximate evaluation of these results showed that, for an electrically insulated drill, the life increased by a factor of less than two in the machining of type-VNS-5 stainless steel, but by a factor of more than five when type-VT5 alloy was being cut. Observations on the dynamics of drill wear on the flank face, made with a type-BMI large toolmaker's microscope, showed convincingly that in the drilling of the titanium alloy the tool is destroyed far more rapidly when the circuit is closed than when the drill is operated with the insulating bushing.

Before describing the main series of tool-life tests, we must remark that there exist several methods of opening the thermoelectric circuit: applying an insulating coating to the drill shank, using drills with conical plastic shanks, using a conical steel adapter with a nylon-6 shell on the inside, insulating the workpiece from the fastening device with plastic inserts, and so on. All these methods have a beneficial effect on tool life and on the quality of the machined surface. However, if the advantages and drawbacks of the several types of electrical insulation are compared, it turns out that in practical terms it is far from indifferent in which way the tool-workpiece-machine circuit is opened.

A plastic cone can be viewed as a damper of mechanical vibrations; its use, regardless of its electrically insulating properties, lowers the roughness of the machined surface and extends the tool life.[251–254] But according to the available information, the use of a plastic shank lowers the stiffness of the drill–holding-device–spindle system, thus causing somewhat greater tool wear.[255] For this reason, a bonded drill chuck (the same part in all trials) was subsequently used to open the thermoelectric circuit. The circuit could be closed artificially through a low-resistance shunt (Fig. 4-12).

The following metals were machined: types-VNS-5 and -EI654 stainless steels, type-EI643 constructional steel, and type-VT5 titanium alloy. The drilling machine was a model 2A135. Since the materials supplied to the enterprise were checked on arrival, no additional chemical analysis was done.

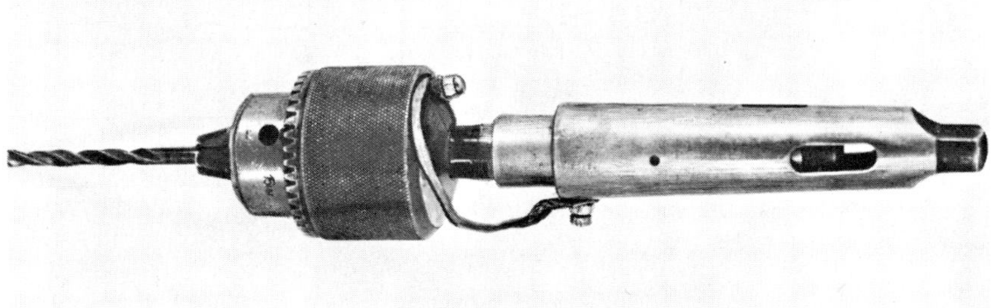

Fig. 4-12. View of the chuck with drill and adapter bushing connected by a single-strand copper conductor.

For each metal type, all the specimens were forged from a single bar, taking on the shape of rectangular plates. The thickness of each plate was selected to go with the drill diameter D and, as a rule, equaled two to three times D. After shaping, the workpieces were heat-treated; after heat treatment the materials VNS-5, EI654, EI643, and VT5 were found to have mechanical properties almost the same as in the as-supplied condition.

Tool-life tests were run on batches of drills, made of type-R18 high-speed steel and with diameters of 5.1, 7.2, 10, and 15 mm. The criterion of dullness was again optimal wear on the flank face at the drill periphery. In each specific case, some fixed number of drilled holes corresponded to the optimal wear δ_0; thus, it was possible to select drills for uniform wear rate. Tests of the efficiency of the GAO method were restricted to those drills for which the wear value δ at a given number of drilled holes did not fall outside fixed upper and lower limits: $\delta_0 - \Delta \leqslant \delta \leqslant \delta_0 + \Delta$, where $\Delta \approx (0.1-0.2)\,\delta_0$. The tool was ground with strict observance of the geometry (back relief angle of $12°$ at the periphery, point angle of $2\varphi = 127 \pm 3°$).

The effect was verified in the presence of a cutting fluid (5 percent aqueous solution of self-emulsifying oil).

The results of this investigation appear in Figs. 4-13 to 4-16 as log-log graphs of tool life T versus cutting speed v, feed s, and diameter D. In each figure, part

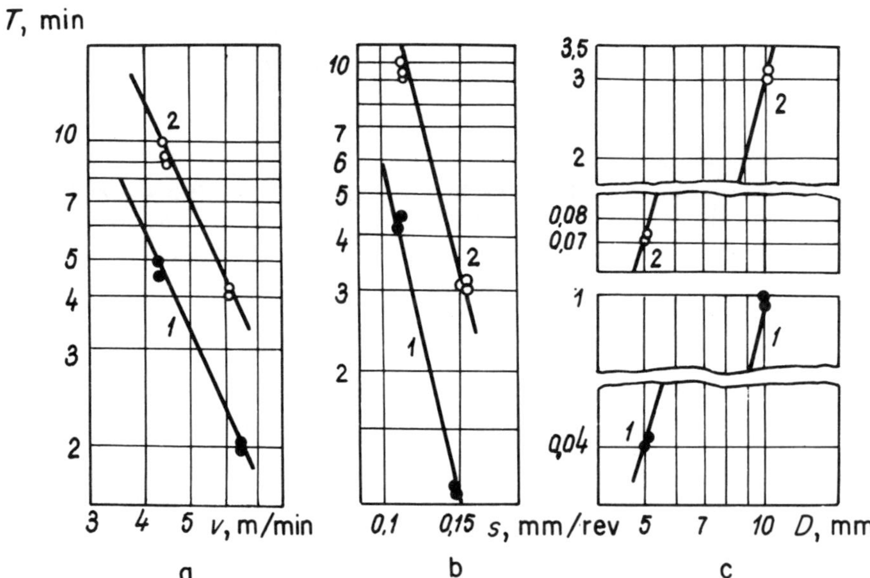

Fig. 4-13. Effect of electrical insulation of drills made of type-R18 steel in the cutting of type-VNS-5 steel ($\delta_0 = 0.4$ mm): (1) closed thermoelectric circuit; (2) open circuit.

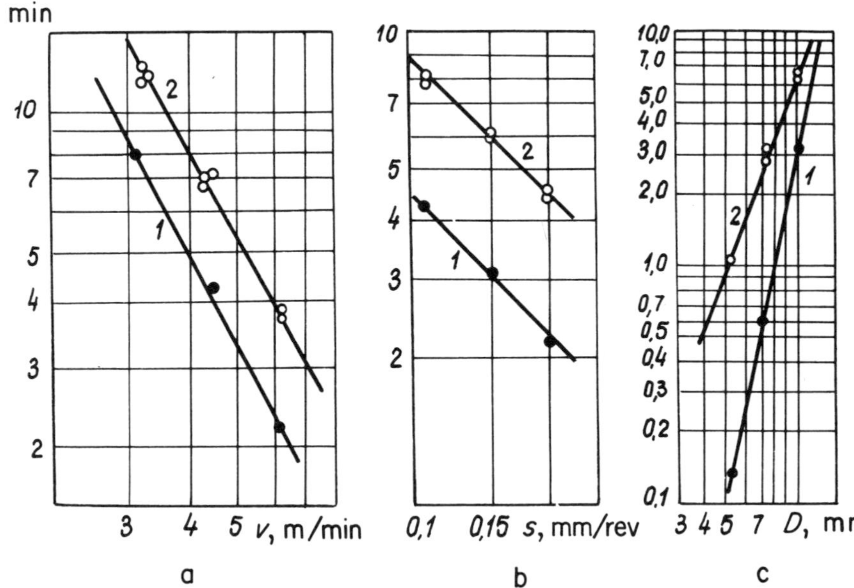

Fig. 4-14. Effect of electrical insulation of drills made of type-R18 steel in the cutting of type-EI654 steel (δ_o = 0.3 mm): (1) closed thermoelectric circuit; (2) open circuit.

(a) has s = 0.11 mm/revolution and D = 10 mm; (b) has v = 4.4 meter/min and D = 10 mm; and (c) has v = 4.4 meter/min and s = 0.15 mm/revolution.

Table 4-2 gives the ratio of open-circuit drill life to drill life with the drill shorted to the machine, for each of the materials machined.

In the cutting of a given alloy, the effect of electrical insulation of the tool is different for different drill diameters D. For example, when D increases, the ratio T_o/T_c decreases sharply when the workpiece is type-EI654 steel, almost doubles when it is type-VT5 alloy, and remains practically unchanged when it is type-VNS-5 steel. This is an exceedingly curious fact, and we will return to it in Section 5.3.

Expressing the results as the slopes of T-v, T-s, and T-D lines made it possible to compare the cutting-speed formulas for different states of the thermoelectric circuit (Table 4-3).

Table 4-2.

Material	T_o/T_c
VNS-5	2–3
EI654	1–7
EI643	1–4
VT5	3–6

Table 4-3. Formulas of the type $v = f(T, s, D)$ for various states of the tool-workpiece-machine thermoelectric circuit.

Type of Material	Ultimate Strength σ_b (kg/mm^2)	External Circuit	
		Closed	Open
VNS-5	130	$v_c = 0.004D^{1.7}/T^{0.36}s^{1.7}$	$v_o = 0.006D^{1.7}/T^{0.36}s^{1.7}$
EI654	73	$v_c = 0.004D^{2.72}/T^{0.58}s^{0.69}$	$v_o = 0.057D^{1.77}/T^{0.58}s^{0.69}$
EI643	190	$v_c = 0.042D^{0.21}/T^{0.33}s^{1.84}$	$v_o = 0.370D^{0.13}/T^{0.33}s$
VT5	75	$v_c = 0.210D^{0.27}/T^{0.58}s^{1.9}$	$v_o = 0.195D^{0.66}/T^{0.58}s^{1.9}$

From these formulas, it follows that in operation with an insulated holding device, the cutting regime can be forced if the condition $T_o = T_c$ is observed. For example, with VNS-5, $v_o/v_c = 1.5$.

We conducted an additional series of trials to determine the value of K_e in the machining, without cutting fluids, of only the titanium alloys, which have very low antifriction properties.[256] We obtained the data of Table 4-4.

On the basis of what has been said, it was concluded that opening the tool-workpiece-machine thermoelectric circuit can extend the lives of high-speed drills, and thus make it possible to increase the productivity of cutting operations.

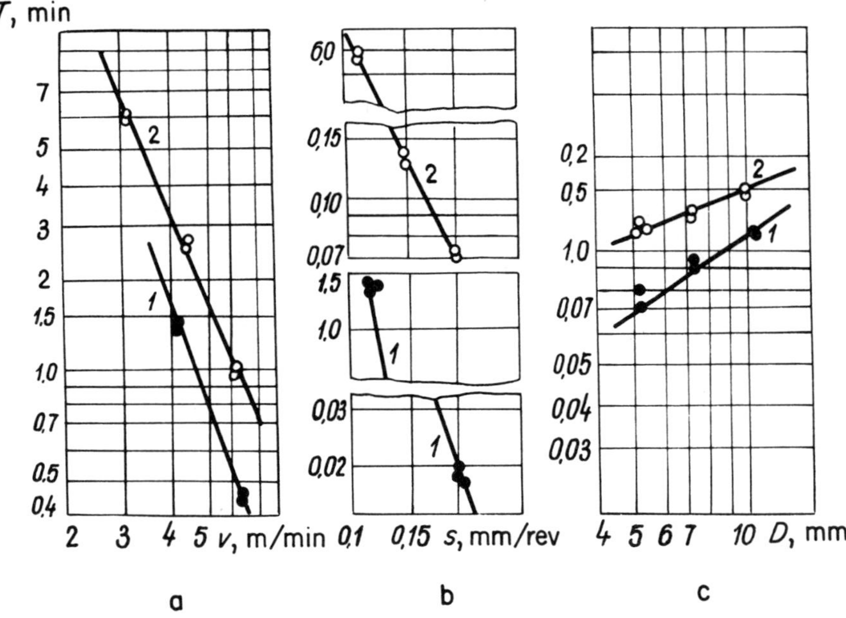

Fig. 4-15. Effect of electrical insulation of drills made of type-R18 steel in the cutting of type-EI643 steel ($\delta_o = 0.2$ mm): (1) closed thermoelectric circuit; (2) open circuit.

Table 4-4.

High-speed Steel	Titanium Alloy	T_O/T_C
R12	VT3	4
R12	OT4-1	4.5
R18	OT4-1	3
R18	VT6	2.5

4.3. THE AC COMPONENT OF THERMAL EMF AND ITS ROLE IN THE STUDY OF THE DYNAMICS OF CONTACT PROCESSES

While pointing out the errors of Dubinin,[110] the author has at the same time emphasized that an energetic approach to the study of the sliding of metals, along with Dubinin's proposed method for recording the ac component of the potential difference between rubbing surfaces, deserves serious attention.[26]

Korobov[245] and the author[26] independently adapted that method to the cutting process.

Repeating Dubinin's incorrect computations in application to the cutting of metals, Korobov found that the electrical potential at the surfaces of microcontacts reaches values of more than 1 kV. This error led to the quite unwarranted

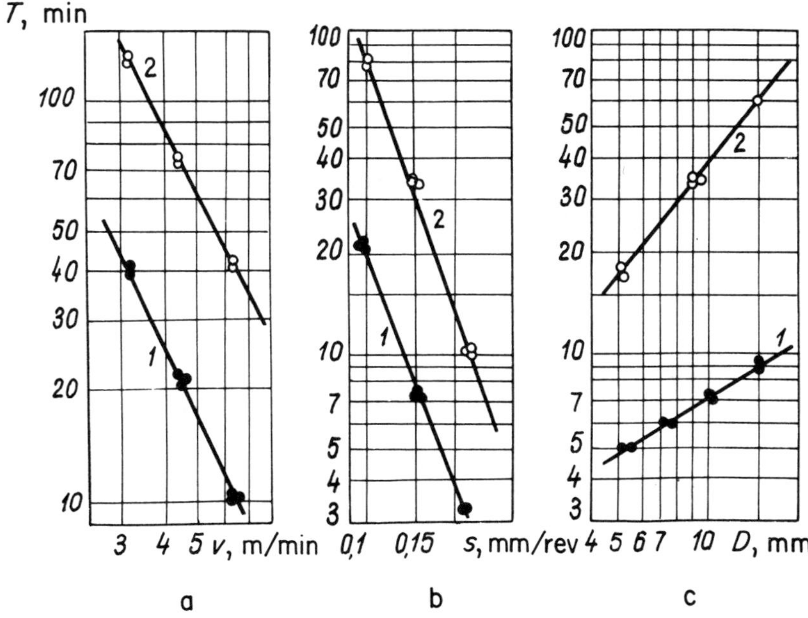

Fig. 4-16. Effect of electrical insulation of drills made of type-R18 steel in the cutting of type-VT5 alloy (δ_O = 0.2 mm): (1) closed thermoelectric circuit; (2) open circuit.

supposition that the discharges between the metal plates of the arbitrary capacitor are sparks. In subsequent publications, Korobov probably took into account the critical remarks that had been made about the method he used for the calculations (see Section 2.2). In 1968, he wrote,

> In the cutting of metals, thermoelectric potentials with continuously varying frequency and amplitude are generated. The potential values fluctuate between 0.25 and 2.5 mV depending on the machining conditions. ... At a cutting speed of $v = 25$–60 meter/min, potential fluctuations of low frequency (100–300 Hz) and small amplitude (0.4–0.6 mV) are typical; at $v = 200$–300 meter/min, the amplitude rises to 2–2.5 mV and the frequency increases to 5000 Hz.[257]

Important results of the study include the observation that the ac component of $\mathcal{E}_t$ is a minimum when the tool geometry is optimal and a maximum when the tool wear is greatest. The article exhibits a clear tendency to stress the dynamics of thermoelectric processes; for example, it assumes that the effect of cutting speed on the amplitude and frequency of the potential fluctuations has to do with the change in the character of the temperature and dynamic influences at sites of thermionic emission. But all of this has the aim of justifying the possibility of concentrated tool wear on the flank face resulting from electrical erosion in microscopic impulse discharges.

The author's investigations, which are based on recording the ac component of thermal emf, have had and continue to have a different purpose. Right from the start we considered recording this component as above all a method of studying the dynamics of contact processes, since the electrical-erosion type of wear cannot clearly predominate in the operation of a metal-metal couple without vibrational rebounds of the indenter (tool). Only in finishing work, when the very small cross section of the cut means that conditions are favorable for the complete destruction of contact during vibrational bumps, is significant erosion allowable. This result agreed with Korobov's conclusions, but on altogether different theoretical grounds (see Section 5.1). Figure 4-17 clearly illustrates the rebounding of the tool (type-T15K6 alloy) when it collides with asperities on the surface of the rotating workpiece (type-45 steel).

Having determined that the character of the potentialograms obtained with a

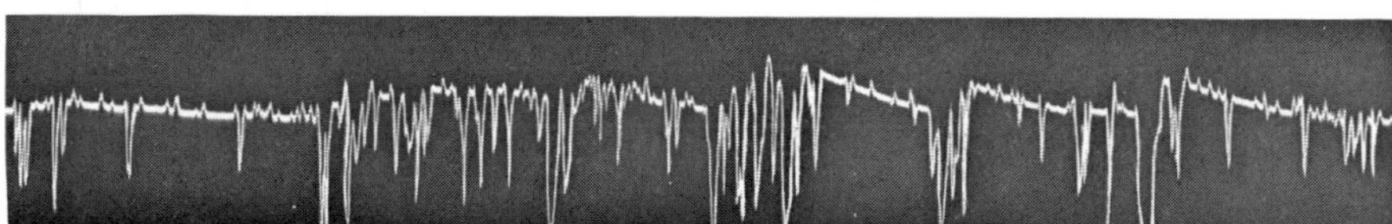

Fig. 4-17. Rebounds of the tool (type-T15K6 alloy) upon collisions with asperities on the surface of a rotating workpiece (type-45 steel).

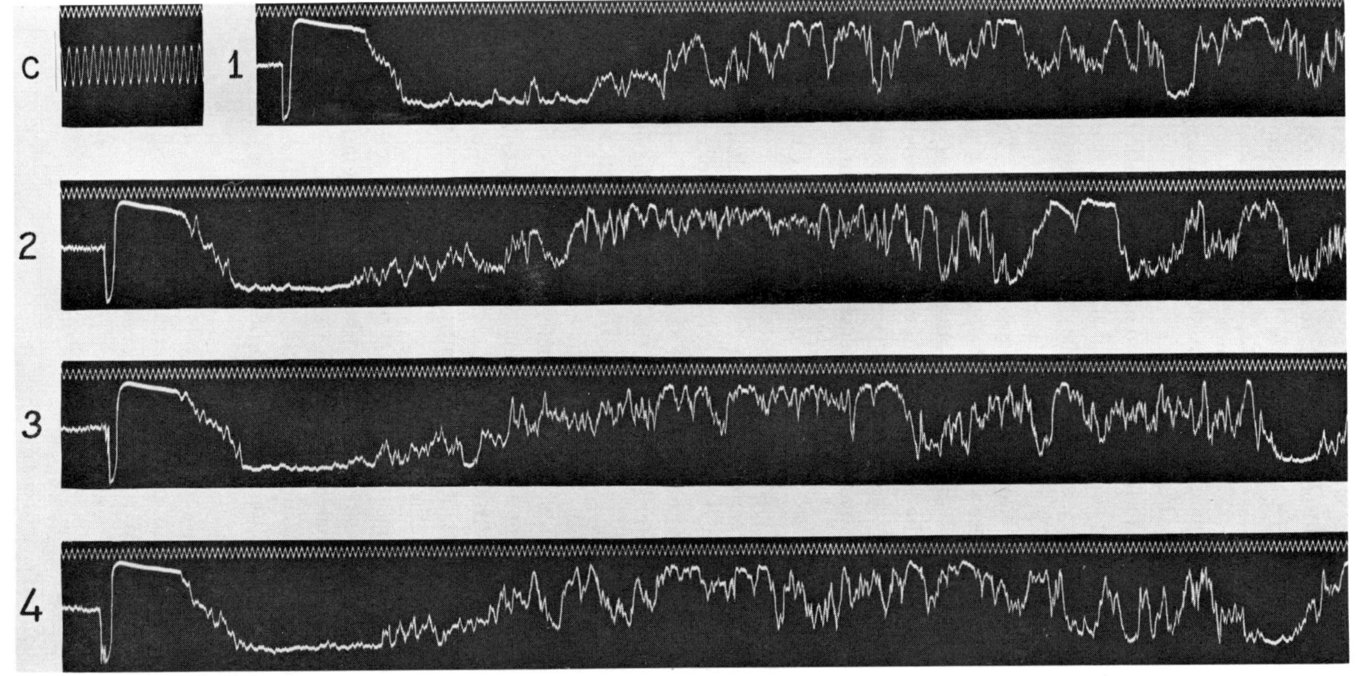

Fig. 4-18. Repetitiveness of the character of curves illustrating the variation in potential difference between tool (type-R18 steel) and work-piece (type-45 steel) after four successive cuttings-in in planing. Cutting conditions: $v = 14$ meter/min; $s = 0.6$ mm per double pass; $t = 1$ mm. Timer frequency $f_t = 500$ Hz; C is the calibration signal (20 μV).

type-UBP1-01 biopotential amplifier* and a loop oscillograph was reproducible for given machining conditions (Fig. 4-18), the author attempted to analyze the effect of cutting regimes on the ac component of $\mathscr{E}_f$. He obtained the following results:[26, 27]

• Recording of this component can be considered a method of gaining information on the tool wear rate (Fig. 4-19).

• This component is practically independent of feed and cutting depth for rough and semifinishing work on metals (Figs. 4-20 and 4-21).

• The amplitude-frequency spectrum of the fluctuating potential difference is a function of cutting speed and reproduces the amplitude-frequency spectrum of vibrations in a system generating electric charges (Fig. 4-22). The low-frequency vibrations of the workpiece show up most distinctly at small v. As the cutting speed increases, it is as if the workpiece-machine subsystem lost sensitivity to forced vibrations, while the tool-support subsystem remains quite susceptible to vibrations. At large v, quasiperiodic tool vibrations appear, at an increased or even a high frequency, with all the consequences that that entails.

An overall analysis of the records obtained in friction and cutting showed that the pulsation of the potential difference between rubbing surfaces is not just a result of the variation in capacitance of the frictional capacitor. Because of the vibrations and the discreteness of the contact, the real area of contact fluctuates continuously. The internal resistance of the seat of emf will alternately rise and fall, the more so if, in sliding with jumps, the freedom of the slider to carry out its normal motions is only partly limited. The superimposition of overtones on the fundamental frequency of the pulsations has two causes: (1) discharges in the form of mixed electron emission (chiefly field emission from the juvenile surfaces), and (2) the exceptional heterogeneity of the surfaces, which consist of alternating sections with differing electrical conductivity. Structural transformations, destruction of order in the crystal lattice at temperature flashes, variation in the microgeometric profile of the surface, oxidation of the metal, destruction of the thin lubricant films, and so on—all these factors are reflected at once in the contact resistance value, the more so when there is vibration-induced fretting.[258] In addition, probable reasons (specific to the cutting process) for fluctuations in $\mathscr{E}_f$ include inhomogeneity of structure and physical properties in the work material, the formation and breaking of a built-up edge, the cyclic nature of chip formation, and so forth.[240]

Quite obviously, the author's experiments imply that it is desirable to perform a harmonic analysis of the spectrum of fluctuations of $\mathscr{E}_f$ and to discover how the spectral characteristics depend on the conditions of the contact interaction, the predominant form of wear, and so on. This also follows from the publica-

---

*The amplifier uses a differential circuit.

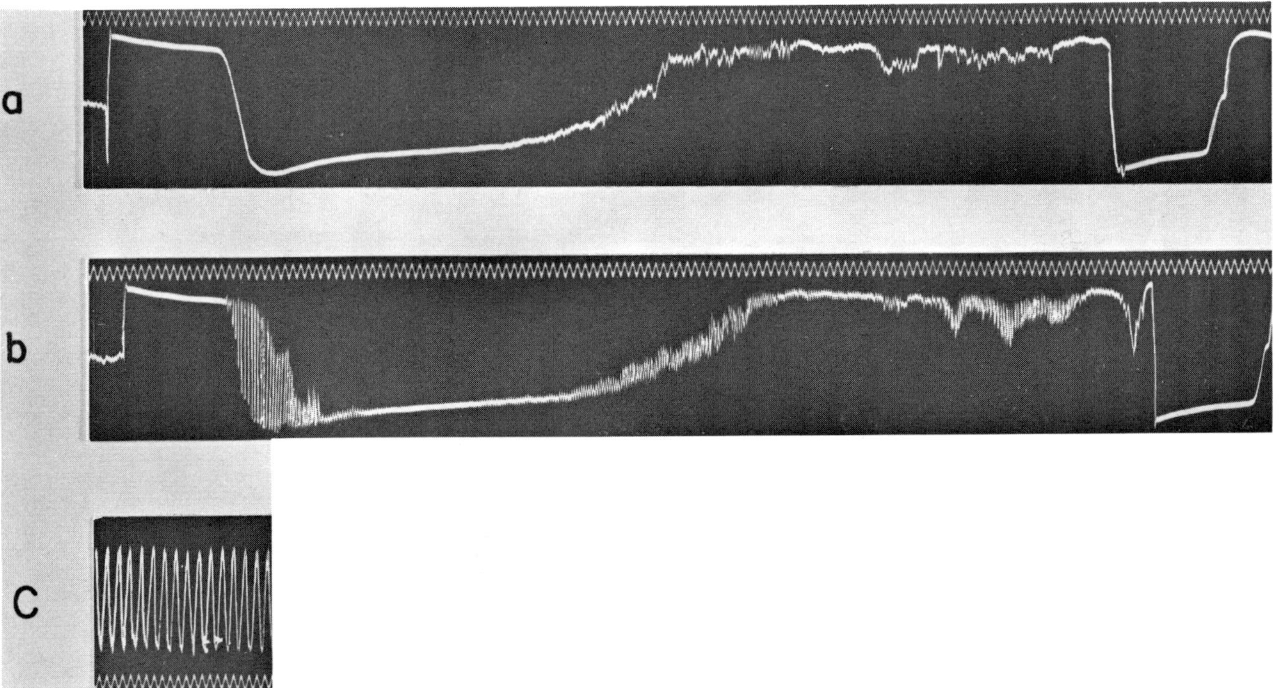

Fig. 4-19. Alternating component of the potential difference between tool (type-R18 steel) and workpiece (type-18Kh2N4VA steel) in planing with a sharp tool (a) and after the tool has become dull (b). Cutting conditions: v = 23 meter/min, s = 0.6 mm per double pass, t = 1 mm. Timer frequency f_t = 500 Hz; C is the calibration signal (20 μV).

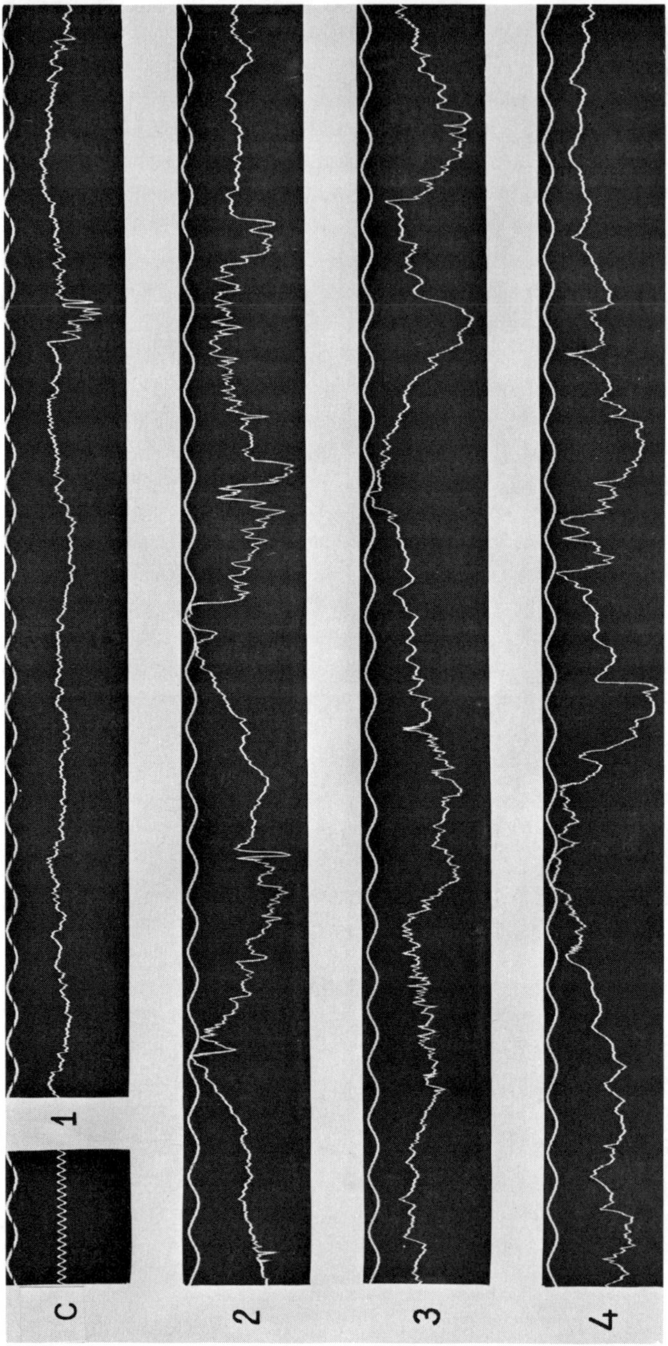

Fig. 4-20. Effect of feed s on the character of the variation of potential difference between the tool (type-T15K6 alloy) and the workpiece in the longitudinal turning of copper ($v = 45$ meter/min, $t = 1$ mm). Feed equals (1) 0.088, (2) 0.164, (3) 0.328, (4) 0.5 mm/revolution. Timer frequency $f_t = 50$ Hz; C is the calibration signal (20 μV).

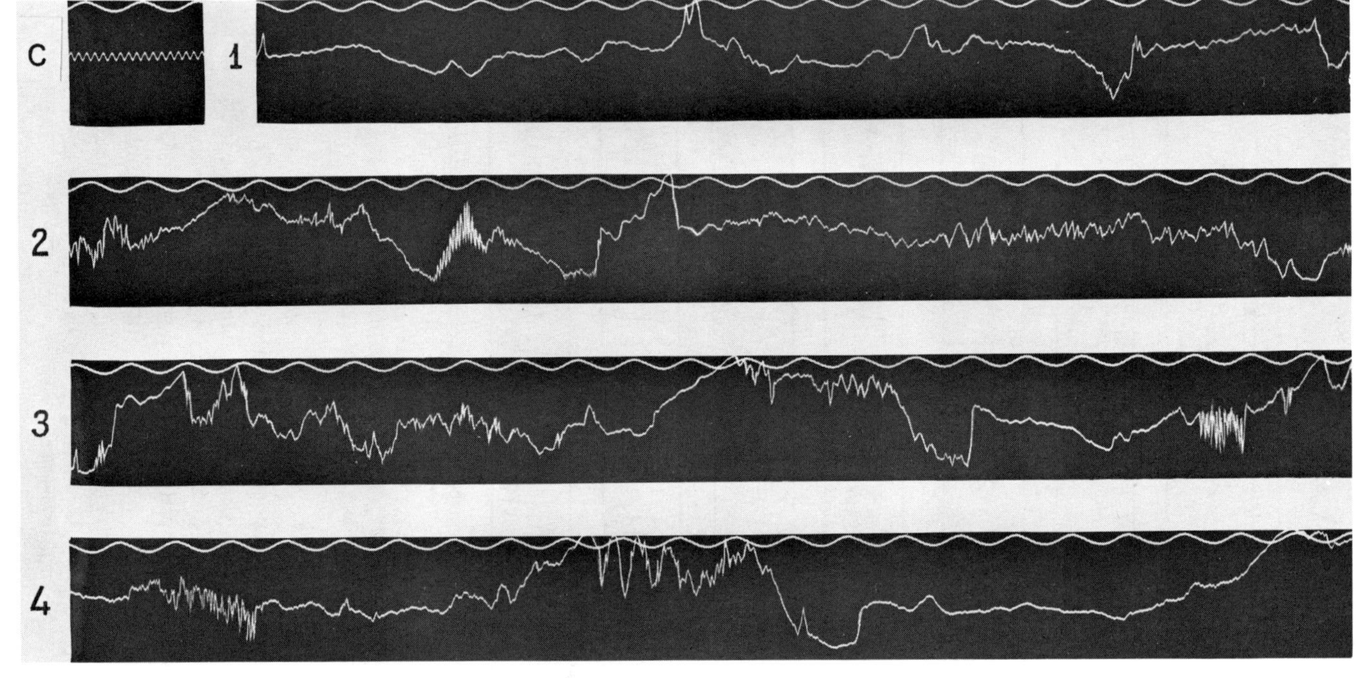

Fig. 4-21. Effect of cutting depth t on the character of the variation of potential difference between the tool (type-T15K6 alloy) and the workpiece in the longitudinal turning of copper ($v = 45$ meter/min, $s = 328$ mm/revolution). Cutting depth equals (1) 0.5, (2) 1, (3) 1.5, (4) 2.5 mm. Timer frequency $f_t = 50$ Hz; C is the calibration signal (20 μV).

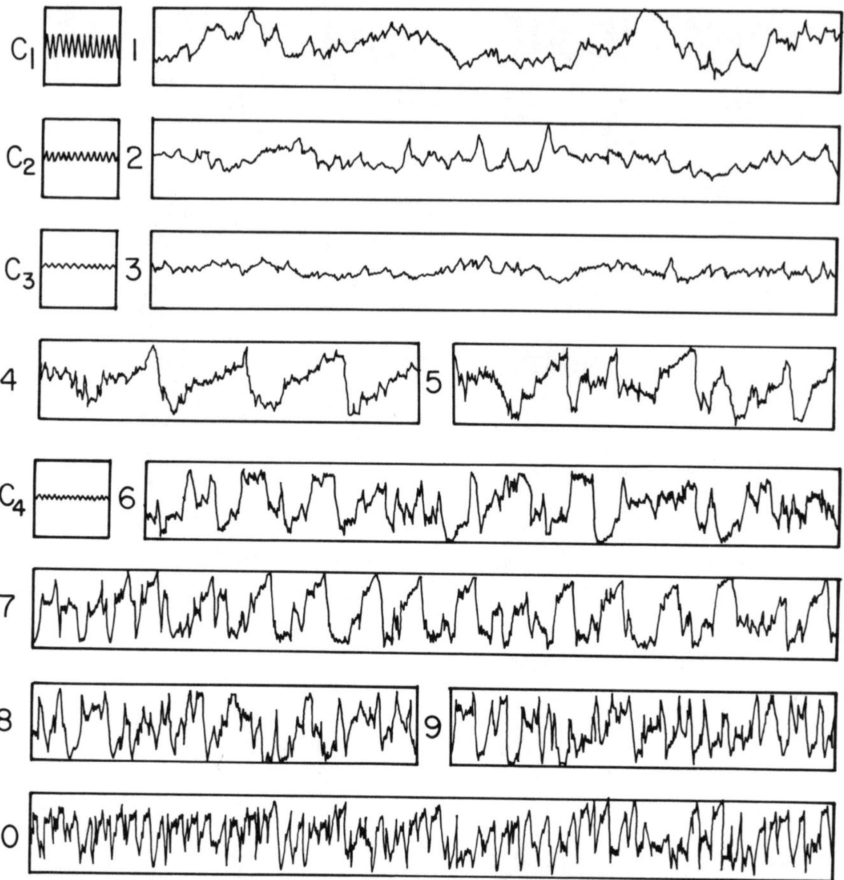

Fig. 4-22. Effect of cutting speed v on the ac component of the potential difference between the tool (type-T15K6 alloy) and the workpiece in the longitudinal turning of type-45 steel ($s = 0.328$ mm/revolution, $t = 1$ mm). Cutting speed equals (1) 4, (2) 8, (3) 13, (4) 34, (5) 55, (6) 86, (7) 110, (8) 137, (9) 172, (10) 220 meter/min. Timer frequency $f_t = 500$ Hz; C_1, C_2, C_3, and C_4 are amplifier calibration signals (20 μV).

tion of Kostetsky and Zaporozhets (mentioned in Chapter 2), which is devoted to a harmonic analysis of the spectrum of fluctuations of the frictional force.[118] Dikushin has given the following formulation of the problem as a whole: "It is probably desirable to investigate another possibility, that of using the natural tool-workpiece thermocouple formed in friction to determine, even if only approximately, the machinability of a material and to measure the rate of dimensional wear of the tool. The overall goal is to automate the establishment of an optimal regime of cutting."[259]

In fact, Kretinin has extended the author's research in this direction.[21,260] Since the ac components of the thermal emf and cutting forces are ergodic stationary random functions, it was possible to sharply limit the number of realizations of them necessary for calculating the spectral densities $S_e(\omega)$ and $S_p(\omega)$ that describe the frequency composition of the fluctuations in $\mathcal{E}_t$ and the force P_y at various stages of the dulling of the cutting tool. The curves obtained (Figs. 4-23 and 4-24) show quite clearly how the frequency distribution of energy varies and how much the intensity of oscillations is reduced when cutting fluids are used. Computation and normalization of the corresponding cross-correlation functions showed that in dry cutting the fluctuations in $\mathcal{E}_t$ and P_y are determined mainly by the same factors: the cross-correlation coefficient for their ac components reaches values of ~ 0.76, whereas in cutting with irrigation by a cutting fluid the value decreases to 0.3–0.4.

In the study of the spectrum of the ac component of thermal emf, an attempt was also made to detect frequency bands correlated with the tool wear rate.[260] The cross-correlation coefficient exceeded the 99 percent significance level for many frequencies; Kretinin considers this fact to support his idea of the tool wear process as a collection of "individual events of destruction taking place at a statistically determined frequency and rate."[260]

Although it is known that statistical criteria can prove no hypothesis, the result obtained is very important, for it indicates that the new method of studying tool-life relations in cutting (friction) is "not rejected." The method can be used, in

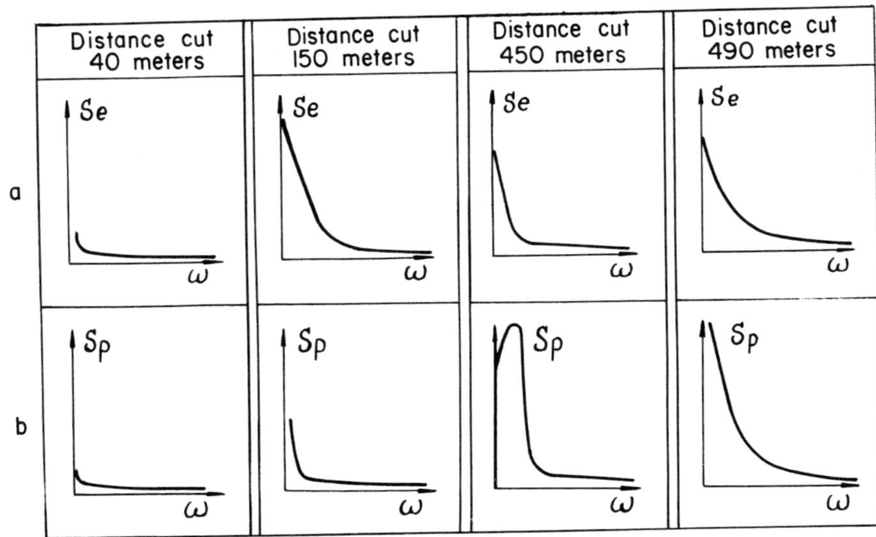

Fig. 4-23. Changes in the fluctuational spectrum of thermal emf (a) and cutting force (b) as wear increases on a cutter made of type-T5K10 alloy in the dry turning of type-40Kh steel.[21]

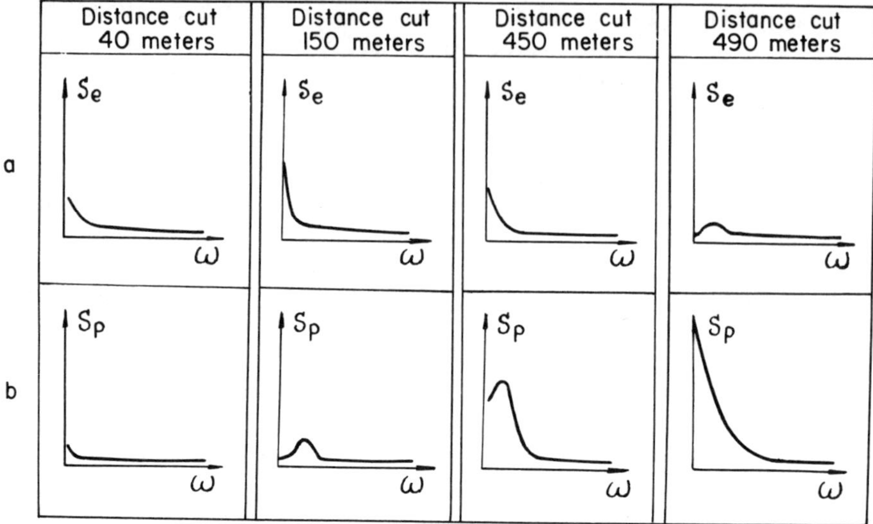

Fig. 4-24. Changes in the fluctuational spectrum of thermal emf (a) and cutting force (b) as wear increases on a cutter made of type-T5K10 alloy in the turning of type-40Kh steel under flooding with a 30 percent emulsion of type-NGL-205 lubricant.[21]

particular, to assign cases of wear to the several types of wear and to investigate the reasons for the "tool-life paradox" (the nonmonotonic variation in the function $T[v]$). One of the reasons, in our view, is the resonance-selective mechanism of energy dissipation in friction (cutting), which results in the appearance of a frictional absorption spectrum (see Sections 1.6 and 1.7). As an automatic source of information on the character of processes in the cutting zone and on the degree of dulling of the cutting edge, the tool-workpiece thermocouple is a superb feedback transducer. Its use, together with a spectrum analyzer for the appropriate frequency range, may make it easier to solve problems of automatic control of manufacturing processes involving metal cutting.

We will not stop to discuss the physical aspects of this problem. We just remark that the existence of fluctuations in the number and mobility of charge carriers, the real area of contact, and the effective surface area from which emission (frictional flicker noise) takes place implies a mathematical analogy between our problems and the study of electrical noise in electron tubes and semiconducting instruments.

In conclusion, we should again emphasize that lateral forces, the work done by which determines the emf generated in friction and cutting (even when there is no water-based cutting fluid), differ in their physical natures (Section 2.3). For this reason, in all the cases of contact interaction under consideration, it would probably be more correct to speak of the integral emf of the frictional system and of its fluctuational spectrum. This is all that we can do when cutting fluids with electrolytic properties are used.

4.4. INFORMATION CONTENT OF POTENTIALOGRAMS FOR CUTTING IN ELECTROLYTIC MEDIA

The ideas developed in this section relate to the penetrating power of cutting fluids, and reflect some results of the author's research aimed at evaluating the actual penetrating power by using a prototype of the method described previously (see Section 3.3).

A cutting fluid is a versatile material whose use drastically improves the economic and productive indices of metal cutting. Almost all industrial cutting fluids now used can be divided into three categories: anhydrous oils, nonreactive and active, based on fatty and mineral oils; self-emulsifying oils; and synthetic fluids (that is, true or colloidal aqueous solutions). The last group of cutting fluids has come into wide industrial use fairly recently. Synthetic fluids, which are excellent coolants, often contain combinations (mixtures) of components such as (a) amines and nitrites (inorganic and organic) to prevent corrosion; (b) nitrates to stabilize the nitrites; (c) phosphates and borates as water softeners; (d) soaps and wetting agents for lubrication and to lower surface tension; (e) compounds of phosphorus, chlorine, and sulfur to form chemical lubricant films; (f) glycols as binders; (g) herbicides to increase bacteriological resistance; and so on.[261] In order to give suitable recommendations on the selection and preparation of cutting fluids—that is, in some measure to predict scientifically the mechanism by which the cutting medium will act in any concrete case—it is necessary to know the composition of the fluid and its physical and chemical activity; to allow for the character of the machining operation and for cutting conditions; and to have available reliable information on the behavior and properties of the workpiece material in chip formation.

There is no direct connection between the present research and questions related to one of the principal functional properties of cutting fluids, namely, cooling power. Nonetheless, it should be emphasized that, with regard to beneficial effects of a fluid on the temperature conditions in the friction of surfaces, lubricating action can also be regarded in part as a cooling function of the cutting fluid. On the other hand, the direct extraction of heat (which depends on such properties of the fluid as the specific heat, heat of vaporization and thermal conductivity) works against the softening of metallic soaps and the development of rotational isomerism in the surfactant molecules, lowers the desorption probability of these molecules, and thus increases the lubricating ability of the medium. Thus, the fluid cools while lubricating and, conversely, lubricates while cooling.

While the layer being cut experiences large plastic deformations and high zonal temperatures, and when the pressure on the tool is high (it may reach 30,000–40,000 kg/cm^2 for special super-strength alloys), the penetrating power of the cutting fluid becomes a matter of first importance. This property of a fluid is usually associated with the ability of the lubricant to reach the communicating portions of the space between chip and tool. The property is in no way similar to the bulk propagation of lubricant molecules through the internal microcapillary system of "inherent" microcrevices and nucleated microcracks that appear in the deformation process; nor does it resemble the penetration of the distorted crystal lattice of a solid by atoms and ions resulting from the thermal and catalytic degradation of molecules. At the same time, there are rather weighty reasons[262–265] to suppose that, even in the presence of three-dimensional compression in the region of primary deformation, particles of surfactants and of chemically active liquids may rapidly become part of the external layer of the transitional plastic zone (from the bead ahead of the chip) and migrate into the bulk of a polycrystalline metal by irregular diffusion along lattice imperfections (hollow dislocational nuclei or block boundaries). This "molecular-atomic permeation" of a solid results from the capillary suction of the cutting fluid, because of thermal-diffusion processes on the internal phase surfaces and electrostatic attraction between particles. Only those particles of low-molecular-weight or ionic compounds with dimensions not exceeding those of the admitting ultrapores can penetrate into a metal being deformed; the mobility of the particles depends on their polar character and on the electrostatic charge of the ions in the solid. Latyshev's conclusions—namely, that the effective internal-lubricant action of chlorine compounds (carbon tetrachloride, barium chloride, calcium chloride, etc.) is due to the small atomic radius (1.07 Å) and the large mobility of chlorine ions*—seem to us quite in order.[143]

Under severe conditions of cutting, when the lubricant goes from the liquid to the gaseous (vapor) state, the prominent role of getters becomes evident. The rate of gas adsorption is determined by the chemical activities of the getters in relation to oxygen, hydrogen, nitrogen, carbon dioxide, and so on, and by their ability to dissolve the gases or retain them on their surfaces. It was mentioned above that tantalum, titanium, niobium, and several other metals are getters. For example, oxygen, sulfur, and halogens act on tantalum at elevated temperatures; the properties of titanium as a getter† probably explain the fact[267] that, as the content of titanium carbides in a solid alloy rises, the built-up edge on the tool rake face disappears and the roughness of the machined surface decreases.‡

The mechanism by which lubricant penetrates to the near-tool layer of the chip and the near-chip layer of the tool, like that of the impregnation of the third body formed in the immediate vicinity of the cutting edge as a result of seizing of the metals, is completely similar to that described above. It should be kept in mind, however, that for the most part a cutting fluid can penetrate a

---

*Cassin and Boothroyd report the decomposition of neutral CCl_4 with the liberation of chlorine in the cutting zone.[265]

†Macklin and Jankee mention the high solubility of oxygen in titanium.[266]

‡We remark in passing that the addition of solid lubricants to the workpiece (e.g., of free lead or of metal sulfides) may severely degrade the functional properties of aerospace materials. An exception, from the standpoint of the desirability of metallurgical impregnation, is tungsten pressed and sintered to 85 percent of theoretical density. Face-centered cubic metals are used as "fillers" in porous tungsten.[268]

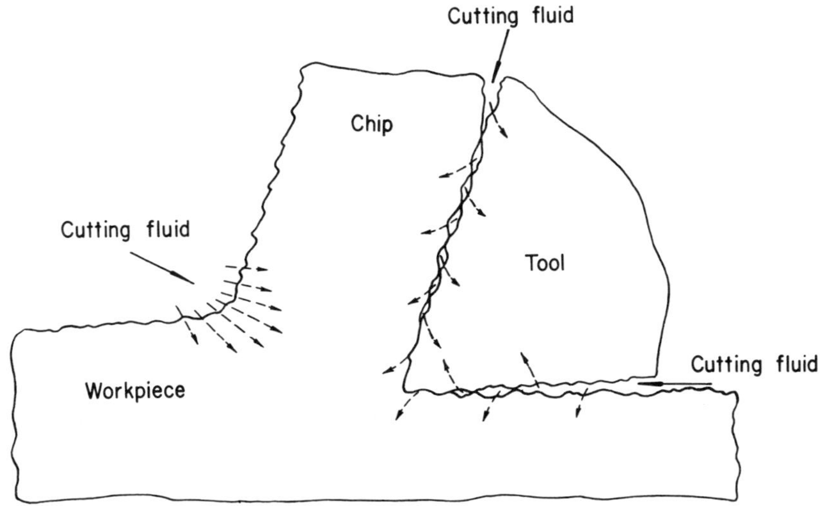

Fig. 4-25. Paths by which cutting fluid moves toward the cutting zone.

given volume of the metal only after it has reached the cavity between the tool and the chip (Fig. 4-25). The propagation of lubricant along the rubbing boundary surfaces is due to all the same capillary forces, electrostatic attraction, and thermal motion of ions, atoms, and molecules. But now events of external (adsorptive) adaptation of surfactants, the formation of cationic and anionic layers on the contacting surfaces,[142] the growth of well-defined chemical films, and other factors take on prime importance. As the lubricant starts to penetrate the crevice-shaped space, coming from the place where the chip withdraws and curls away from the tool rake face, the lubricant figuratively soaks through a series of filters differing in porosity, the pore size decreasing with approach to the cutting edge (down to a separation of a few angstroms). Naturally, it is for this reason that, in such selective filtration, the large molecules of typical surfactants (organic acids and alcohols) may stick at a rather substantial distance from the tool edge, even when the surface ions have been solvated and a disjoining pressure has appeared.* As for the probability of flood lubrication of the whole rake face, there is quite a convincing argument that vacuum suction of cutting fluid into the cavity between tool and chip occurs because of the vibrations of the cutting tool and breaks in the built-up edge on its surface.[271, 272] The suction conditions noted contribute to the development of some physical properties of cutting fluids, including the surface tension, viscosity, and density.

On the basis of the foregoing, the author remarked that great care is needed in the selection of criteria making it possible, in the best case, to estimate the abili-

---

*B. V. Deryagin has made circumstantial studies of the disjoining effect of thin liquid layers.[269, 270]

ties of lubricants to reach the interface between the contacting metals under conditions known ahead of time.[26] It would evidently be rash to call these criteria the penetrating power of a liquid or to consider any one of them as anything like a constant of the cutting fluid. The author considers, for example, that the electrolytic current density in the presence of a porous partition[143] can be only tentatively treated as an indicator of the penetrating powers of synthetic lubricants used in metal cutting.

The question of the mechanism by which cutting fluids penetrate into areas of tool-chip and tool-workpiece contact becomes less difficult to answer if information of the following types is contained in the experimental data: "film" or "no film," "continuous film" (in the sense that no contacting metallic islands are present) or "film with breaks," and so on (of course the initial conditions of the trials should be kept constant).

So that this information could be obtained, the integral emf $\mathcal{E}_f$ was supplied to the input of a type-UBP1-01 differential amplifier, but now in cutting with cutting fluids. The fluids chosen were aqueous solutions of halogen derivatives and mixtures of these with a surfactant containing a hydrophilic group. Isopropyl alcohol was used to activate the cutting fluid; this alcohol had exhibited greater wetting ability than glycol or glycerin. The materials machined were copper and types-45, -18Kh2N4VA, and -OKhNZM steel. The cutting tools were made of type-R18 high-speed steel or type-T15K6 titanium-tungsten-cobalt alloy. Table 4-5 makes it possible to arrange the galvanic couples of workpiece material and tool in order of emf value when the circuits are completed through a large resistance.

The amplifier will be open if its sensitivity matches the value of the ac component of $\mathcal{E}_f \approx \mathcal{E}_t$ in the case of direct metallic contact, but it will close rapidly (cutoff of the signal [Fig. 4-26]) when $\mathcal{E}_f \longrightarrow \mathcal{E}_g$. That is, the metal surfaces make contact through a film of electrolyte-indicator and films of chemical com-

Table 4-5. Potential difference between tool and workpiece immersed in an aqueous electrolyte solution (mV).

		Electrodes*		
Electrolyte	Concentration (%)	Type-R18 steel–copper	Type-R18 steel–type-45 steel	Type-R18 steel–type-OKhN3M steel
HCl	1	200	−25	20
HCl	5	210	−28	30
NH_4Cl	1	190	−110	−60
NH_4Cl	5	220	−80	−20
NaCl	1	140	−44	−20
NaCl	5	150	−56	10

*Forward polarity of the galvanic cell: tool (type-R18 steel) is anode. The external circuit has a resistance of 1 kΩ.

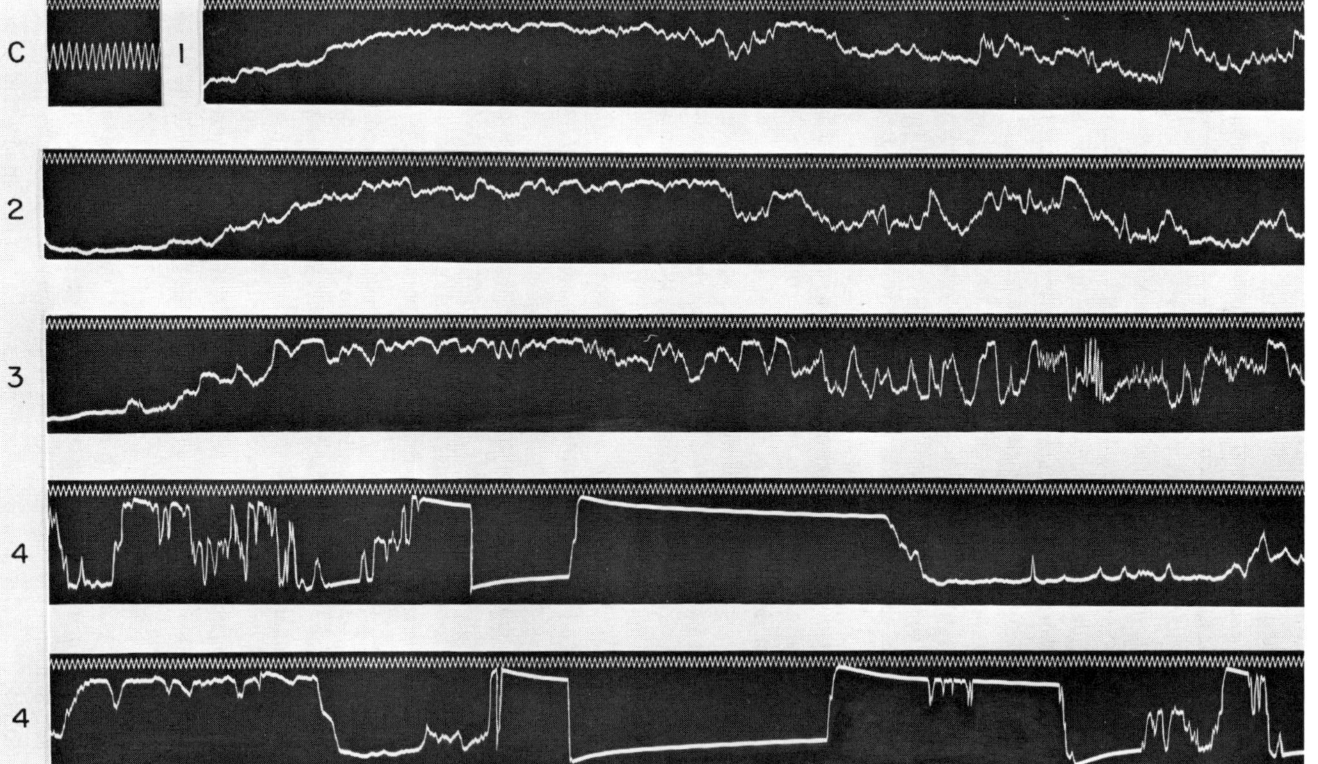

Fig. 4-26. Oscillograms of the alternating component of the potential difference between the tool (type-R18 steel) and the workpiece (type-18Kh2N4VA steel), in planing in air and in various aqueous media (irrigation): (1) air; (2) 15% solution of C_3H_7OH; (3) 15% solution of NH_4Cl; (4) 15% solution of a mixture of C_3H_7OH and NH_4Cl. Timer frequency $f_t = 500$ Hz; C is the calibration signal (20 μV). Cutting conditions: $v = 10$ meter/min; $s = 0.3$ mm per double pass; $t = 1$ mm.

pounds that have formed on the metals, with resistances significantly larger than those of the metallic a-spots. Probably it is these films that carry the main load; but they have two exceedingly active aids, without which the stability of the films would be far less, even in the cutting of plastic copper. We have in mind, first, that solvated molecular clusters oppose the closer approach of the rubbing surfaces and, second, that methylene chains have great axial elasticity if the isopropyl alcohol molecules are considered as rather stiff atomic complexes.

To prevent misunderstandings, one shortcoming in the method should be pointed out. This shortcoming concerns planing experiments and has to do with the fact that the shortness of the workpiece and the resulting short distance cut were not taken into account at the proper time. While the trials were still in progress it became apparent that, of the two kinds of signals—those corresponding to transitions $0 \rightleftarrows \mathcal{E}_f (\mathcal{E}_g)$ before cutting in, and after the tool withdraws from the workpiece in the case of irrigation with an electrolyte, and those generated by the actual events of appearance and disappearance of the contact (we remember that the contact potential difference attains values of ~ 1 V)—the first kind does not block the amplifier nearly as much as the second does. For this reason, when planing was carried out in air or with the use of an isopropyl alcohol solution and when this circumstance was not taken into account, only the last sections of the records (which, incidentally, display clearly the moment when the tool withdrew from the workpiece [Fig. 4-27, parts 1–3]) conveyed any useful information. The situation began to repeat itself in trials with electrolytes, but could be eliminated if the amplification channel was connected immediately after the tool cut in, when the information-obscuring signal had probably had time to die out because of the short relaxation time (Fig. 4-27, parts 4–6).

It should also be emphasized that the nature of the potential discontinuity that blocks the intermediate amplifier remains an open question. According to Bibikov and Turik, for example, water and water vapor are as good as the best semiconductors with respect to thermoelectric properties, or more precisely to the value of the Seebeck coefficient.[273] As for the duration of the signal cutoff, specific features of the operation of the amplifier do not allow this duration to be identified completely with the "lifetime" of the continuous screening films. Moreover, when similar experiments are being conducted, it is necessary to bear in mind the possibility that the components of the integral emf may compensate one another, and consequently that the integral emf may become inverted.

Independent of these remarks, the potentialograms obtained in planing when the cutter was vibrating intensively make it possible to ascertain two points:[18, 113]

1. The ability of a cutting fluid to penetrate to the contact areas is directly linked with dynamic phenomena in the machine-tool-workpiece system.
2. The addition of a monohydric alcohol (C_3H_7OH) to aqueous solutions of electrolytes increases the metallophilic character of the media and their wetting and lubricating abilities.

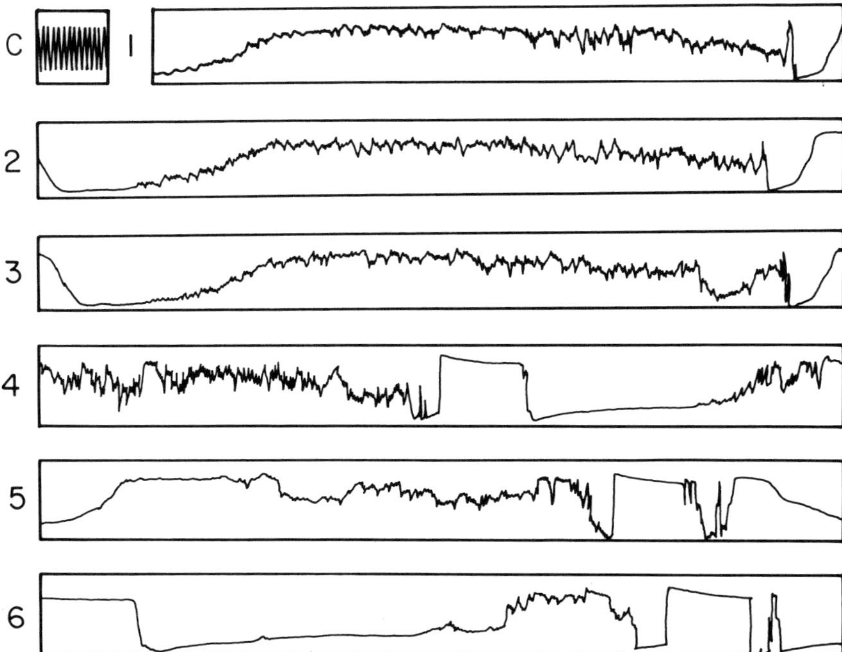

Fig. 4-27. Potentialograms taken in the planing of copper in air and in various aqueous media (irrigation) with a tool made of type-R18 steel: (1) air; (2) 1% solution of C_3H_7OH; (3) 15% solution of C_3H_7OH; (4) 1% solution of NH_4Cl; (5) 15% solution of NH_4Cl; (6) 15% solution of a mixture of C_3H_7OH and NH_4Cl. Timer frequency f_t = 500 Hz; C is the calibration signal (20 μV). Cutting conditions: v = 23 meter/min, s = 0.6 mm per double pass, t = 1 mm.

From a practical standpoint, the results of planing experiments indicate the applicability of the new method not in the case of conventional, continuous cutting, but rather in vibrational, intermittent cutting. The method has the aim of selecting cutting-fluid formulas and optimal regimes of harmonic oscillations of the tool (forced vibrations).

However, in continuous cutting too, recording the ac component of the potential difference between the tool and the workpiece makes it possible to obtain information on the degree to which the cutting fluid participates in contact processes.[18, 27] It has been clear to the author that the extremal spikes observed on the curves taken in longitudinal turning (potentialograms 3 and 4 in Figs. 4-28 and 4-29) are also connected with some peculiar factors that are characteristic of the cutting process and that bring about a drastic and cyclic anomaly in the behavior of the real area of contact. As the records showed, the frequency of the extremal spikes in the machining of copper correspond precisely to the number of revolutions per second of the workpiece. Hence, it followed that, in the case

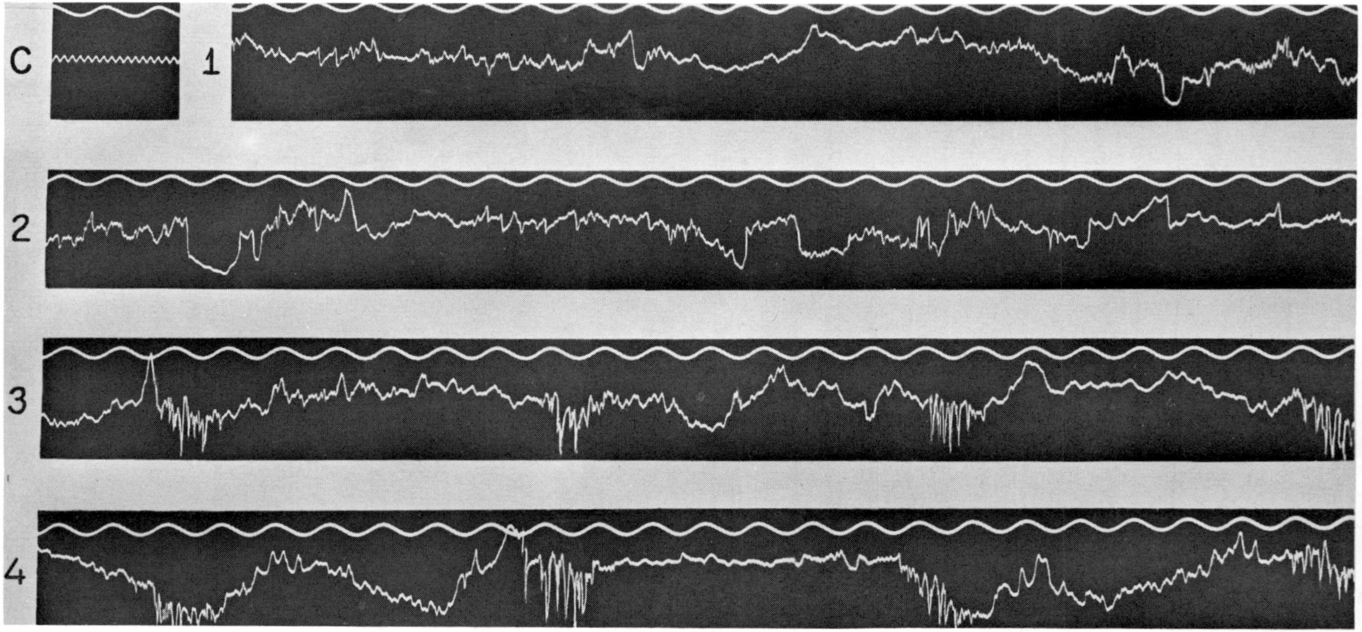

Fig. 4-28. Potentialograms taken in the longitudinal turning of copper with a tool made of type-T15K6 alloy, in various aqueous media (irrigation): (1) 1% solution of NaBr; (2) 1% solution of NaI; (3) 1% solution of NaBr + 10% solution of C_3H_7OH; (4) 1% solution of NaI + 10% solution of C_3H_7OH. Timer frequency f_t = 50 Hz; C is the calibration signal (20 μV). Cutting conditions: v = 45 meter/min, s = 0.328 mm/revolution, t = 1 mm.

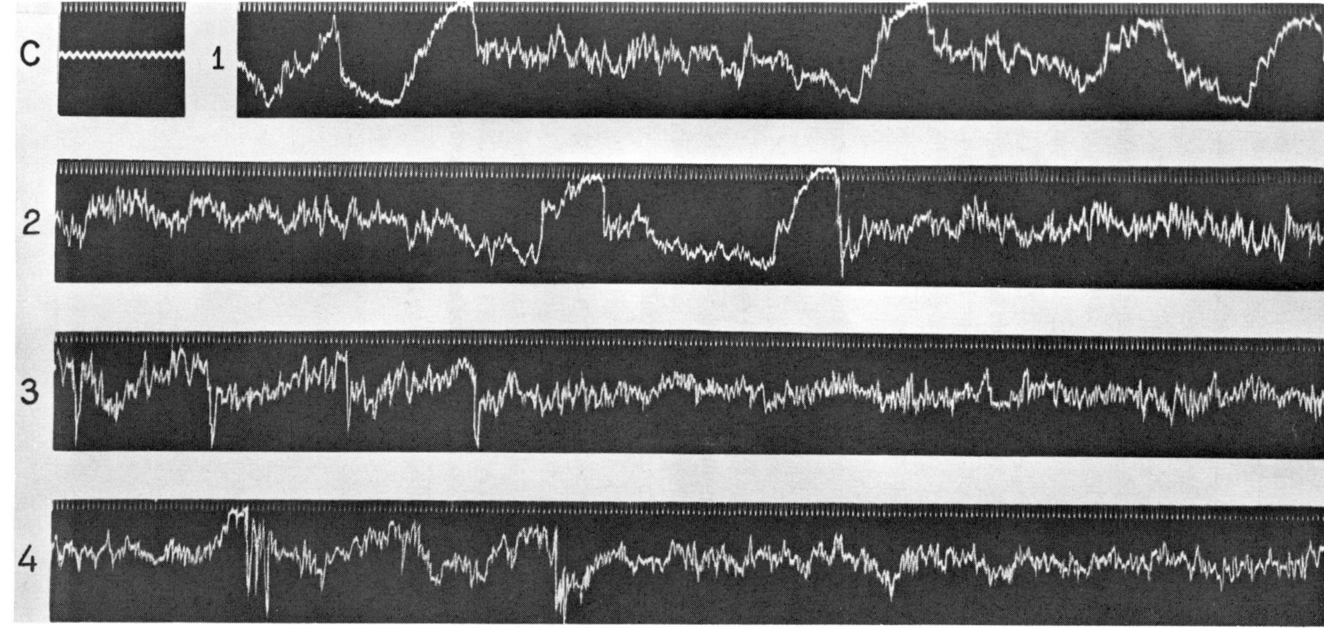

Fig. 4-29. Potentialograms taken in the longitudinal turning of type-45 steel with a tool made of type-T15K6 alloy, in various aqueous media (irrigation): (1) 1% solution of NaCl; (2) 1% solution of KCl; (3) 1% solution of NaCl + 10% solution of C_3H_7OH; (4) 1% solution of KCl + 10% solution of C_3H_7OH. Timer frequency f_t = 500 Hz; C is the calibration signal (20 μV). Cutting conditions: v = 55 meter/min, s = 0.328 mm/revolution, t = 1 mm.

under consideration, the mechanism by which lubricant penetrates to the metal-metal interface has a direct relation to vibrational motions: the cutting fluid or its vapor moves toward the cutting zone when the low-frequency vibrations of the workpiece and the high-frequency vibrations of the tool do not match in phase and amplitude. But there are, it seems to us, grounds for believing that the periodic breaking of the built-up edge on the tool rake face may be such a peculiar factor, bringing about drastic weakening of the metallic contact in the cutting of type-45 steel. This breaking, by forming a vacuum, results in the practically instantaneous sucking of cutting fluid into the cavity between the tool and the chip (see the potentialograms).

After the fluid has been sucked in, the low-frequency fluctuations in $\mathcal{E}_f$ ($\mathcal{E}_t$) weaken markedly. This effect, along with the lubricating action of the cutting fluid, which alters the character of the tool vibrations, can be linked with at least two other causes: first, the presence of an electrically conductive intermediate phase contributes to the rapid neutralization of electric charges generated on the plates of the frictional capacitor and, second, the electrolytic medium "stabilizes" the internal ohmic resistance of the seat of emf.

Once again, experimental results have confirmed the great adsorbability of linear polar molecules with an active center at one end.

The author and Teplov examined the question of whether it is possible to make an automatic search for optimal cutting-fluid compositions with the help of an adaptive system in which the cutting zone itself is simultaneously the information source and the feedback transducer.[19] The idea for the creation of such a system and even the proposed methods of realizing it (in general form) were based on the same physical groundwork as underlay the idea of developing instruments for automatically controlling the cutting process (this was touched on in the preceding section). Really and truly, we are dealing with links in one and the same chain of investigations, whose initial element is the recording of the ac component of the integral emf.

5

Theoretical and Experimental Substantiation of the Destructive Function of Thermoelectric Current in the Cutting of Titanium Alloys

5.1. CRITICAL REVIEW OF IDEAS ON THE MECHANISM BY WHICH THERMOELECTRIC CURRENT AFFECTS THE WEAR OF RUBBING COUPLES AND METAL-CUTTING TOOLS

Experimental material accumulated up to the present indicates that the destructive function of thermoelectric current can be linked with processes differing in character. The complexity of the mechanism by which thermoelectric current affects wear lies precisely in the fact that the mechanism is determined by the totality of these processes. However, under concrete conditions of contact interaction, as a rule only one of the processes prevails; for this reason, one proceeds to examine one by one the types of wear—for example, electrical-erosion, electrical-diffusion, and so on—caused by the passage of a thermoelectric current through an interface between metals.

Having studied how the natural state of the thermoelectric circuit of a frictional couple and the connection of an external source to this circuit are reflected in the wear of test pieces, Gordienko and Gordienko suggested that the thermo-electric current generated in the friction of metallic solids, like the current supplied by the external source, exerts on the rubbing surfaces an erosive, destructive effect, evidently similar in nature to the effect seen in electrical-discharge machining of metals.[22]

It has already been remarked that fundamentally incorrect theoretical reasoning led Dubinin[110] and Korobov[245] to the conclusion that spark and corona discharges could bring about molecular subdivision of the indenter (cutter) material.

Dubrov and coworkers maintain a similar point of view,[23,274,275] except that they lay special stress on the thermal nature of erosional processes. On the basis of a thermal emf on the order of 1–15 mV, a machine resistance on the order of tenths to tens of ohms, and finally thermoelectric currents in the range of tenths to hundreds of milliamperes, they hold that with respect to the cutting of metals, "One can speak of low-voltage, low-power discharges in very small spark gaps." Of course, once the validity of this thesis is granted, it makes sense to go on and discuss the generally unexceptionable version in which the liberation of additional thermal energy in contact microvolumes leads to the appearance of a a defect layer, because the carbides and the bonding material have far different coefficients of linear expansion. Incidentally, this same version of Dubrov, in the light of ideas developed by him on electrical-erosion tool wear, was upheld by Makarov and Kolenchenko.[276]

Without exception, all the publications of Dubrov and coworkers on this question either state directly or imply two basic points at the outset:

It is assumed that the cutter-workpiece contact can be considered intermittent (this refers not to disruption of the whole workpiece-tool contact, but to the continuous succession of spots of contact between the workpiece and the cutter that moves along it). By analogy with ordinary electrical contacts, it is assumed that wear of the cutting tool in the cutting (friction) of metals results from discharges occurring on microcontacts in the cutting zone.

Evidently, Dubrov and coworkers did not take into account that the destruction of a multipoint electrical contact at its individual points, even at the maximum speed attainable in practice, can in no way be equivalent to the interruption of the current flowing through some contact areas or other before they begin to move apart. Hence, even if specific features of frictional seats of emf are not touched on, it follows that there are no grounds for the conclusion those workers drew—namely, that the causes and character of electrical-erosion wear in cutting are the same as "in the switching of any electrical circuit"[274]—and trials aimed at modeling this type of wear with the help of an electric arc[275] are all the more unjustified.

According to Bobrovsky, the increase in tool life upon the opening of the thermoelectric circuit cannot be explained on the basis of reduced electrical-erosion wear of the surfaces.[111,277] Bobrovsky, however, misinterpreted the author's position, and only after appropriate additional clarification (in personal correspondence) did he come to refer to the author's "amendment" to the hypotheses that had appeared following the prior work of Dubinin.

What follows is a reproduction, with some insignificant additions, of what the author had written about earlier and had reported at a scientific seminar in the Institute of Mechanical Engineering.[26,27]

Suppose, for example, that the sliding of a chip is accompanied by rebounds of the tool from the workpiece; the condition of complete loss of contact between them is for the present not obligatory.* The amplitude of the vibrations is such that "oxygen starvation" of the rake and flank faces comes to a stop, and the portions of these faces situated near the cutting edges are enabled to "breathe," so to speak; that is, to draw in air coming into contact with the open contact areas. Obviously, the tendency of metals to form oxides, and in many cases nitrides as well, takes effect practically instantaneously. Centers of embryonic films, products of a short-lived oxidation reaction, begin to appear on the clean surface of the heated alloy. But the embryonic chemosorbed (or adsorbed) film cannot develop as a continuous film if the field inside the film is of the order of 10^8 V/meter (which involves breakdown of the film).

In high-frequency vibrations of the tool (indenter), the electric field strength evidently attains the values necessary for initiation of a self-maintained discharge in the "silent" form. The time in which the distance between plates of the frictional capacitor increases to 50 Å is a fairly small fraction of the total rebound time. This means that, at the earliest stage of separation of the surfaces, not nearly all the charges of the electrical double layer will have time to recombine through surface conductivity (back currents) or the tunnel effect (as field emission of electrons). Neutralization of some of the unlike charges may go by a gas-discharge mechanism, the more so when the interface has a large electrical resistance or the contact is completely broken.

Contact charging is especially marked in friction between a metal and a nonmetal (see Chapter 6). If no charge leakage occurs from a solid surface, and if the concentrated total charge is increasing, we rightly expect a discharge, even in spark form. Consequently, the possibility of erosion damage to materials in microscopic spark discharges must be taken into account in boundary friction between metal surfaces in the presence of lubricant films with pronounced dielectric properties.[113] Here it is very important to note that a reduction in the wear rate of rubbing couples under these conditions is now accomplished not by opening but, on the contrary, by the *closing* of the electrical circuit between the

---

*Instances of turning accompanied by severe vibrations ("visible chattering") are reported, in particular by Barrow and Spenser.[225]

principal elements of the couple;[186] therefore, the opening of the thermoelectric circuit in cutting is, from this standpoint, unsuitable for flood-lubricated elements in the kinematic chain of the machine.

The situation becomes just the opposite, however, if another cause of the excitation of an electric field in the crevice-shaped gap is kept in mind. We are speaking of complete breaks in the current-carrying thermoelectric circuits in vibrational rebounds of the indenter from the element conjugated with it. Realization of conditions of this type is especially distinctive for intermittent cutting. Under such conditions, the potential possibility of electrical microerosion of the surfaces is created by the emf of self-induction $\mathscr{E}_s$, which is induced in the external thermoelectric circuit as a result of breaking of this circuit. It is quite probable that, upon rapid breaking of the circuit,

$$\mathscr{E}_s = r\,\frac{\mathscr{E}_t}{R}\,e^{-\frac{r}{L}\tau} \gg \mathscr{E}_t \tag{5-1}$$

at the instant immediately following the beginning of the break ($r \to \infty, \tau \to 0$).

The inductance of the tool-machine-workpiece circuit was determined from the resonant frequency of the circuit shown in Fig. 5-1, or with the help of a high-frequency meter, type E12-1. It turned out that $L < 10^{-5}$ H. The energy of the circuit is so small that it can be absorbed by a miniature capacitor.

As a rule, circuit inductances in friction and cutting of metals are very small. Therefore, under ordinary conditions, when there is no highly resistive third phase in the gap of the frictional capacitor, the probability of spark discharges is low. However, if such discharges do take place (when mechanical relaxational vibrations are present in frictional systems), then blocking of thermoelectric currents may prove desirable, for the reasons stated below.

Erosion of the very thin oxide films that act as lubricant frees sections of the surface for direct metallic contact. The heat evolved in the narrow conductive channel causes the material to soften, and sometimes even to melt. If electrical breakdown is repeated at a high rate on the same spots, destruction of the metal

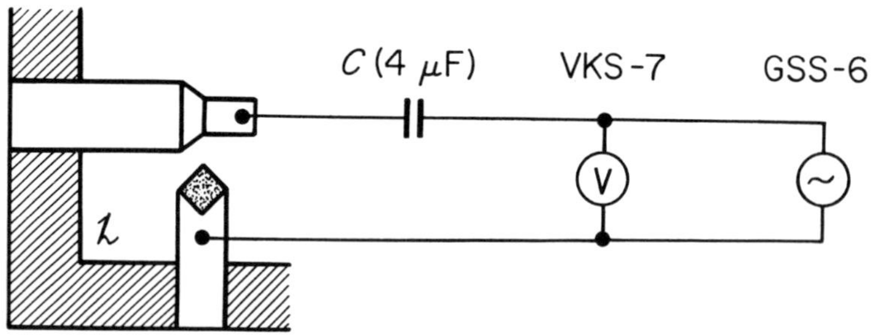

Fig. 5-1. Oscillator circuit for determination of the inductance of the external portion of the thermoelectric circuit.

in the oxidation process leads to cratering wear.* Films of organic lubricants may be destroyed too. The erosion process restricts the access of the nonconductive cutting fluid to the crevice-shaped space between phases. If the electrolytic medium is sucked in, then as a second-class conductor it prevents a dangerous type of discharge. It is rather remarkable that not only man-made (synthetic) liquids are able to limit the overvoltages in the zone of friction (cutting). Particles of self-emulsifying oil, for example, becoming ionized on being atomized, also weaken the electric field in the gap between the separating contact areas ("fog" effect).

In conclusion to the discussion of the questions raised in the preceding, it should be emphasized once again that not only the causes and types of discharges, but also the ways of combating them in friction, are quite different and sometimes mutually exclusive.

It has already been remarked that the effect of thermoelectric currents on cutting-tool wear was first observed by Axer and Opitz. Measurements of flank wear of carbide-tipped cutters have shown that compensation of thermal emf or opening of the external circuit (insulation of the cutter from the slide) may result in tool life extended over that in conventional cutting conditions. According to Opitz, the introduction of weak currents into the cutting zone makes it possible to control oxidation processes at points of contact between tool and chip and between tool and workpiece, thus influencing the rate of electrochemical wear of the cutting edges.[24] The rate of formation and growth of the oxide layer decreases if the thermoelectric current is not channeled through the machine.

The author thinks it pertinent to state at once that, in the publications of Opitz and coworkers that are known to him and that date from after the London conference of 1957, the point presented here was not developed. What is more, both this point and the desirability of practical utilization of the proposed method have been placed in doubt, since the continuing research of the German specialists has demonstrated the possibility of a detrimental effect of thermoelectric compensation on the formation of coatings (chemical reaction products) retarding tool wear.†

Systematic investigations of the way in which thermoelectric currents and currents injected into the cutting zone affect the rate and character of oxidative tool wear have already been carried on for a number of years by Rizhkin and coworkers.[25, 244, 278, 279] Sharing the views of Opitz and Axer, these researchers have from the very start been guided by data on the conditions of oxidation of contact areas in cutting (friction) and the lubricating ability of oxide films as a function of thickness, composition, and properties.[267, 280–282] The fundamen-

---

*This type of wear is not considered to be linked with the formation of macroscopic craters and grooves on the working faces of the tool.
†This information became known to the author through a copy of a letter from W. Koenig (Aachen, Federal Republic of Germany) to R. G. Markosyan.

tal element in their reasoning, which is based on the Wagner theory of high-temperature oxidation,[283] is the notion that the migration of metal cations (Fe^{2+}, W^{6+}, CO^{2+}, etc.), oxygen anions, and electrons through the primary oxide film is accelerated (on closing of the thermoelectric circuit, "aiding" connection of source) or slowed down (on opening of the thermoelectric circuit, "opposing" connection of source) by the change in potential difference (electric field strength) between the "cationic" and "anionic" plates of the film; this change can be due to an external emf or to the emf of the natural thermocouple. Accelerated diffusion of ions of the tool material toward the interface between oxide and atmospheric oxygen brings about continuous growth in the thickness of the oxide films and causes the films to take on the properties of a well-defined oxide, which is known to be far easier to remove from the surface of the metal substrate. Because the result of the destruction of these films is the denudation of juvenile surfaces, Rizhkin and coworkers propose considering thermoelectric current as a stimulator of adhesional processes on contact areas;[279] this idea is in complete agreement with the conclusion that the author drew earlier, on the basis of other theoretical premises and published experimental data.[26,27,30]

Zhilin, remarking that the wear of a carbide-tipped tool is caused by several processes occurring simultaneously (mechanical scratching by abrasive, adhesion, corrosion, solid-phase diffusion, contact-reaction melting, plastic flow), also touches on the question of how thermal emf's and external currents affect the corrosion component of integral wear, whose contribution rises sharply at temperatures of $600°$ to $900\,°C$.[284] Specific features of gaseous corrosion (treatment in air) under the conditions of occurrence of thermoelectric phenomena are explained on the Mott-Cabrera model; but the explanation differs little from Rizhkin's approach to the problem. As for the pecularities of electrochemical corrosion (in cutting with cutting fluids) under the same conditions, a simple enumeration of these peculiarities makes any forecasting undesirable that goes beyond the considerations set forth in the previous chapters. The most interesting portion of the publication cited[284] is, in our view, the following explanation of why series VK alloys are more vulnerable to corrosion than series TK alloys:

An isolated oxygen atom has the valence-electron configuration s^2p^4. As a rule, oxygen exhibits acceptor properties relative to its partner in a chemical compound, and tends to form a stable s^2p^4 electron configuration. Tungsten carbides, together with a high statistical weight of atoms with stable d^5 configurations, have a relatively low percentage of delocalized electrons, so that it is difficult for s^2p^6 configurations to form or for stable chemical compounds, which prevent further oxidation, to appear.

Atoms of titanium, which forms part of the composition of type TK alloys, display donor properties with respect to oxygen, forming stable compounds with increased statistical weight of atoms with stable s^2p^6 configurations; these configurations hinder the diffusion of metal and oxygen atoms through the film and thus reduce the rate of oxidation.

This assumption seems quite logical and justifiable—something that cannot as yet be said for the author's attempt to connect the effect of thermal emf's and external currents on the rates of corrosion processes with the "dependence in principle" of the formation of electron configurations on the electric field strength resulting from the superposition of fields.

Bobrovsky denies the possibility that thermoelectric current can have a substantial influence on the rate of oxidative tool wear;[111] he has conducted special trials on shallow milling of type-18Kh2N4VA steel with a large-diameter one-piece cylindrical cutter made of type-R18 high-speed steel. Having ascertained that electrically insulating the tool does have an effect, and supposing that prolonged residence of the cutter teeth in air would make it impossible to see the effect, even though it were caused by the reduction in oxidative wear as a result of opening of the thermoelectric circuit, the author came to disagree with the hypothesis developed by Rizhkin.

It is not difficult to see that Bobrovsky contradicts himself in the present instance. Having particularly set up conditions under which the effect of thermoelectric current on the rate of oxidative wear would be veiled, the author uses this very fact to deny the indicated effect that had been considered the object of his investigation. In our judgment, the results of Bobrovsky's experiments have made it possible to say with assurance only that other reasons exist for the destructive function of thermoelectric current, independent of the one just examined.

It is doubtful too whether one could agree with Bobrovsky's remark that it is incorrect to model the process of oxidation, "which can take place in cutting, by oxidation in a furnace."[111] This means of modeling (it has been used, in particular, by Opitz, Axer, and Rizhkin) is supported by data on the similarity between, on the one hand, the phase composition of the surface layers of the rubbing specimens and, on the other, the composition of oxides formed on the surfaces after the specimens have been heated in a furnace to the same temperature.[285] (The data were obtained in an investigation of the oxidation of frictional materials by high-temperature X-ray diffractometry in the temperature range of 20° to 1,100 °C.)

The intervention of thermoelectric current during the oxidation of the tool may indeed be active and significant, if oxidative wear is one of the leading forms of wear in cutting. We consider the most convincing proof of this to be experiments demonstrating that electrical phenomena do not in any way affect the life and surface roughness in cutting under an inert gas (argon) that prevents the appearance of oxide films.[229]

Now we will turn to some ideas on the role of electrical diffusion in cutting-tool wear.

Hehenkamp[28] was the first to attempt to explain the destructive function of thermoelectric current in cutting from the standpoint of mass transfer caused by these currents, that is, electrical transfer. Referring to the lack of published data on electrical diffusion in solid alloys, and all the more in the oxide films that appear on them in cutting, he considers it possible that particles belonging to various components of the solid alloy will countermigrate. Results of this migration will include changes, for example, in the concentrations of carbon, tungsten, and cobalt; this will be reflected at once in the mechanical properties of the alloy. According to Hehenkamp, electrical-diffusion wear of a tool should go especially fast just at those places where the thermoelectric current density attains the highest values. In order to estimate the scale of mass transfer involved in the directed migration in an electric field of the ions making up the framework of the crystal lattice, he suggests making use of the following formula expressing the volume of transferred matter V when a thermoelectric current I_t has flowed for a time τ:

$$V = BI_t\tau \tag{5-2}$$

The quantity B, which characterizes the rate of electrical transfer at a given current, has a value of $\sim 10^{-7}$ mm^3/A-sec according to the highly approximate calculations of Hehenkamp; even this small value is too high, as he himself recognizes.

Having seen no difference in the wear of insulated and uninsulated carbide-tipped cutters in the turning of steels, and having established that the machine resistance goes up to 200–1,000 Ω in operation, Hehenkamp concluded that a resistance of this size in the external circuit is equivalent to electrical insulation when the thermal emf acting in the circuit is ~ 10 mV, and thus, according to Equation 5-2, no effect will be gained through opening of the thermoelectric circuit with the weak current flowing in it. At the same time, Hehenkamp did not retreat from his hypothesis that additional tool wear is possible as a consequence of electrical transfer, for in his judgment "purely electrical" wear occurs on the passage of large short-circuit thermoelectric currents, localized in the immediate vicinity of the interface, through which they circulate. He attempts to use changes in the value and the character of the distribution of internal thermoelectric currents to explain the increased tool life detected by him when an outside source was connected to the cutting zone.

In fact, Bobrovsky[111] constructs his reasoning on the same physical base, with just one difference: the notion of a redistribution of components of the interacting solids in the electric field, which "may lead to weakening of the structure of the tool material and to the creation of more favorable conditions for its mechanical destruction," is now extended to the external thermoelectric circuit, where, according to Bobrovsky's data, the operative currents are as a rule measurable in milliamperes. Apart from general—and, for the most part, well-known —results taken from numerous publications on diffusion in elementary binary systems, Bobrovsky presents not one single direct proof that his propositions are correct, referring to the limited capabilities of the modern experiment and, it seems to us, undervaluing the tracer method (autoradiography). Let us look at data obtained in an investigation of electrical diffusion in binary alloys, keeping in view ultimately that any quantitative estimate in the present case serves as an instrument only of purely qualitative analysis.

It is well known that carbon atoms have the highest mobility in solid alloys in a steady electric field. In iron-carbon alloys, all these atoms are in the ionized state; under the action of ponderomotive forces and also because of entrainment by the electron wind, they advance in a front in the direction toward the cathode, attempting to leave the anode region altogether.[286] The radioactive carburized zone was seen to move without losing its symmetry in electrical transfer of carbon in iron, nickel, cobalt, titanium, tantalum, and tungsten.[287] However, it follows from Lebedev and Guterman's work[288] that an estimate of the magnitude of the displacement occurring over several hours at a temperature of $\sim 1{,}000\ ^{\circ}\mathrm{C}$ yields values on the order of 10^{-5} mm.

Partial speeds of electrical transfer, in centimeters per second, are $\sim 10^{-8}$ for chromium in a Ni-Cr alloy at 950–1,100 $^{\circ}\mathrm{C}$;[289] $\sim 10^{-8}$-10^{-7} for silver and zinc in Ag-Zn alloys at 550 $^{\circ}\mathrm{C}$;[290] and $\sim 10^{-7}$-10^{-6} for tungsten and iron in a solid solution of tungsten in iron at 900–1,150$^{\circ}\mathrm{C}$.[291] Regardless of how closely these speeds are linked with the migration of ions formed when the atoms dissociate by

$$\mathrm{Me(atom)} \rightleftarrows \mathrm{Me(ion)} + e,$$

and how closely they are linked with the motion of neutral atoms re-formed by ions that have stayed for very short times in electron-trap sites, these data are far from convincing us that electrical diffusion (which can be eliminated by opening of the tool-workpiece-machine circuit) can be serious competition to thermal diffusion in weakening the structure of the cutting-edge material, even though electrical diffusion takes place in a force field while thermal diffusion is free. Instead, the following conclusion, drawn by Avakov and Markosyan on the basis of an even more detailed discussion of the question,* suggests itself: "The cutter fails through mechanical wear significantly short of the time required for 'purely' electrical wear to make even the slightest contribution to overall wear."

It should also be emphasized that contact electric fields at the interface can either promote or hinder the diffusional mass transfer.[72] Diffusion, whatever its nature, is made difficult in friction and cutting by the continuous variation in the number and distribution of contact spots where metallic bonds arise. As for a high-speed-steel tool, in Loladze's judgment, it hardly suffers diffusional wear, since the maximum temperature at which it retains its shape is below the temperature at which marked diffusion begins.[242]

*In a manuscript entitled "Investigation of the effect of thermoelectric and thermomagnetic phenomena on wear properties and life of cutters" (Leninakan, 1972).

All this does not, however, mean that the author is ignoring the role of electrical-diffusion processes in cutting-tool wear. On the contrary, as his earlier publications have suggested,[27,30] these processes may predetermine the strength of the cohesion joint between the tool and the workpiece, with all the consequences that that entails. But the next section will give the reader a closer acquaintance with the author's hypothesis.

Any discussion of thermoelectric cooling of the tool, which is brought about by the passage of relatively weak currents through the region where the metals interact,* should begin by dividing the existing ideas about the role of the Peltier phenomenon in cutting into false and true (in physical fundamentals).

Galey managed to approximately double the life of type-R18 steel drills in the machining of type-45 steel by introducing into the cutting zone an external current opposite in sense to the thermoelectric (natural) current; he achieved the same result in the grinding of high-speed tungsten and molybdenum steels. He explained these effects by the withdrawal of heat from the interaction zone—by heat transfer from tool to workpiece through the Peltier effect.[293]

Bobrovsky, in attempting to clarify to what extent the cooling "resulting from the Peltier phenomenon" is able to influence the extension of tool life,[111] entirely shares Galey's judgment on the heating (cooling) of the tool by current upon aiding (opposing) connection of a source. On this basis, he builds up all his reasoning on the role of the Peltier phenomenon in cutting. In particular, he compares the abilities of the "method of breaking the thermoelectric circuit," the "compensation method," and the "countercurrent method" to secure a tool-life advantage by causing absorption of Peltier heat in the contact areas.

Of the hypotheses "about the physical nature of the effect exerted by thermoelectric current on the principal indices of the friction and cutting process" on which Dubrov and coworkers[229] analyze experimental data, they formulate one in the following way: "Thermoelectric current flowing in the tool-workpiece contact is a cause of additional heat liberated in the contact, as a result of which the tool life is shortened." Furthermore, Dubrov and coworkers state directly that by this additional heat they mean Peltier heat; the possibility that they discuss, of additional cooling of the hot junction through breaking of the thermoelectric circuit, intensification of this cooling by the insertion of countercurrent, and so on, is not in doubt at all.

Evidently under the influence of Dubrov's publications, Shulga starts from the same considerations.[294]

There is no need to prove that Galey,[293] Bobrovsky,[111] Dubrov and coworkers,[229] and Shulga[294] were incorrect in the way they understood the essence of the manifestation of the Peltier effect in cutting (friction), and we omit the discussion of results obtained with the use of ideas contradicting the Second Law

*M. T. Galey was the first to take up this question in print.[292]

of Thermodynamics. At the same time, we can point to a series of studies differing from those mentioned primarily in taking a correct approach to the actual statement of the problem.[230, 238, 243, 295–297]

Tyushev analyzed the possibility of using natural thermoelectric cooling by the Peltier effect to limit the temperature rise and stabilize the temperature in friction.[295] The following values of the Peltier coefficient P (in units of 10^{-3} J/C) were found for various pairs of materials (the arrows indicate the sense of current for which the contact is cooled): antimony ⟵ lead, 0.8; copper ⟵ aluminum, 1.7; copper ⟵ lead, 2.4; type-55 steel ⟶ copper, 2.8; lead ⟵ constantan, 9.2; constantan ⟶ copper, 11.0; steel ⟵ constantan, 12.0; bismuth ⟶ copper, 16.0; lead ⟵ bismuth, 21.5; antimony ⟵ bismuth, 45.0; steel ⟵ graphite, 150.0. The publication examined the fundamental principles of temperature control in graphite-metal contact couples, where, as experimental data have shown, thermoelectric phenomena can most efficiently lower the temperature of the rubbing surfaces and contribute to prolonged operation of the frictional couple without the use of lubricant.

Yakunin and Yakubov studied the effect of thermoelectric current on the lives of one-piece high-speed cutters made of type-R18 steel in the longitudinal turning of type-St3 steel under oxygen and in air.[243] The external source was a chromel-alumel thermocouple heated so that its emf was 14–16 mV. The results of the life tests confirmed that there is a close interrelation between the manifestation of thermoelectric effects, on the one hand, and, on the other, the effect of the medium in which cutting takes place and the properties of the oxide films that form. In cutting under oxygen, it was determined that the connection of the external source aiding (opposing) the cutter-workpiece natural thermocouple may lead to a significant increase (decrease) in life, the peak life being displaced toward higher (lower) cutting speeds. Yakunin and Yakubov proceeded to the physically justified conclusion that the result seen is explained by lowering (raising) of the contact temperature of the intensively oxidized cutter and chip surfaces by the Peltier effect. They also came to the following important conclusion, but unfortunately left it without reinforcement by elementary calculations: "The part of the heat flux that is due to the presence of the thermoelectric current* is quite considerable, and the temperature field in the tool and the tool life may depend to a significant degree on the direction of this current."

It follows from the experiments of Yakunin and coworkers that the thermoelectric current has not only a destructive but also a creative function due to the Peltier effect. This convinces us that it was correct to assume that the opening of the thermoelectric circuit is not desirable under certain conditions of cutting.

According to Yakunin and coworkers, the thermal effect of a thermoelectric current, on which the properties of the screening films depend, becomes appar-

---

*Here, as before, we mean the connection of an external thermoelectric source.

ent in different ways, depending on the "temperature-speed" conditions of cutting.[230]   As the cutting speed increases, the temperature on the contact areas automatically rises, because plastic deformation processes go more intensively. Therefore, as Yakunin and coworkers suppose, the temperature may exceed the optimal value from the standpoint of realizing conditions in which the thermoelectric phenomena exert an especially marked effect on tool life. According to Usmanov and Yakunin's data, "If the effect on increased life in an oxygen medium is to be maintained on a changeover from one range of cutting speeds to another, the feed should be changed so as to keep the product of cutting speed and feed constant."[298] As an example of the reconstruction of conditions favoring the appearance of the effect, Yakunin and coworkers consider a change from one set of conditions ($v = 104$ meter/min, $s = 0.1$ mm/revolution) to another (80 meter/min, 0.13 mm/revolution) under the condition $v \cdot s \approx$ const (longitudinal turning of type-45 steel with a cutter of type-R18M steel, cutting in air).[230] Connecting a chromel-alumel thermocouple (emf $\sim$ 15 mV) with the artificial thermoelectric current agreeing in sense with that from the natural current re-resulted in a marked rise in tool life, expressed in the K_e values before ($\sim$ 2.3) and after ($\sim$ 1.5) the changeover.

The arguments of Yakunin and coworkers,[230] according to which the action of thermoelectric currents on tool life has to do with thermal phenomena, are supported to some extent by the results of Hashimov and Yakunin's investigation.[299] If connecting the external source really does change the temperature on the rubbing surfaces, then the effectiveness of a thermoelectric current should be expected to vary, depending on whether such a temperature change does or does not contribute to an approach to current-sensitive temperature-speed conditions of cutting. This proved to be true when the authors began artificially setting up conditions with different initial temperatures on the contact areas, by varying the tool rake angle γ (from $0°$ to $35°$). The effect of a current introduced to the cutting zone from an external source was especially strong at some $\gamma = \gamma^*$ ($\sim$ 25–27°) and considerably less perceptible for $\gamma < \gamma^*$ and $\gamma > \gamma^*$. Since qualitatively similar results were obtained in trials with different media (air, oxygen), the authors concluded that, under certain conditions of cutting, the thermal effect of thermoelectric currents may prevail and the change in temperature, although slight, may lead to a significant change in the life of a tool with rake angle $\sim \gamma^*$.

The approach set forth here is in close contact with Makarov's hypothesis[300] that the optimal cutting temperature for a concrete combination of tool material (carbide-tipped) and work material (material of construction) is invariant with respect to cutting conditions, tool geometry, and other external conditions.

Ladakina and Gufan in their time narrowed the question of the upper limit of the thermoelectric-current power evolved in the tool-workpiece contact zone, emphasizing too the obvious (for steady-state processes) dominance of the

Peltier heat, which can be withdrawn into the external portion of the thermo-electric circuit, over the Lenz-Joule energy.[297] As they supposed, the very small values of the power developed by the natural seat of thermal emf in the cutting zone ($\sim 10^{-5}$–10^{-4} W) point to the necessity of using "some quite specific mechanisms" to explain the effect of thermoelectric currents on cutting-tool wear.

In order to give an experimentally grounded answer to the question of the role of natural thermoelectric heating of the cutting zone, Avakov and Markosyan used a roundabout method to solve the problem, consisting in a wear comparison between two identical, symmetrically operating cutters to which an external seat of emf is connected (Fig. 5-2).[238] The thermocouples formed by either cutter and the workpiece and connected in opposing senses should be exactly symmetrical; this was checked ahead of time with a millivoltmeter used as a null indicator. The trials showed that the different senses (relative to the interfaces) of the currents in the two contact zones do not give rise to differences in the wear of type-VK8 alloy blades in the machining of type-OT4 titanium alloy when the current varies in the range of 0.05–0.2 A (v = 122.5 meter/min, s = 0.1 mm/revolution, t = 0.5 mm). That is, regardless of whether Peltier heat is liberated or absorbed on the tool contact areas, this heat is not very important under the given conditions of cutting.

In our judgment, the author's attempt can be considered methodologically irreproachable only in relation to steady-state processes. It does not make clear the interconnection of thermal and electrical effects in cutting in the presence of the inevitable dynamic phenomena that actually predetermine the nature of tool wear.

Now let us go on to a more detailed exposition of the essence and specific features of the processes that bring about the direct connection between thermo-

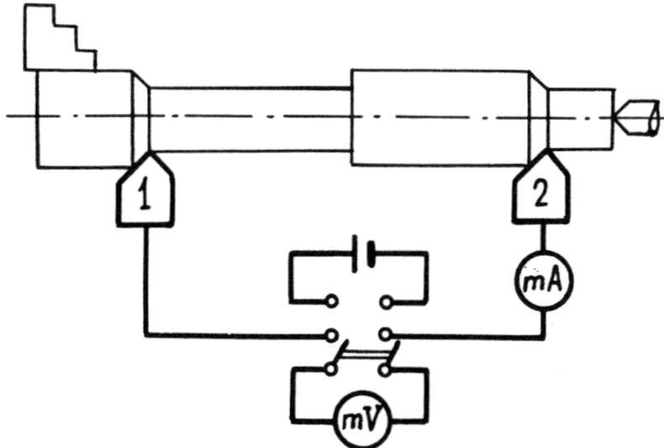

Fig. 5-2. Arrangement for clarifying the role of the Peltier effect in the natural thermoelectric cooling of the cutting zone (after A. A. Avakov and R. G. Markosyan).

electric phenomena in friction (cutting) and the mechanism by which metals and alloys in contact interaction are destroyed.

5.2. THERMOELECTRIC CURRENT AS A STIMULATOR OF SEIZING AND DIFFUSIONAL SINTERING OF MATERIALS

According to several publications cited earlier,[26, 27, 30] it is natural to link the idea of thermoelectric-current surges (which as a rule are disorganized in character) with the idea of jumplike surmounting of the energy threshold of seizing. We have in mind Semyonov's hypothesis of seizing,[301–305] expressed as its author expressed it and not in the way it has been interpreted or even misinterpreted in several other publications.

In Semyonov's judgment, accomplishing the seizing of metals requires communicating to them some activation energy drawn from the work of plastic deformation. Comparing the interface obtained through crystallization from the liquid state or through recrystallization with the interface formed in the seizing of like metals, Semyonov notes[303] that the latter surface has the higher energy, and supports his idea with a passage from Smit:[306] "The transition from one to another structure cannot be accomplished if the atoms do not acquire an additional energy, called the activation energy. In polymorphous transformations, this energy barrier is rather high."

This point of view seems plausible to us. It is doubtful whether one can agree with Likhtman's judgment[307] that plastic deformation plays an auxiliary role, consisting only in creating conditions for the appearance and development of juvenile contact areas, whereas the seizing itself is supposed to provide the whole excess of free surface energy accompanying this process.

The author holds that even in an ideal vacuum, when we are completely rid of oxides and sorbed films that screen the fields of free valences and prevent the development of molecular coherence forces, the formation of new metallic surfaces, which are defects extended in two dimensions, inevitably involves a drastic change in the energy states of the surfaces, under the action of the internal pressure of the electron gas. The association of atomic systems "pushed apart" by this pressure, including systems with differently oriented lattices, is actually made easier by mechanical-thermal activation.

On the other hand, we do not finally intend to dispute that the work of deformation may significantly exceed the activation energy needed for the surmounting of the energy threshold of seizing.[308] Depending on specific features of the interaction of the solids, the liberation and absorption of energy in contact microvolumes goes by various routes, stimulating not only seizing but also diffusion, twinning, phase transitions, and other atomic rearrangements of a thermal-fluctuational nature; here it is especially important that thermal fluctuations play a decisive role, not only in the secondary processes just mentioned but, rather, even in the primary events of deformation and destruction of the metals, events that are directly connected with the evolution of the defect structures of the metals.[309–311] From this standpoint, incidentally, it is rather easy to explain the physical picture of the productive method of hot machining of difficultly machinable materials (cutting of metals heated to high temperature),[312] which is especially effective for the utilization of powdered ceramic materials.

The external manifestation of seizing is scratching and transfer of material in rubbing couples. With the help of an electronic probe, it was observed that some 85 percent of the contact area of the tool may be covered by a thin layer of work material, in which structural transformations are going on.[313]

As Ainbinder remarks, couples of unlike but readily weldable metals differing in hardness "possess, as a rule, incomparably better antifriction properties than couples of like metals."[314] However, we do not see any contradiction here either, if we speak of the desirability of making up frictional systems, including tool-workpiece couples, in accordance with the criterion of clear mismatching of the chemical potentials of the metals, with the aim of assuring plasticization of the surface layer of one of the elements in contact interaction (Section 1.9). The notion of an electronic mechanism for the plasticization of contacting solids also strips all the mystery away from the questions of why a ranking of the principal properties of metals in an evaluation of their cuttabilities showed the work function to be the most informative property.[315]

Rikalin and coworkers divide the process of formation of a firm bond between unlike substances into two stages:[316] a preparatory stage of drawing together as joint plastic deformation takes place, on account of the surface diffusional motion of atoms and the like, when the primary role is that of electrostatic interaction processes between particles; and a final stage, leading to the appearance of a strong joint, during which the decisive role belongs to quantum processes of electron interaction. Regardless of whether these quantum processes are equivalent to the formation of a metallic bond (pure metals, solid solutions), a stable bond of covalent type (contact of metal with metalloid, semiconductor, intermetallic compound, etc.), a coordination-covalent bond (e.g., between donor and acceptor oxides) or whatever, the occurrence of the processes requires activation of the state of the surfaces by energy in a certain quantity communicated in the form of heat (thermal activation), as energy of elastoplastic deformation (mechanical activation), or through various types of irradiation (radiative activation).

As we see, Semyonov's "energetic hypothesis of seizing" has been refined and developed on the basis of contemporary quantum-mechanical ideas.

The points of importance to us cited by Rikalin and coworkers[316] are, first, considerations having to do with the strong acceptor properties of titanium, which are due to the presence of two unfilled electron shells, and, second, the fact that if a significant portion of the chemical bonds are ruptured in both the contacting substances within limited periods of time, then the probability increases that atomic residues will capture foreign bonds (under conditions when paired-electron bonds form as a result of the translation of valence bonds).

Now we have come right up to the question of participation by thermoelectric currents in the events of seizing of metals in friction (cutting).

We write the equation of thermoelectric energy balance for the external circuit under steady-state conditions of the frictional interaction:

$$I_t^2 R_m - \alpha T_1 I_t + I_t^2 R_o + \alpha T_2 I_t = 0. \tag{5-3}$$

Here $I_t^2 R_m$ is the Joule-Lenz heat evolved in the hot junction at temperature T_1; $\alpha T_1 I_t$ is the Peltier heat absorbed from the same junction ($\alpha T_1 = P_1$); and $I_t^2 R_o$ and $\alpha T_2 I_t$ are the Joule-Lenz heat and the Peltier heat evolved in the cold junction ($\alpha T_2 = P_2$).

Depending on the value of R_o, now the greater and now the lesser part of the thermal energy (work of deformation) is transformed into electrical energy. In fluctuations of R_o, relaxational thermoelectric processes arise; these, together with elastoplastic deformation and dispersion, are responsible for the continuous liberation and absorption of energy in contact microvolumes. A consequence of the thermoelectric-current surges upon sharp drops in R_o is the rapid pumping of energy from the hot to the cold junction; this effect reflects the tendency of the system to even out the temperatures T_1 and T_2 and to lower the current. The external circuit (machine) plays the role of a thermoelectric radiator in relation to the hot junction (cutting zone). Here we see a fairly graphic confirmation of Le Chatelier's principle, which states that any external influence moving a system away from equilibrium stimulates processes in the system that tend to weaken the results of this influence.[317]

In connection with what has been said, it should be remarked that a general frictional couple often gives the impression of being a "thinking" system whose behavior indicates that it is trying to go to the state of least energy consumption (minimal perturbation), either by adapting to the character of the external influence or else at the cost of its own destruction as a "protest" against loads and speeds that it finds out of proportion.

If the case of the cutting of metals is considered, we may think that an arbitrary heater and cooler have an infinitely large specific heat, with $T_1 \gg T_2$. Then

$$Q_{P1} = \alpha T_1 I_t \approx \frac{\mathcal{E}_t^2}{R_m + R_o} \tag{5-4}$$

while

$$Q_{D1} = I_t^2 R_m = \frac{\mathcal{E}_t^2 R_m}{(R_m + R_o)^2}. \tag{5-5}$$

It is clear that $Q_{P1} > Q_{D1}$ and the quantity of heat absorbed from the hot junction, $Q_1 = Q_{P1} - Q_{D1}$, will depend on the ratio of resistances of machine and cutting zone, as Fig. 5-3 shows.

It is not difficult to see that fluctuations in the machine resistance cause additional fluctuations about T_1 of the temperature on the contact areas of the tool; that is, a particular kind of temperature fluctuation that should give an even more nonuniform character to the thermal motion of the atoms, this nonuniform character being expressed in energy (thermal) fluctuations. Under these conditions, as it is natural to suppose, the probability increases that foreign bonds will be captured as a result of quantum processes of interaction of atomic electron shells. The energy fluctuations continually change the configuration of the equipotential surfaces of the external molecular fields of the interacting condensed phases. The more frequent and the more different these changes are, the greater the odds are on these configurations resembling each other closely; we consider coincidence of the configurations the basic indication that the interface between crystal lattices has disappeared. Consequently, thermoelectric current (more precisely, relaxational processes connected with the passage of that current at variable R_o) stimulates the events of seizing.

Relaxational processes of this type in the internal thermoelectric circuits play an even more important role; unfortunately, because of their microscopic dimensions, these circuits do not permit instrumental intervention from outside. For one of the group of comparatively large, local eddy thermoelectric currents that flow, for example, right in the cutting zone, Equation 5-3 takes on the form

$$i_t^2 r_i - \alpha T_1 i_t + i_t^2 r_k + \alpha T_2 i_t = 0. \tag{5-6}$$

While the processes going on in the external thermoelectric circuit correspond to only the right-hand branch of the curve in Fig. 5-3 ($R_o/R_m > 1$), relaxational processes in the internal circuits are described by both branches of a similar curve. Passes through the fluctuating maximum of such a curve (from $r_k/r_i < 1$ to $r_k/r_i > 1$) signify in fact that the hot and cold junctions have changed places (after an intermediate leveling out of the temperatures T_1 and T_2).

Let us estimate roughly the thermal power dissipated by a local thermoelectric current in the cold junction. If we take several values for $r_i = r_k$, then for $T_1/T_2 \sim 2$ and $e_t \sim 1$ mV we obtain for $Q_2 = Q_{P2} + Q_{D2}$ the approximate values shown in Table 5-1. These would seem to be very low values. If, however, we allow for the fact that power on the order of 10^{-4}–10^{-2} W is dissipated in very small contact microvolumes, then we can no longer neglect the heat effect of thermoelectric currents, which leads to a rise in temperature and softening of the contact spots. The power per cubic millimeter of the thermoelectric current may amount to several watts, quite an appreciable value and sufficient, in our judgment, for the thermal activation of seizing. Looped thermoelectric eddy currents

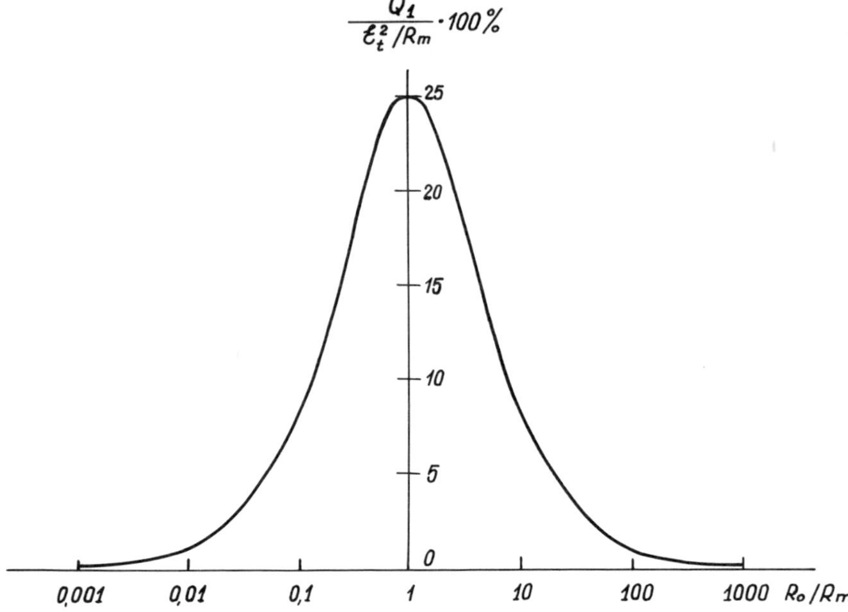

Fig. 5-3. Effect of the ratio between load resistance of the tool-workpiece thermocouple and internal resistance of the thermocouple (logarithmic scale) on the relative quantity of heat withdrawn from the cutting zone in natural thermoelectric cooling. The short-circuit power of the thermocouple ($\mathcal{E}_t^2/R_m$) is taken as 100%.

may promote local plastic deformation on precisely those microscopic portions of the surface where the energy barrier has not yet been surmounted. Because of the thermoelectric relaxational processes, thermal cycles are imposed on the repeatedly overstrained asperities; these cycles bring about "adhesive attack."* It is under just these conditions that the avalanche formation of "welded junctions,"* which causes jamming in sliding, can take place.

*F. P. Bowden's terminology.

Table 5-1.

$r_i = r_k$ (ohms)	$Q_2 = Q_{P_2} + Q_{D_2}$ (watts)
~1	~10^{-6}
~0.1	~10^{-5}
~0.01	~10^{-4}
~0.001	~10^{-3}
~0.0001	~10^{-2}

It must be supposed that the drastic fluctuations in contact-spot resistance are linked in the most intimate way with processes of oxidation of the rubbing surfaces and with the properties of the films that form on them. Screening films of oxides, which guard the surfaces against seizing, differ rather greatly in their electrical properties (see Fig. 2-34), mechanical properties (hardness, brittleness), and, especially important, the character of the adhesion bond with the substrate and of the degree of adaptation to the crystalline relief of the substrate surface.[318] Destruction of the oxides in local constrictions is like short-circuiting of the thermoelectric seat of emf across low-resistance metallic a-spots. Here the thermoelectric current density is especially large, and here both the Peltier energy (brought from the hot junction where a temperature flash has just occurred) and the Joule-Lenz energy are concentrated. In turn, the thermoelectric current facilitates "healing" of the damaged oxide and affects the mechanism and growth rate of newly formed film.

The external and internal portions of the thermoelectric circuit are linked. Therefore, disconnecting the external circuit leads to a redistribution of thermoelectric currents in the internal circuits; this event must be reflected in the conditions of the contact interaction and in the rate of destruction of the rubbing solids. This behavior is probably one reason for the change in shape of the wear areas on the flank faces of drills, which has been observed several times by various researchers after the tool-workpiece-machine thermoelectric circuit was opened.

From the notion of chaotic thermoelectric-current surges, it follows[30] that in the cutting of metals the transfer of the diffusing species due to electron-hole entrainment of activated ions proceeds nonuniformly. This fact is probably very important, regardless of the mechanism by which electrical diffusion affects cutting-tool wear; that is, regardless of whether the "drawing off" of the migrating impurity brings about a weakening of the structure of the tool material,[29,111] or the "injection" of diffusing species through the arbitrary interface between lattices* results in welding of the surface layers.[27] The author held, however, that in the presence of fluctuations of the electric field, which can cause a "jerky" severalfold change in the diffusion flux density, the second assumption in particular is not without heuristic potential.[30] Now we can speak of this with even more confidence. The ideas just set forth concerning the scale of electrical mass transfer, which is distinctly insufficient to allow electrical-diffusion weakening of the tool-material structure to be set apart as a special type of wear (see Section 5-1), leave almost no doubt that it is more correct to fix our attention not on electrical-diffusion wear as such, but on "gusts" of electron-hole "wind" as one of the principal causes of the strengthening of the adhesion joint, and thus of cohesive wear. Consequently, not only thermal but

---

*After the appearance of metallic bonds.

also electrical fluctuations are under consideration; because of the latter, we do consider thermoelectric current as a stimulator of the diffusional sintering of materials.

In the light of ideas about the effects of thermoelectric currents on the deformational-thermal activation of atomic-molecular rearrangement processes (hardening and resoftening, recrystallization, dissolution and precipitation of excess phases, allotropic transformations, etc.), it is necessary to take special account of the heterogeneity of structure of real solids. Cyclic heating of contact microvolumes as a result of thermoelectric relaxational processes increases the nonuniformity of the thermal motion of atoms, that is, intensifies the energy fluctuations. According to the kinetic conception of strength,[310] this effect leads to accelerated accumulation of breaks in interatomic bonds in regions of structural stress concentrators, arrangement of the breaks one beside another, and localization of the damage to the solids through the formation of submicroscopic, then microscopic, and finally macroscopic cracks. It is known, incidentally, that considerable internal stresses arise in a solid solution of tungsten carbides in cobalt (this solution is the cementing material in heat-resistant cutting blades made of group VK alloys) as a result of the significant difference between the coefficients of thermal expansion of tungsten carbide and the cobalt phase.

Barrow and Spenser attempted to discover how important a contribution "thermoelectric wear" makes to overall tool wear, in life tests on lathes with different electrical resistances.[225] They concluded that the effect of thermoelectric phenomena on tool life can be important only when machines with very low resistances are used; in the great majority of cases, insulation of the tool "does not increase the reliability of the life tests." In other words, if there is additional wear due to the passage of thermoelectric currents, then it is usually so slight as not to change the life value outside the range over which the experimental data are scattered for all other causes. As Barrow and Spenser suppose, the reliability of life-test results is determined by a dominating factor such as the dynamic characteristic of the machine-tool-workpiece system.

Levin and coworkers are conducting a systematic study of the effect exerted by the resistance of a given system to vibration on the life of the cutting tool. They have explained the increase in life of drills with short-circuited conical plastic shanks as due to the damping, vibration-suppressing action of the shank,[251] although in more recent work[253] Levin points out the desirability of using a drill design developed by him (together with I. V. Gogolev), not only to weaken the destructive effect of high-frequency mechanical vibrations on drill life but also to limit the destructive action of thermoelectric currents regardless of its mechanism.

The effect of high-frequency power impulses arising in cutting on the elastically stressed elements of the cutting tool and on the stability of the geometric tool parameters has formed a subject for special investigation.[319] It has been found that, for blades made of hardened tool steel and for the working portions of, for example, twist drills, the stability of shape is determined by the amount of influence of "two principal factors: elastic stress and high-frequency power impulses arising in cutting."* In studying the hyperfine structure of the Mössbauer spectrum of ^{57}Fe nuclei in the ferromagnetic iron of the hardened tool steel,† Levin and Nikitina concluded,

> The phenomenon of residual deformation of elastically stressed polycrystalline metals under the influence of anharmonic power impulses is accompanied by intensification of the directed diffusional transfer of atoms of carbon and other impurities, the transfer being due to the variation in internal energy of the solid being deformed.

If we turn to the ideas just set forth, the following remarks should be made about the publications mentioned here[225, 253, 319, 320] that touch to some degree on the dynamics of the contact processes:

● Mechanical vibrations in the cutting zone cannot be isolated from thermoelectric relaxational processes; thus, both factors always have a joint effect on the scatter of experimental life values. The accuracy of life forecasting depends on the law of distribution of life values.

● The high-frequency impulse application of external force affects the variation in shape of a cutting edge with elastically stressed interatomic bonds. This effect has a simple, elegant explanation in the light of ideas about (1) the decisive role played by thermal-fluctuation events of atomic-molecular rearrangement in the development of deformation processes and (2) the directed character of these events in a mechanical stress field.[321] Directed diffusion of atoms, which Levin and Nikitina detected by nuclear gamma-resonance spectroscopy, was evidently only a secondary one of these rearrangements (a "concomitant" rearrangement, as Levin and Nikitina properly characterized it).

● The conditions of dynamic loading of the contact areas, in which the thermoelectric relaxational processes take place, are in themselves especially favorable for seizing.[322]

● It is altogether probable that the increase in tool life is caused by precisely the aggregate effect of the electrically insulating holding device, which acts as a damper of mechanical vibrations, attenuates the relaxational processes in the internal thermoelectric circuits, and eliminates processes of this type in the external circuit.

While according to Bobrovsky's data[111] the destructive function of thermoelectric current is stronger when the machine resistance is lower, Budzynski,[296] who maintains a diametrically opposed position on the role of thermoelectric

---

*The blades were put through model tests under appropriate conditions of static and dynamic loading.

†The gamma source was metastable ^{57}Co diffused in Pd (W_γ = 14.4 keV).

currents in cutting, recommends that machines be manufactured with relatively small R_o.

From what has been said, it follows that not so much the value of R_o but its stability is important. However, because fluctuations in the machine resistance are in fact inevitable, it is doubtful whether one can agree that it is desirable to aim at increasing the thermoelectric currents in the external portion of the circuit.

In addition to what has been said, we note that the use of lubricants is one effective way of attenuating the high-frequency vibrations (see Section 2.4), that is, of weakening the mechanical-thermal activation of frictional systems. What is more, cutting fluids bring the system back from the energy threshold of seizing through cooling of the contact areas and the screening action of the lubricant films, which prevent the formation of a joint between lattices when the crystallographic match is close. For this reason, the effect of electrical insulation of the tool either decreases markedly or almost entirely disappears when cutting fluids are used.

The more pronounced the relaxational processes in friction and cutting, the more actively the natural thermoelectric currents perform in stimulating the formation of an adhesion joint between the metals. Information on the character of the relaxational processes comes from the spectrum of the ac component of the integral emf $\mathscr{E}_f(\mathscr{E}_t)$ (see Section 4.3). It follows that the study of these spectra from the standpoint of the ideas set forth above may be very fruitful.

5.3. SOME FACTORS PREDETERMINING THE EFFECT OF ELECTRICAL INSULATION OF THE TOOL

The action of thermoelectric current on wear resistance is connected with the features of relaxational, oxidative, diffusional, and other processes that certainly affect one another. This action manifests itself in different ways, depending on many factors or properties characterizing the frictional system. The author earlier attempted to show how, by taking account of several of them, it was possible to see ahead of time in what instance opening the external portion of the tool-workpiece-machine thermoelectric circuit would be a useful measure.[30]

We are aiming for minimal thermoelectric current flowing through the conductive contact spots for given dimensions of the spots. That is, other conditions remaining equal, the thermal emf causing this current $\mathscr{E}_t = \alpha\Delta T$ should be a minimum, where α is the specific thermal emf of the given couple and ΔT is the temperature difference between the hot and cold zones of the natural thermocouple.†

We remark, incidentally, that selection of materials with allowance for their thermoelectric properties can evidently yield even more of an increase in cutting-

---

†In the following, we will use the symbol * to denote static parameters of the analogous artificial thermocouple.

tool life than does insulation of the tool. As a matter of fact, by insulating the drill, broach, cutter, or the like, we disconnect only the external portion of the tool-workpiece-machine thermoelectric circuit. The optimal combination of tool and work materials ($\alpha \longrightarrow 0$) sharply limits the thermoelectric currents both in the external circuit and in the internal thermoelectric circuits right in the cutting zone.

The quantity α, expressed through the absolute Seebeck coefficients, equals $\alpha_{tool} - \alpha_{work}$ when the tool and work materials have the same conduction mechanism, but $\alpha = \alpha_{tool} + \alpha_{work}$ if the workpiece-cutter thermocouple is a combination of electron and hole conductors. (Here and in the following, what is meant is the predominant type of conduction in the alloys.) It follows that if the numerical values of α_{tool} and α_{work} are not much different, then the second variant is most suitable for verifying the effect of opening the external thermoelectric circuit.

The alloy conduction mechanism can be assessed from the percent content and types of intrinsic conduction of the alloying elements[234] (see Tables 2-4 and 2-5). As for the values of the absolute Seebeck coefficients, their determination in ordinary engineering alloys does not make practical sense. The explanation for this is not just the exceptional complexity of the experimental problem itself (the Thomson heat must be measured over a very wide temperature range)[112] but also the exceedingly poor reproducibility of the results. The quantity α^*_{work} (or α^*_{tool}) will vary greatly with the test-piece material—from delivery to delivery within a given grade, from forging to forging, bar to bar, even from section to section of the workpiece itself.

In order to compare the thermoelectric properties of alloys, let us attempt to find the specific thermal emf's α^* for couples with some standard (the same in all cases) "reference thermoelectrode." Here the value of α^*, also called the relative Seebeck coefficient, should "automatically" take account of both the concentration and the sign of charge carriers in each of the contacting conductors.

Cutting tools made from a single grade of material are used in the machining of many metals. For this reason, it is quite logical to manufacture the reference thermoelectrode from high-speed tool steel or a hard alloy.

Figure 5-4 presents curves of the emf $\mathcal{E}_t^*$ of the artificial thermocouple versus the hot-junction temperature t^*; the curves were obtained for a group of work materials being heated in couples with type-R18 steel. (A tubular electric furnace, a Pt-Rh standard thermocouple, and a type-R300 potentiometer were used.) The temperature of the cold junction was maintained at 18–20 °C.

From the curves it is clear that the value of $\mathcal{E}_t^*$ rises approximately in proportion to the temperature of the hot junction over the range from about 450 °C to about 750 °C; this statement is valid only for alloys with n-type conduction (Table 5-2). In the temperature range below about 400 °C, the function $\mathcal{E}_t^*(t^*)$ behaves in a rather complex way; this behavior may be linked with the predomi-

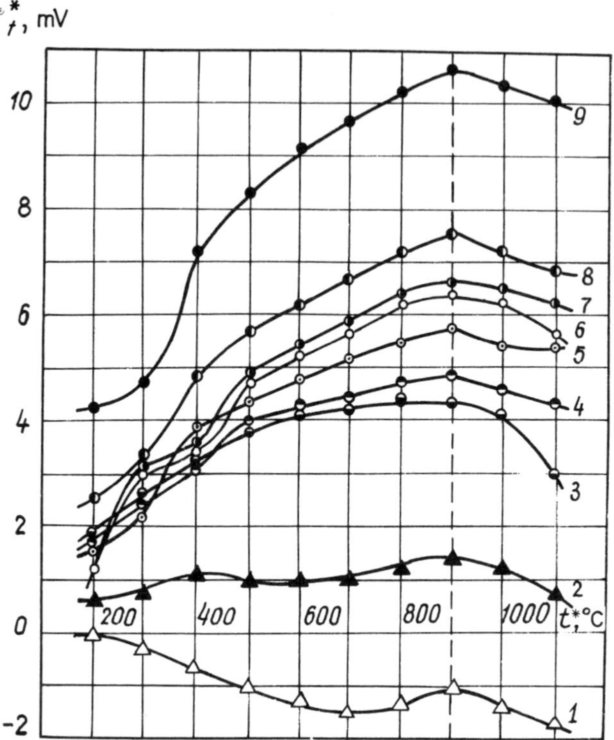

Fig. 5-4. Thermal emf of the artificial cell versus the temperature of the hot junction in the heating of a number of materials coupled with type-R18 steel: (1) VNS-2; (2) VNS-5; (3) EI703; (4) EI654; (5) 30KhGSA; (6) OT4-1; (7) SN-3; (8) Kh18N10T; (9) VT5.

nant effect of semiconducting oxide ingredients retained on contact areas of the hot junction. A distinctive feature of all the curves, regardless of the type of alloy conduction, is the presence of a clearly defined maximum; and it is most curious that eight of the nine extremal points correspond to a temperature of 800 °C. This latter fact, it seems to us, is a consequence of polymorphous transformations taking place in the type-R18 steel. This very steel was the invariant component of the experimental thermocouple, and a temperature of 800 °C is an allotropic phase-transition point for it.[256]

Now it becomes clear that the thermoelectric properties of work materials cannot be compared by the specific thermal emf's in couples with the tool material, because of the nonlinearity of the function $\mathscr{E}_f^*(t^*)$—that is, the coefficient $\alpha^* = \alpha^*[t^*]$.

However, alloys can be ranked in a sort of thermoelectric series by their abilities to yield a Seebeck effect in conjunction with type-R18 steel (see Table 5-2).

Table 5-2.   Thermoelectric series for a group of alloys paired with type-R18 steel.

Alloy Grade	Predominant Type of Conduction	$\mathcal{E}^{*}_{800\,°C}$ (mV)	$\epsilon^{*}_{R\,18}$ (μV/degree)
VNS-2	p	−1.00	−1.25
R18	p	0	0
VNS-5	p	1.39	1.75
EI703	n	4.43	5.55
EI654	n	4.80	6.00
30KhGSA	n	5.69	7.10
OT4-1	n	6.45	8.05
SN-3	n	6.60	8.25
Kh18N10T	n	7.48	9.35
VT5	n	10.68	13.35

To this end, it is necessary to use as the criterion for comparison not α^{*} but the value $\mathcal{E}^{*}_{t_{tr}}$ corresponding to the transformation temperature t_{tr} or, still more conveniently, the value $\epsilon^{*} = \mathcal{E}^{*}_{t_{tr}}/t^{*}_{tr}$. This ratio makes it possible to define the coefficient ϵ^{*} as the specific thermal emf of the artificial thermocouple at the transformation temperature. In the case under consideration here, $t^{*}_{tr} = 800\,°C$, so that

$$\epsilon^{*}_{R18} = 1.25 \cdot 10^{-3}\,\mathcal{E}^{*}_{800\,°C}\ \mu V/degree. \tag{5-7}$$

Among the thermophysical properties of the work material, the thermal conductivity commands especial interest. According to various sources in the literature, it is possible to trace what a profound influence this property exerts on the character of heat exchange and on the overall heat balance in the cutting of metals. As Danielyan has clearly shown,[323] the smaller the coefficient of thermal conductivity, the more nearly adiabatic is the process of plastic deformation. The propagation velocity of heat energy in the bulk of the workpiece determines the mean temperature of the near-tool layers of the chip and of the near-chip layers of the cutter, the temperature of the cutting zone as a whole, and consequently the values of ΔT and $\mathcal{E}_{t}$. To some degree associated with the participation of thermoelectric currents in events of seizing and diffusional sintering of the materials is the dependence of machinability on thermal conductivity. We will indeed attempt to clarify this dependence, after making a digression, which, it later will be easy to see, is entirely justifiable.

One of the main problems in optimizing cutting conditions is to determine correctly the machinabilities of materials. From the standpoint of cutting down expenditures of materials and time, indirect methods of evaluating the machinability are very promising. These methods involve considering machinability as a function of the physicomechanical, physicochemical, and metallurgical properties of the metal. The most complete set of such properties is known only for pure metals; therefore, it was for pure metals that a satisfactory formula for the machinability index[315] could be obtained on the basis of tabular data from the literature. This formula, however, is completely inapplicable to the determination of machinability in steels and alloys; the same publication,[315] incidentally, brings up this very point. For these materials, unfortunately, many of the properties that could be determined ahead of time for pure metals cannot be stated. A steel or alloy of a given grade may differ considerably in its properties from delivery to delivery. Moreover, even within a single blank the properties may vary substantially from layer to layer. Different heats and semifinished products as supplied have different previous histories, and each workpiece is nonuniform; these facts result in uneven machinability of material within a given grade and have a particularly detrimental effect on normal operation of automated processes, causing equipment failure.

It does not follow, however, that indirect methods cannot be applied to the evaluation of machinability in steels and alloys. Many investigators have already found more or less accurate functional relationships between the index of machinability and such material properties as hardness, effective ultimate strength, thermal conductivity, surface tension at melting point, and so on. The author pursues the same goal,[324] deriving a formula for the index of machinability through the use of a specially developed statistical method, which is set out in Appendix 2.*

The steel grade book issued by the Scientific Research Institute of Information on Machinery Construction lists values of the index of cuttability K_v by a tool made of type-R18 steel, for 163 materials used in machinery construction; the hardness HB or its range[†] for 154 materials; and the coefficient of thermal conductivity at various temperatures (λ_{100}, λ_{200}, . . . , $\lambda_{t^\circ C}$) for 65 alloys. On the basis of data on the steady-state temperature fields in the cutting zone and on the maximal temperature on the contact areas,[215, 325–331] an equivalent thermal conductivity (λ) was computed for each alloy with the help of an electronic digital computer. This equivalent thermal conductivity was such that, on the condition that it is independent of the work-material temperature, the material affects heat withdrawal to practically the same extent as when the thermal conductivity varies from layer to layer.

Having this information at hand, we find an analytical expression for the response function:

$$K_v(\text{HB}, \lambda).$$

---

*For the reader already familiar with the method, it will be useful to remark that, if the symbols used are the arbitrary ones that have appeared widely in the literature, there is no need to alter them to indicate the degree of determinacy of the random function.

[†]In this case, the true hardness is taken as the value of HB in the center of the range.

Table 5-3 gives the distribution of materials with respect to machinability over selected hardness ranges.

We exclude three alloys (entries in boxes) from further consideration as lying outside the most probable range of the distribution. We find interval-by-interval estimates for the mathematical expectations of HB and K_v (Table 5-4).

Let us look into the character of the relation obtained (Fig. 5-5). For HB $<$ 175, the behavior of the points is so irregular that it is very difficult to establish any general law governing it. In the range HB $\geqslant$ 175, the disposition of the points distinctly suggests a linear relation; smoothing of this dependence by the method of least squares yields

$$\bar{m}_{K_v} = 3.735(367.5 - \text{HB}) \cdot 10^{-3}. \tag{5-8}$$

The numerical values of the index of machinability can be reduced to a given value of hardness (HB* $= 233$; $\bar{m}_{K_v} = 0.5$) by the formula

$$K_{v\,(\text{HB})} = K_v \frac{0.5}{3.735(367.5 - \text{HB}) \cdot 10^{-3}}. \tag{5-9}$$

It is not difficult to see that the conversion from the function $K_v(\text{HB},\lambda)$ to the function $K_{v\,(\text{HB})}(\lambda)$ is completed.

A comparison of graphs on which all pairs of values of the random quantities were expressed as points showed that

$$D[K_{v\,(\text{HB})}(\lambda)] < D[\lambda(\text{HB})], \tag{5-10}$$

so that we can agree to consider the factors λ and HB independent.

The distribution of the remaining 47 materials with respect to $K_{v\,(\text{HB})}$ in the selected ranges of λ is shown in Table 5-5.

We find estimates of the mathematical expectations of the quantities λ and $K_{v\,(\text{HB})}$ (see Table 5-6).

Table 5-3.

	HB				
K_v	100–150	150–200	200–250	250–300	300–350
0–0.2	—	—	5	4	3
0.2–0.4	☐1	9	14	8	1
0.4–0.6	—	17	27	4	—
0.6–0.8	—	15	10	1	—
0.8–1.0	—	13	4	—	—
1.0–1.2	—	4	2	—	—
1.2–1.4	3	1	—	—	—
1.4–1.6	5	—	—	—	—
1.6–1.8	1	☐1	—	—	—
1.8–2.8	—	—	—	—	—
2.8–3.0	—	☐1	—	—	—

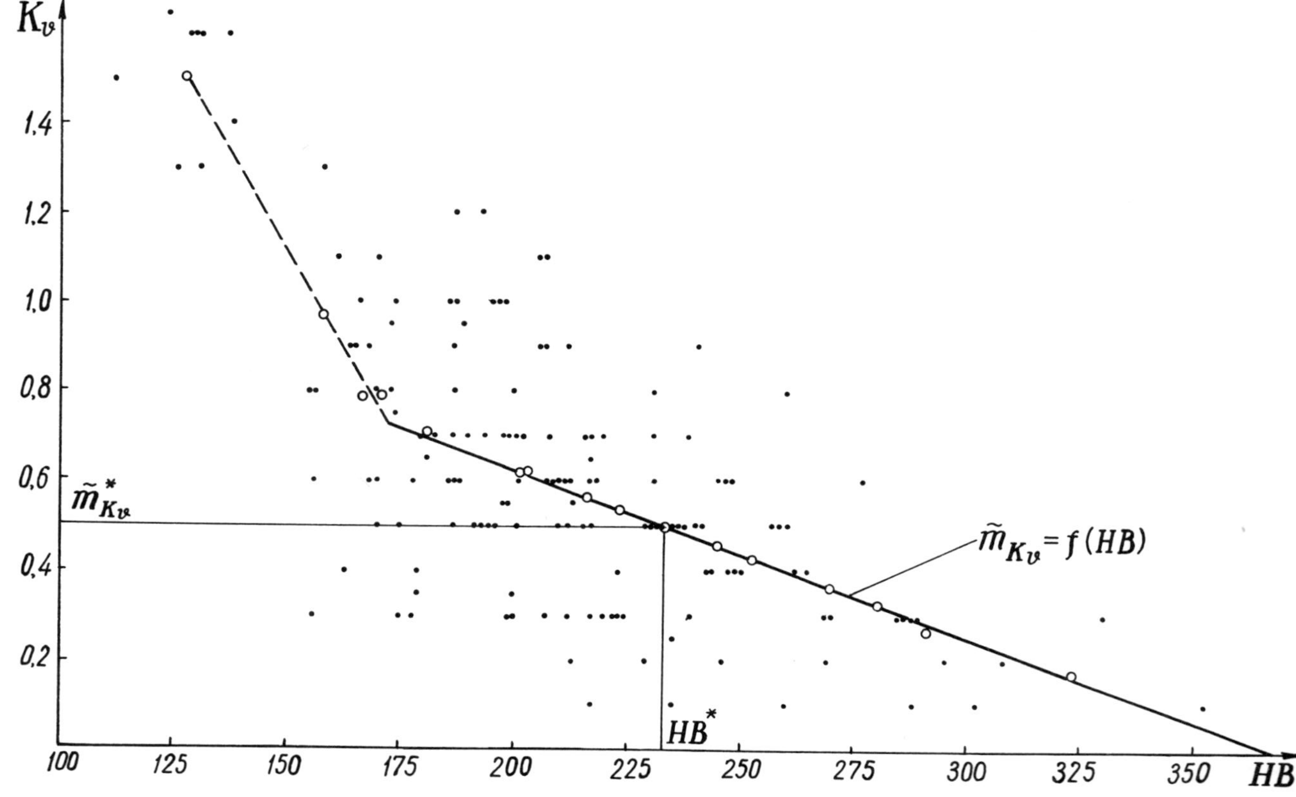

Fig. 5-5. Graph of $K_v = f(HB)$.

Table 5-4.

HB	Number of Alloys	$\tilde{m}_{HB}$	$\tilde{m}_{K_v}$
100–150	9	128	1.505
125–175	30	158	0.970
135–185	32	167	0.790
137–187	37	171	0.790
150–200	59	181.5	0.708
175–225	73	201.5	0.620
200–250	62	223.5	0.533
225–275	37	245	0.460
250–300	17	270	0.364
275–325	8	291	0.268
300–350	4	323	0.175
150–250	121	203	0.620
175–275	110	216	0.565
200–300	79	233.5	0.500
225–325	45	253	0.425
250–350	21	280.5	0.330

The relation obtained is expressed best in base-ten logarithms (Fig. 5-6), for which the equation has the form

$$\tilde{m}_{K_v(\text{HB})} = \log 0.1065\lambda. \tag{5-11}$$

But, in accordance with the proposed method, we may assume that the following holds with accuracy sufficient for the subsequent analysis:

$$K_{v(\text{HB})} \approx \log 0.1065\lambda, \tag{5-12}$$

then

$$K_v \frac{}{3.735(367.5 - \text{HB}) \cdot 10^{-3}} \approx \log 0.1065\lambda$$

and, finally, we write

$$K_v \approx 7.5 \cdot 10^{-3} (365 - \text{HB}) \log 0.1065\lambda. \tag{5-13}$$

Table 5-5.

$K_{v(\text{HB})}$	λ (kcal/meter-hour-degree)			
	11–19	19–27	27–35	35–43
0–0.2	4	—	—	—
0.2–0.4	9	5	1	—
0.4–0.6	—	8	7	3
0.6–0.8	—	—	3	4
0.8–1.0	1	1	—	—
1.0–1.2	—	1	—	—

Table 5-6.

λ (kcal/meter-hour-degree)	Number of Alloys	$\tilde{m}_\lambda$	$\tilde{m}K_{v(HB)}$
11–19	13	15.54	0.215
19–27	13	23.38	0.414
27–35	11	32.49	0.534
35–43	7	38.93	0.624
11–27	26	19.46	0.315
19–35	24	27.55	0.469
27–43	18	35.00	0.569

The values computed by this formula for the index of machinability were compared with the experimental K_v values taken from the steel grade book. The calculation covered only 24 alloys, for each of which the source gave either an accurate value of the Brinell hardness number, HB with a $\leqslant$ sign, HB with a $\geqslant$ sign, or a range of hardness with a width of Δ HB $\leqslant 10$. (For the remaining 20 materials, Δ HB > 25.) The results obtained (Table 5-7) show that the departures of calculated from experimental data are of the same order of magnitude as in the case of a special investigation conducted on 15 steels to find how the

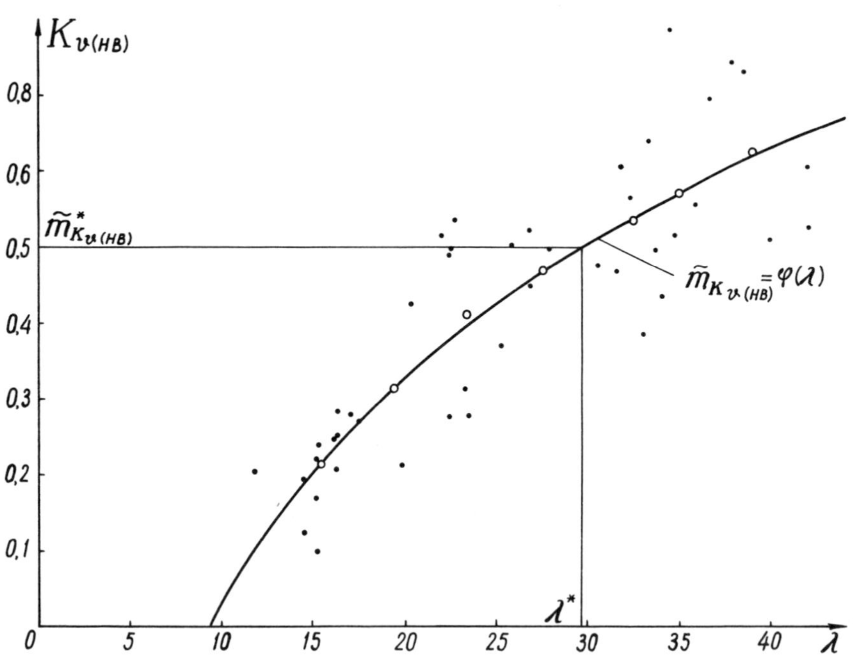

Fig. 5-6. Graph of $K_{v(HB)} = f(\lambda)$.

Table 5-7.

No. Alloys	Distribution of Alloys with Respect to Deviation between Calculated and Experimental Values of the Index of Machinability (%)				
	0–10	10–15	15–25	25–50	50–100
24	10	4	7	2	1
15	9	4	1	1	—

effective ultimate strength and thermal conductivity of the materials affected machinability.[332]

If the numerical values of $K_{v\,(HB)}$ are referred to the same value of the coefficient of thermal conductivity ($\lambda^* = 29.7$; $\bar{m}^*_{K_v\,(HB)} = 0.5$), then we obtain the distribution of 44 alloys with respect to $K_{v\,(HB,\,\lambda)}$ shown in Fig. 5-7.

Now we have data at our disposal for seeking a third factor whose application, together with HB and λ, would make it possible to refine the final formula and ultimately would contribute to solving the problem of optimizing cutting conditions in the modern technology of mechanical treatment. Analysis of the unaccounted-for scatter of points (of realizations of the function) suggests that one such factor might be the crystallographic match between the atom-electron systems (of the tool and work materials). The need for further correction to the formula for the index of machinability was most clearly expressed in the case of nickel-based alloys containing titanium and aluminum, which are poor conductors of heat.*

Here it is pertinent to mention how the criterion of efficiency of the GAO method as a function of increasing drill diameter behaves differently in the machining of type-VT5 titanium alloy and of type-EI654 and type-VNS-5 steels (see Section 4.2). It is not out of the question that the low thermal conductivity of the titanium alloy and its predisposition to seizing actually brought about an increase in $K_e(D)$, despite the improvement in the conditions of heat withdrawal from the cutting zone through the tool. Nor is it unreasonable that, with respect to the influence of the latter factor on the effect of electrical insulation, type-EI654 steel proved just opposite to type-VT5 alloy, with type-VNS-5 steel occupying a middle position between them.

It is well known that hexagonal closest-packed (hcp) metals are less apt to seize and have lower coefficients of friction than bcc and fcc metals (for example, Semyonov[305]). If thermoelectric currents are able to intervene in the very processes of atomic rearrangement, hastening their onset, then polymorphous trans-

---

*From the standpoint of the occurrence of thermoelectric relaxational processes in friction and cutting, especial interest attaches to the fact that titanium-nickel alloys contract on heating and expand on cooling.

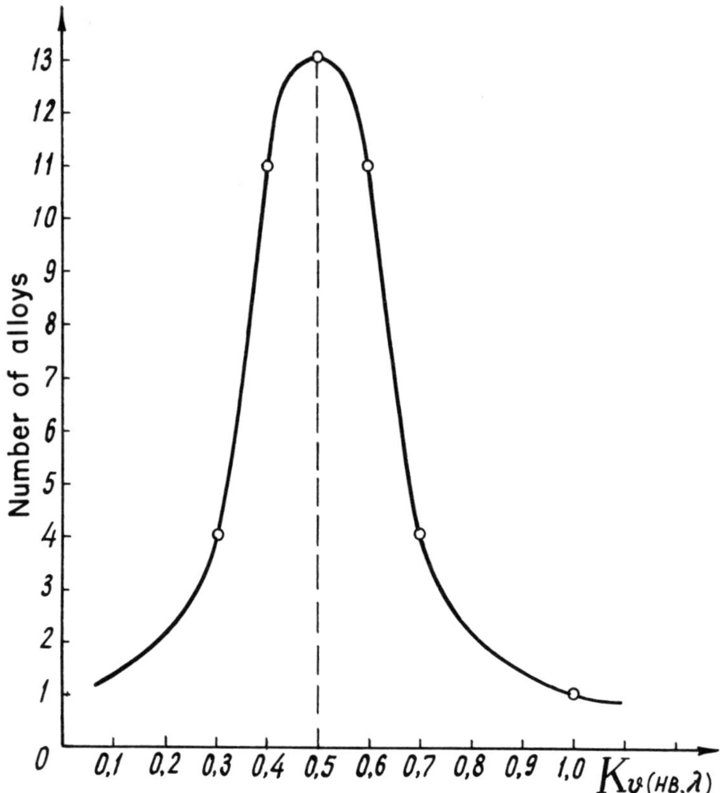

Fig. 5-7. Distribution of alloys over $K_{v(HB, \lambda)}$ with a mode of 0.5.

formations, which move in the direction of realization of conditions favorable for seizing (hcp $\longrightarrow$ bcc, fcc), should promote this intervention and intensify its external manifestation.

Now the factors of primary importance that predetermine the consequences of opening the external thermoelectric circuit should include the crystallographic match between atom-electron systems, their temperature-dependent allotropy, and the thermoelectric and thermophysical properties.

5.4. ROLE OF THERMOELECTRIC PHENOMENA IN THE FORMATION OF A COHESION JOINT AND MACHINING OF TITANIUM ALLOYS

From the ideas developed above, it follows that when type-VT5 titanium alloy is machined with type-R18 high-speed steel, the effect of electrical insulation of the tool should make itself quite apparent. This conclusion is supported concretely by the following points:

(1) the combination R18-VT5 represents a *p-n* transition;

(2) the greatest value obtained for the coefficient $\mathcal{E}^*_{R18}$ is that for type-VT5 alloy;

(3) titanium alloys, including VT5, have low thermal conductivities;

(4) in the machining of type-VT5 alloy, the temperature on contact areas of the tool rises to 800 °C or higher, as indicated by the presence of the maxima on curves 3 and 4 in Fig. 4-2;

(5) the crystal structures of the elements making up type-R18 steel are predisposed to unite with the similar structure of β titanium, which appears in type-VT5 alloy upon an allotropic transformation (Table 5-8);*

(6) the elements with which type-R18 steel is alloyed dissolve in titanium (V to an unlimited extent, W and Cr with the formation of intermetallic compounds;† all of these are isomorphic to β titanium, so that as they diffuse into the VT5 alloy during cutting they can hasten the appearance of the high-temperature allotropic modification in the alloy);

(7) electrical transfer of carbon into type-VT5 alloy lowers the plasticity of the near-tool layer, at the same time contributing to the loss of tool hardness.

Figure 5-8 shows photomicrographs illustrating the effect of electrical insulation of drills as applied to the combination of tool and work materials under consideration (R18-VT5). In the case of a closed external circuit, adhesion and sintering phenomena are incomparably more pronounced, and fit of the tool takes place far more rapidly, than in operation with an insulating device.

The photomicrographs of Fig. 5-9 are a no less graphic confirmation of the destructive function of thermoelectric current under similar conditions. When the external thermoelectric circuit is closed, under selected operating conditions it is possible to drill only one or two holes before the drill becomes firmly welded to the titanium and breaks off (Fig. 5-9[b]). For a similar drill, but one that is electrically insulated, the tendency to seizing and sintering with the titanium is hardly detectable after the drill has made the same number of holes (Fig. 5-9[c]). The tool continues to perform, and only after it has drilled seven to eight holes does it become welded to the workpiece and chip (Fig. 5-9[d]).

Let us now dwell on the changes in the structure of the high-speed steel. In the cutting of type-VT5 alloy, the number and size of special carbides scattered "on the field" of fine martensite needles (compare Figs. 5-9[a] and [b]) are seen to decrease. If this reduction is linked with diffusion processes, then the transfer of diffusing species should weaken when it is due only to the temperature gradient and there is no electric force field (compare Figs. 5-9[b] and [c]). The drift

---

*Being an α stabilizer, aluminum raises somewhat the temperature of the dimorphous $\alpha \longrightarrow \alpha + \beta$ transformation of titanium in type-VT5 alloy.

†For the solubility of tungsten carbides in titanium carbides, see Kazakov's book.[333]

Table 5-8. Characteristics of the crystal structures[a] of metallic elements in type-R18 high-speed steel and type-VT5 titanium alloy.[56, 256]

Atomic No.	Element	Allo-tropic Form	Range of Stability (°C)	Lat-tice Type	Coor-dina-tion No.	Lattice Constant a (Å)	Nearest-Neighbor Distance d (Å)	Content in Alloy (%)	
								R18	VT5
13	Al		Up to 660	fcc	12	4.04	2.86	—	4.0–5.5
22	Ti	α	Up to 882	hcp	12	2.95	2.91	—	Base
		β	882–1,660	bcc	8	3.31 (900 °C)	2.87 (900 °C)		
23	V		Up to 1,700	bcc	8	3.03	2.63	1.0–1.4	—
24	Cr		Up to 1,830	bcc	8	2.88	2.49	3.8–4.4	—
26	Fe	α	Up to 910 1,400–1,535	bcc bcc	8	2.86	2.48	Base	—
		λ	910–1,400	fcc	12	3.63 (910 °C)	2.57 (910 °C)		
74	W		Up to 3,410	bcc	8	3.16	2.73	17.5–19.0	—

[a]Except for two cases (temperatures in parentheses), values of a and d are given for room temperature, i.e., 18–25 °C.

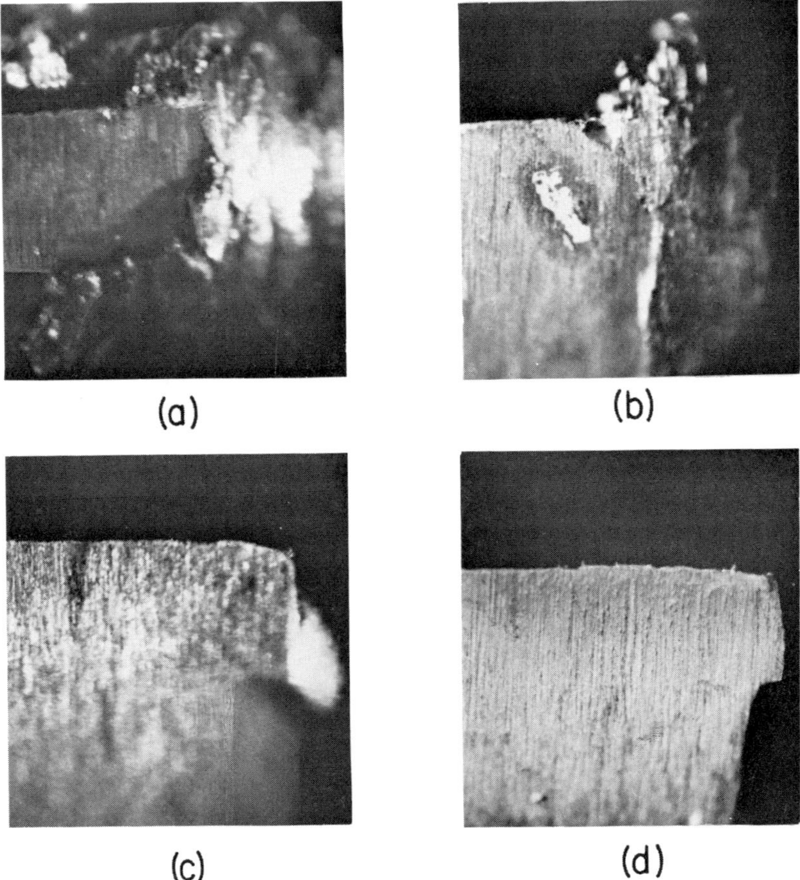

Fig. 5-8. Adhesion of chip to tool (left) and flank wear of drills (right), for various states of the tool-workpiece-machine thermoelectric circuit (30X): (a, c) external circuit closed (after 7 holes); (b, d) open (after 7 and 28 holes, respectively). Workpiece thickness, 25 mm; drill diameter, 10 mm. Cutting conditions: v = 8 meter/min; s = 0.10 mm/revolution.

of ions toward the titanium alloy may be accompanied by growth of the granules, whose sizes are especially large in the very boundary layer of type-R18 steel (Fig. 5-9[d]). Very probably what is happening here is ordinary coagulation of the special carbides. However, the absence of any such coagulation in closed-circuit operation makes it possible to look at this process in an entirely different light. Let us suppose, for example, that the intense effect of electron-hole entrainment guarantees not only that migrating ions will be transferred through the interface, but also that they will be transported farther into the bulk of the work alloy. It goes without saying that transportation of this type is impossible if the

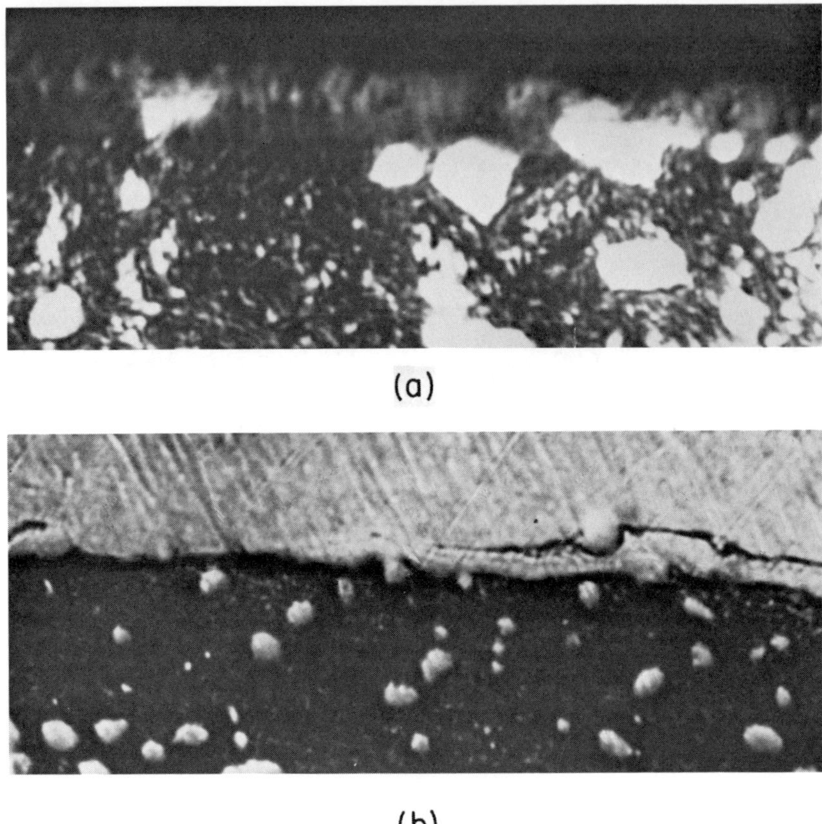

(a)

(b)

Fig. 5-9. Photomicrographs of drills, in cross section perpendicular to the cutting lip (1,500×): (a) new drill after grinding; (b) after drilling of two holes with closed circuit; (c) with open circuit; (d) after drilling of seven holes. Cutting conditions: v = 12.5 meter/min; s = 0.11 mm/revolution.

external thermoelectric circuit is open. In that case, the diffusing species would "saturate" the boundary layer of the titanium; this would lead to retardation of the ions and thus would make thermal diffusion of alloying elements out of the type-R18 steel more difficult.

The experimental material obtained confirms the validity of the hypothesis that thermoelectric current stimulates seizing and diffusional sintering of metals.

The lives of high-speed drills insulated from the machine by a special holding device were observed to increase by a factor of more than three, not only in the machining of titanium alloys (VT3, VT5, OT4-1, etc.), but also in that of stainless steels containing titanium. However, although we have thought of the drilling of titanium alloys as one of the most probable areas where the effect should

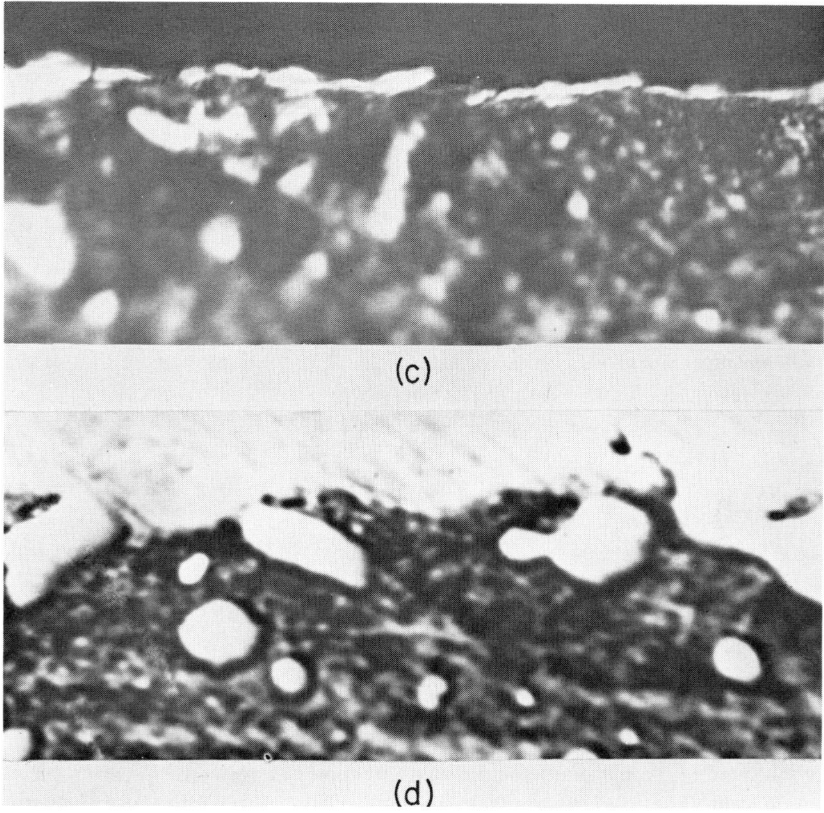

Fig. 5-9. (continued)

appear, the effect has proved unstable even in this concrete case that would seem so appropriate for its practical utilization.[334] The thermoelectric current flowing through the tool-workpiece-machine circuit has a destructive effect only under certain machining conditions; the effect depends on the totality of the factors examined above, which in themselves may exert an even more profound, and at times a dominant, influence on the cutting process. One-time trials have shown, for example, that, in contrast to the results already presented (Section 4.2), the use of a bonded chuck in the drilling of type-0Kh18N10T steel yields an advantage in the life of a high-speed tool only under some intermediate machining conditions, when the cohesion of chip to drill is markedly weakened after electrical insulation of the drill. Changing over to relatively low or high cutting speeds involves either a significant weakening or the actual absence of the effect. In addition, the use of a bonded chuck extends the lives of only individual ones, but always the same ones, of the experimental batch of drills (Fig. 5-10). For this reason, we may think that the certain conditions mentioned previously are effectively linked with the dynamics of contact phenomena and are set up in the time when the system surmounts the energy threshold of seizing, when fluctuations of the thermoelectric currents are transformed into something like a catalyst for the processes taking place in the contact zone at the atomic level (coupling of crystal lattices, recrystallization, etc.).

The sum total of this kind of consideration and the conclusion that the effect of electrical insulation depends in a direct way on the structural states of the tool and work materials became the principal prerequisites for the conduct of a controlled experiment, which consisted in comparing the K_e values in the drilling of titanium alloys with the equilibrium and quenched microstructures. With regret we must verify that, under selected cutting conditions, the use of a special holding device had little effect in the operation of drills with a cooling emulsion (Table 5-9). However, in the case where the cutting fluid did not prevent encounters between the juvenile surfaces and mechanical vibrations in the system were amplified, the effect of electrical insulation of the tool sometimes became appreciable right away.

Trials on the drilling of type-3V titanium alloy without cutting fluids provide graphic confirmation of the influence of work-material microstructure on the effect of opening the external circuit. Test pieces in the as-supplied condition had an α-phase structure made up of coarse Widmanstätten plates (Fig. 5-11[a]). After quenching from 1,100 °C, the same test pieces acquired the thermodynamically unstable α' phase structure, in thin plates on cooling in air (Fig. 5-11[b]) and in thin needles on cooling in water (Fig. 5-11[c]). The test piece hardnesses were compared by the diameter of the impression of a spherical indenter; pieces quenched in water had the greatest hardness (d = 3.80 mm), those quenched in air a somewhat smaller hardness (d = 3.90 mm), and those in the as-supplied condition the least hardness (d = 4.05 mm). The respective values for the machina-

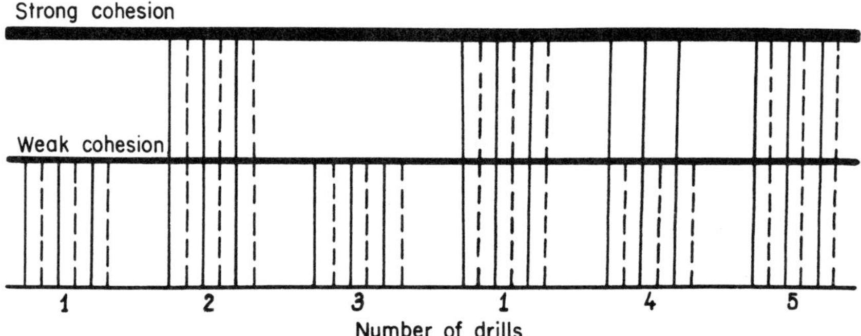

Fig. 5-10. Cohesion of type-R18 high-speed steel to type-OT4 titanium alloy, for various states of the thermoelectric circuit (—— closed, – – – open). Cutting conditions: (1, 2, 3) v = 8 meter/min, s = 0.12 mm/revolution; (1, 4, 5) v = 10.5 meter/min, s = 0.09 mm/revolution. Drill diameter 11 mm; dry drilling.

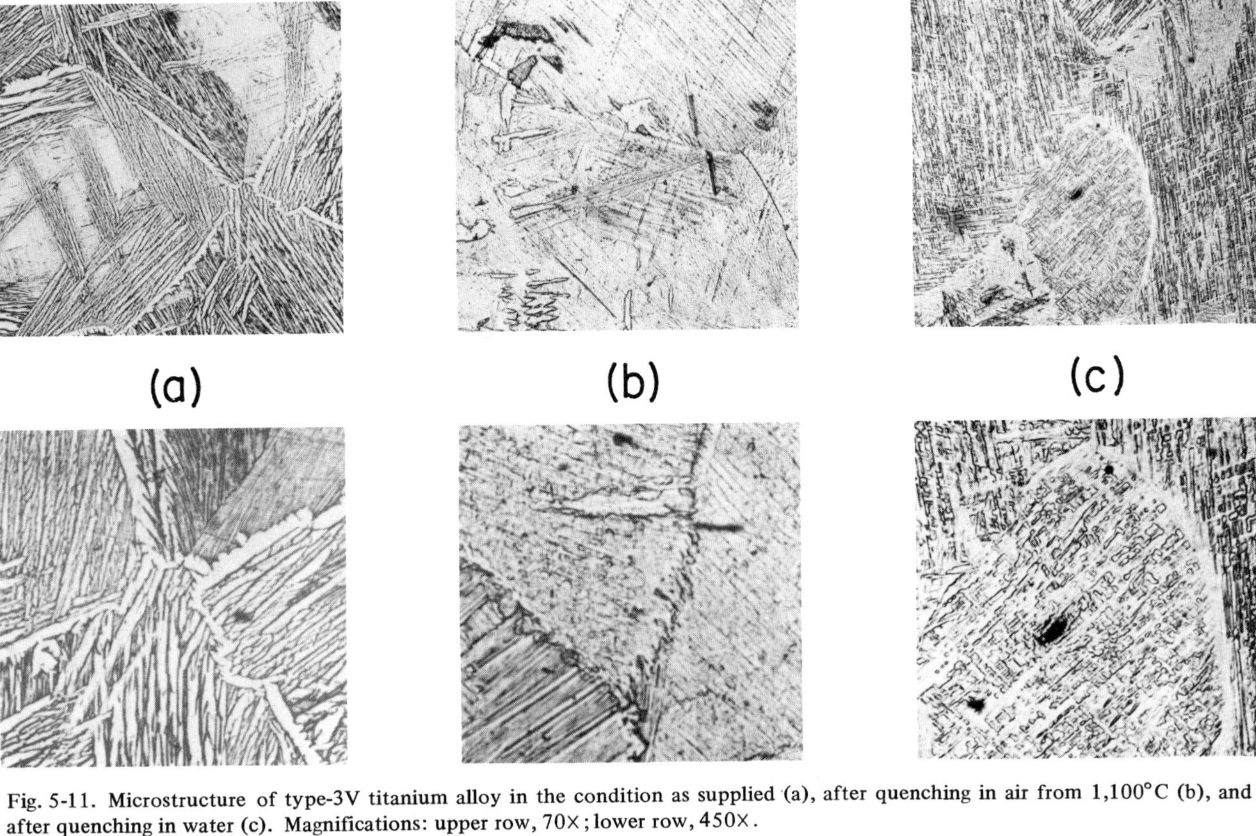

Fig. 5-11. Microstructure of type-3V titanium alloy in the condition as supplied (a), after quenching in air from 1,100°C (b), and after quenching in water (c). Magnifications: upper row, 70X; lower row, 450X.

Table 5-9. Criterion of efficiency K_e for the drilling of heat-treated titanium alloys with cutting fluids.[a]

Alloy Type	Condition as Supplied	Heat Treatment			
		Heating to 800 °C, Cooling in Air	Quenching from 1,100 °C in Air	Heating to 800 °C, Cooling in Air	Quenching from 1,100 °C in Water
OT4	~1	<1	>1	~1	~1
VT3	~1	<1	>1	~1	~2
VT6	~1	~1	~1	>1 (~2)	>1 (~2)
ZV	~1	>1 (~2)	~1 (~4)	<1	~1

[a] Figures in parentheses are K_e values from results of single trials of the given series. Cutting conditions: v = 12.5 meter/min; s = 0.15 mm/revolution; D = 10.5 mm; drilling depth h = 22 mm; 5 percent solution of self-emulsifying oil in water; bonded chuck; type-2A55 radial drilling machine.

bility of type-3V alloy with drills shorted to the machine (closed thermoelectric circuit) were approximately in the ratio 1:2:4. As for the mean number of holes drilled between regrindings without the shorting shunt (open thermoelectric circuit), in comparison with the number drilled with the shunt connected it was practically the same in the machining of the unquenched alloy but roughly twice as large in cutting after quenching. We will attempt to explain this result by the direct intervention of thermoelectric currents in events of interaction between solids, leading to welding of the solids.

As the temperature in the cutting zone rises, phase transformations take place in the contact microvolumes belonging to the tool-workpiece interface. At first these transformations occur in the high-speed steel: bcc martensite is converted to fcc austenite. Then the allotropic $\alpha \longrightarrow \beta$ transition begins in the titanium alloy. As soon as local centers of seizing appear—that is, as soon as metallic bonds (which play the role of tunnels for migrating ions) appear—the diffusing species is injected into the very boundary layer of the titanium alloy. Since some of the elements alloyed in the high-speed steel are β stabilizers for titanium, especially vanadium, these elements initiate the growth of β-phase regions, thus creating favorable conditions for the welding of the metals. Since the energy fluctuations of ions and the diffusion flux density itself depend on the presence of an electrical-diffusion component of the flux, it should be thought natural that the state of the external circuit (closed or open) can be reflected in tool life and the roughness of the machined surface.

What has been set forth here does not mean that Bobrovsky's recommendations[111] on the universal practical utilization of the GAO method are valid; the discussion does point again to the need for further study of problems in the area of the electrophysics of cutting.

6

Triboelectric Phenomena in Metal-Polymer Systems

6.1. INTRODUCTION

It is not the aim of this chapter somehow to characterize briefly the entire complex of phenomena merged under the term "static charging." The study of static electricity is a branch of physics so vast and autonomous that the present monograph need not give any such description; there exists a special literature.[335, 336] The material here is primarily informational and serves to illustrate the role that electrical phenomena play in the operation of frictional systems consisting of a metal and a dielectric. In these systems, in distinction to metal-metal couples, when the plates of a frictional microcapacitor are separated, the charges can be neutralized by a self-maintained gas discharge where an avalanche process develops in the gas gap. Avalanche discharges of this type, which destroy electrical double layers when dielectric films are separated, were detected by Deryagin and Krotova (see Fig. 6-1). At the time, their finding served as experimental confirmation that adhesion forces are electrical in nature (electrical theory of the adherence of films to solids).[35]

If electrical erosion of rubbing surfaces can take place in microscopic spark discharges, then this erosion obviously shares to some extent the traits of the artificial destruction of materials in impulse electrical discharges, which is electrothermal in nature[337, 338] and is widely used in industry. Generalizing on the spark form of electrical discharge, Lazarenko and Lazarenko write, in particular:[184]

Fig. 6-1. Discharge upon tearing of a nitrocellulose film from gelatine (photograph by N. N. Zakhavaeva).[35]

This type of electrical discharge occurs only at very high electric field strengths. . . .

The spark electrical discharge is a non-steady-state physical process lasting 10^{-3} sec or less. . . .

The current and power amplitudes attainable in a spark electrical discharge cannot be achieved by any other means. . . .

The temperature of the spark path is near that of the sun and cannot be obtained in any other controlled process. . . .

The cleaving action of the impulse is so great that all the chemical elements not only in the interelectrode medium but also in the electrodes emit atomic spectra.

As we see, the electrical-erosion type of wear in friction, resembling (if only in miniature) the electrical erosion of metals in the presence of an external source, may be quite dangerous.

Various methods exist for reducing frictional electric charging: increasing the relative humidity, using static neutralizers with radioisotopes, increasing the electrical conductivities of polymer materials by adding special fillers to them, and so on. One of the most recent recommendations for lowering the electric charging and wear of man-made yarns in friction over the guides of textile machinery is to make the guides from porous sintered materials impregnated with antistatic preparations based on mineral oils with surfactant additives.[339]

At the same time, instead of being something to combat, electric charging is viewed more and more often as beneficial. For example, it assures conditions favorable for sign-altering transfer, a special variety of selective transfer, in metal-polymer sliding systems.[340] The phenomenon of frictional charging is, in fact, the basis for the devising of a new class of devices for automatic loading of miniature parts, in particular the crystals of discrete semiconducting instruments, integrated circuits, and optoelectronic devices.[341] Adhesion between these easily-damaged boards and dielectrics charged by friction on a metal has been studied by the author and co-workers,[342] who used a pendulum adhesiometer constructed on the principle of the Deryagin adhesiometer.[35] According to statistical estimates of the measure of adhesion, the mean work of separation $\overline{W}$, dielectrics charged under identical conditions have widely varying abilities to retain miniature parts by electrostatic attraction. As for quantitative estimates of the measure of adhesion on a particular theoretical basis, such estimates cannot yet be given, for the reasons set out in the next section.

6.2. CURRENT IDEAS ON THE MECHANISM OF FRICTIONAL CHARGING

True triboelectric charging occurs only in asymmetric friction between chemically identical materials.[336] This kind of charging involves the propagation of charge at different rates through the interface between the rubbing bodies, which results from the difference in temperature gradient between the bodies.[132] In most cases, however, the charging of bodies in friction is due to "contact charging," and the frictional process itself leads only to an increase in the number of contacting regions. At the same time, it must be taken into account that effects characteristic of the process of friction may exert a substantial influence on the way in which electrons or ions are transferred across the interface, and may even alter the mechanism of contact charging. In what follows, however, the term "triboelectric charging" will be ascribed a more general meaning: It will also refer to the case of contact between the surfaces of two unlike materials.

Only in relation to contact between two metals, between two semiconductors, or between a metal and a semiconductor can the mechanism of charge redistribution be said to be more or less clearly understood. Contact charging in this case is due to the preferential transfer of electrons through the interface from the substance with the lower work function to that with the higher work function. But as for dielectrics, a review of the special literature indicates some uncertainty, and even conflict, concerning the mechanism of their contact charging. Unfortunately, it has become traditional to repeat that there exists as yet no unified concept of the mechanism of triboelectric charging of dielectrics. It is thought, for example, that in the contact of a metal and a dielectric charging is due to the transfer of electrons from the metal to the dielectric and the transfer of positive or negative ions to the metal surface; in the contact of two dielectrics, charging results from the diffusion of charge carriers from one substance to the other. Physical and chemical "contamination" in this case is considered an incidental factor that may distort the phenomenon of charging but cannot alter its nature.

Harper has pointed out the possibility of a different approach to the formation of static charges.[343]   He also remarks that the transfer of ions plays an important role in the charging of solids, but gives preference to those of the mobile ions that have been formed through electrolytic dissociation in the surface films of moisture. There is a suggestion that the rupture of a double diffusional layer of electrolytic ions in the extremely thin water films between two dielectrics (between a metal and a dielectric) is what causes the charging of surfaces being parted one from the other. In other words, the dielectric itself cannot acquire a significant amount of charge in events of the frictional interaction; it can become charged only when it is "soiled" with electrolytic contamination.

Harper divides dielectrics into two classes.[343]   The first comprises "electrophilic" dielectrics (hydrophilic materials), which acquire extremely strong charges through contact with metals (glass, fused quartz, magnesium oxide). The second consists of electrophobic dielectrics (hydrophobic synthetic polymers), which under similar conditions take on a charge that cannot be detected even by sensitive apparatus. This class includes polyethylene, amber, polystyrene, nylon, silicon, poly(methyl methacrylate), poly(tetrafluoroethylene), and other materials. From the standpoint of chargeability, two types of contamination—electrolytic and greasy—correspond to these two classes of dielectrics. Harper remarks that as greasy (electrophobic) contamination is removed from the surface of an electrophilic dielectric, the charge on the dielectric increases. On the other hand, removal of electrolytic (electrophilic) contamination from the surface of an electrophobic dielectric results in a decrease in the charge, ultimately in neutralization. In actual cases, where contamination is complex in composition (that is, includes both greasy and electrolytic components), the effect of cleaning the dielectric on the quantity of charge will be less strongly expressed. Harper emphasizes that electrophilic (electrophobic) character is an intrinsic, inherent property of a dielectric, linked with its exhibiting a hydrophilic (hydrophobic) nature.

Kornfeld recently made an original proposal for the mechanism of triboelectric charging of dielectrics.[344]   The basis for this suggestion was the hypothesis that solid dielectrics (ionic crystals and amorphous and polymeric substances) have a characteristic electric charge (at sites where vacancies of like sign pile up). Besides the investigations of Kornfeld himself,[345, 346] the results from other experimental research have confirmed that such a charge exists. Coehn and Lotz, for example, have demonstrated that electrically charged specimens (tourmaline, carborundum, quartz) can retain their charges upon heating in high vacuum to temperatures that assure no water is present.[347]

According to Kornfeld, compensation of the characteristic charge of a dielectric takes place through selective adsorption of oppositely charged ions. Thus, the physical and chemical contamination on the dielectric surface becomes charged itself. In friction between two specimens, the contamination layers are intermixed and redistributed between the rubbing surfaces. The compensation of the characteristic charges is destroyed, which causes triboelectric charging.

In light of the foregoing, it becomes clear why adhesion forces that are triboelectric in nature manifest themselves in such a complex and diverse way. The functional dependences of these forces are usually studied under conditions allowing reproducible results to be obtained. The effect of the surface charge density, for example, can be taken into account in the case where the rubbing couple dielectric-clean metal is in a monitored atmosphere.[348] But suppose a dielectric is used to make the operating part (conveyor) of a triboelectric charging device. Now, under natural conditions, the surface of the dielectric will be exposed to media with varying ionic composition, and materials used for the body against which the dielectric rubs will include, in particular, steels and alloys. Therefore, it is not nearly always possible to say ahead of time to which of the dielectrics the minature parts will best adhere.

Since this chapter is devoted to metal-polymer rubbing couples, it must be emphasized that the author, in speaking of the adhesion of minature parts, does not refer to metallic parts, such as those made of solder, alone; in fact, not so much to metallic parts as to the principal elements of electronics hardware, namely, semiconductor crystals.

The inequality of dielectrics from the standpoint of their usefulness in the fabrication of conveyors is illustrated by a group in the author's laboratory for the example of the attractive interaction with a substrate of germanium single crystals.[342] Table 6-1 presents the results of this investigation, which are in agreement with some of the ideas just set forth.

As we see, the adhesion of the sheets to any material charged in a couple with brass was greater than that to the material charged by friction against steel.

The dielectrics of Table 6-1 rank in the following order with respect to ability to retain germanium sheets:

> Poly(methyl methacrylate)
> Viniplast
> Poly(tetrafluoroethylene)

Table 6-1. Adhesion between sheets of single-crystal germanium and various dielectrics charged by friction.

Dielectric (specimen)	Measure of Adhesion $\overline{W}$ (erg) When Dielectric Rubs On	
	Steel	Brass
Poly (methyl methacrylate)	0.32	0.56
"Viniplast" (rigid PVC)	0.41	0.44
Poly (tetrafluoroethylene)	0.26	0.27
Polypropylene	0.19	—
Polystyrene	0.14	—

Polypropylene
Polystyrene

Polystyrene has the lowest value of $\overline{W}$; thus, it is no accident that this material is a typical electrophobic dielectric. According to Harper, the charge acquired in triboelectric charging of polystyrene is four orders of magnitude less than that picked up on the surface of crystalline quartz, an electrophilic dielectric. We should note, however, that this same Harper, in whose experiments polystyrene took on charge very poorly, emphasizes the possibility of imparting a high charge to polystyrene films under plant conditions (in the manufacture of capacitors). The author and his colleagues also had to face this inconstancy of the "triboelectric susceptibility" of a material.[342] Measurements on a quadrant electrometer showed, for example, that in the case of a Viniplast specimen the static charges on opposite sides of the specimen, both charged in the same way, may have drastically different magnitudes. A difference in color between the sides, indicating a difference in surface structure, was visible to the naked eye. This fact suggests that when the surface of a solid is treated under certain process conditions, the solid may go from the electrophobic to the electrophilic class of dielectrics, or vice versa. Everything depends on whether the conditions favor the appearance of structural defects, which, according to Kornfeld, contribute to the appearance of a characteristic electric charge on the surface of the dielectric.

Although many experimental facts support Kornfeld's hypothesis,[344-346] it may not be correct to expect, at least for the near future, that some single mechanism can be established for static electric charging. Considering the phenomenon of charging in the most general sense, as the formation and destruction of an electrical double layer—Böning in particular does this[349]—we must allow not only for the tremendous variety of substances (inorganic and organic materials, dielectrics, semiconductors, conducting solids) covered by research on static charging, but even more for the widening variety of conditions under which this complex phenomenon takes place. This is precisely the basis on which we can

Table 6-2. Some laws of triboelectricity.

Frictional Couple	Sign of Charge	
	+	**−**
Two chemically identical bodies	Denser body	Less dense body
Dielectric and fine particles (dust) of the same dielectric	Dielectric	Fine particles
Dielectric and dielectric	Dielectric with higher permittivity	Dielectric with lower permittivity
Dielectric and dielectric	Harder dielectric	Less hard dielectric
Metal and metal	Less hard metal	Harder metal
Metal and dielectric	Dielectric	Metal

Series of Faraday: (+) fur, flannel, ivory, feathers, rock crystal, flint glass, cotton cloth, silk, wood, metals, sulfur (−).
Series of Gezekhus: (+) diamond (hardness 10), topaz (8), rock crystal (7), smooth glass (5), mica (3), calcite (3), sulfur (2), wax (<1) (−).

understand the conflicts among the many studies of contact electricity. Against the background of these conflicts, the laws governing triboelectricity, which constitute a generalization of laboratory and plant experience (Table 6-2), take on special value.

Now we have seen once more that each concrete case of triboelectric charging has its own specific features, which are profoundly influenced by the process of friction itself and by variations in the external conditions.

6.3. EFFECT OF ELECTRIC CHARGING ON TRANSFER PHENOMENA IN METAL-POLYMER FRICTIONAL COUPLES

Studying in parallel the mechanical-chemical and triboelectric phenomena that take place when a plastic slides over a metal makes it easier to understand the way in which plastic films are formed on metal surfaces and, conversely, the way in which metal is transferred to plastic surfaces. Mechanical-chemical effects reduce to three categories: rupture of carbon-carbon and other chemical bonds at sites in the polymer chains where the bond energy is sufficiently high; cracking of chain macromolecules, with the appearance of an anomalously high concentration of free radicals; and participation of these radicals in various chemical reactions.[350–352] The reactions referred to include, first, direct ones "with oxidized and juvenile metal surfaces on sections denuded in friction and wear, yielding compounds $R-(CH_2)_n-CH_2-Me$"[33] and, second, reactions going through the formation of a peroxide radical $R-CH_2-O-O-$ by the addition of oxygen.[353]

It is usually assumed that the rupture of covalent bonds in the molecules is accomplished by mechanical energy supplied from outside, with no contribution from thermomechanical effects. The author does not share this point of view. Instead, following the kinetic concept of strength,[310] he attributes great importance to thermal fluctuations in the initial degradation of overstressed polymer molecules. It is this degradation that gives rise to the first free radicals appearing in polymer interlayers. These thermal fluctuations act as a "trigger" and result in the rapid degradation of a great number of other polymer molecules; this behavior "releases the large quantity of elastic mechanical energy concentrated in the molecules."[310] The evolution of this energy (as heat) contributes, in our view, to the polymerization of macromolecular fragments, which can go through the formation of secondary structures with markedly different physical properties.

Obviously, some dynamic equilibrium between the rates of degradation and formation of polymers (copolymers) corresponds to steady-state friction in a metal-polymer system. This equilibrium is affected in the most direct way by the quantity and sign of charge acquired. Indeed, the sliding of a plastic over a metal is accompanied by large upsets in the electron density at every point of the frictional contact. The transfer of charges through the interface can go in either direction; the plastic will pick up a positive or negative charge in relation to the metal, depending on which direction the transfer takes. For this reason, it is customary to make an arbitrary division of plastics into electropositive and electronegative groups. For example, it is quite possible for electrons to be pumped from the organic compound to the interface, because electrons migrate along the macromolecule chains following the lines of covalent bonds and transfer between neighboring long-chain molecules.[5] The resupply of the near-surface layer of the metal with free electrons depends on the electrical capacitance of the metal specimen and can go on indefinitely, provided the specimen is connected to a practically "bottomless" reservoir of electric charges, such as the earth. The fact that a dielectric surface acquires a potential of some thousands of volts proves that a very intense electric field exists between the plates of a frictional capacitor. This field activates the erosional and diffusional processes and, most important, causes ions of the metal to be plucked out and transferred to an electronegative plastic or, conversely, causes electropositive products of the thermomechanical degradation of the polymer to be transferred to the metal surface, forming a protective film on it. The strength of the resulting film depends on the reactivities of carbon atoms with unsaturated valence bonds, and also to some extent (if the reversible reduction of homogeneous chains and the appearance of new heterogenous molecules are allowed for) on the adhesional activity of the functional groups in the polymer material. The kinetics of mechanical-chemical and triboelectric phenomena is reflected in the behavior of the coefficient of friction, which determines the order of magnitude of the forces that resist sliding and must be applied to the metal-polymer frictional system.

There is some basis for the ideas just set forth, in a short series of publications making the following principal points:

• The spreading of an antifriction plastic on a metal occurs when the plastic acquires a positive electrical charge; if a negative polymer is coupled with a metal, then the polymer surface becomes metallized.[354]

• The wear (measured by weight loss) of a metal or a frictional plastic is least when the plastic contains film–forming additives with high adhering power in relation to the metal.[355]

• The frictional interaction causes a substantial change in the state and properties (electrical conductivity and thermal conductivity) of the near-surface layer of a polymer in contact with a metal.[355, 356] After a sharp drop in the electrical resistance of such a layer (by 4–10 orders of magnitude), a relatively stable regime of friction comes into being.[357] This behavior is evidently linked with the increase in the number of charge carriers per unit volume before the appearance of a stable electrical modification of the surface (the quantity of charge acquired also depends on the surface resistivity).

• Taking curves of the charging current (or the rate of accumulation of charges) is a suitable method for investigating the kinetics of electric charging and its effect on the development of transfer processes in metal-polymer frictional couples.[355, 357]

• Products of the mechanical degradation of polymers migrate under the action of ponderomotive forces of the electric fields of microcapacitors that arise in the friction zone. Replacing the metal with an electronegative polymer (ebonite) does not change the direction in which the products of mechanical degradation of an electropositive polymer (poly[tetrafluoroethylene]) are transferred. If conditions are created artificially under which the charging potential of the positively charged plastic becomes less than the potential of the metal (steel) against which it rubs, then the transfer of plastic to the steel surface comes to a stop.[340]

The most interesting result of the publication last cited (it should be singled out) is an illustration of Yevdokimov and Sanches' assumption about the electrical nature of the driving forces of transfer in a polymer-metal system, for the example of "sign-alternating" transfer.[340] This kind of transfer is due to the ability of a metal "to change the sign of its electric charge in a frictional contact, as a function of the sign of the charge on the polymer." This ability should have the direct result that the products of mechanical degradation of an electropositive polymer, when they are introduced into the frictional zone of a steel-electronegative polymer couple, are transferred in first one and then the other direction, until both conjugate surfaces are completely covered with the products. The continuous regeneration of these coatings (thin polymer films) assures freedom from wear in rubbing couples under conditions where a liquid lubricant cannot be used. Yevdokimov and Sanches found a powerful confirmation of this result in trials with a steel-ebonite couple (charging potential of ebonite, −600 V), where poly(tetrafluoroethylene) (charging potential, +500 V) was used as a lubricant in rotary printing.

Let us remark, finally, that concepts of how electrical phenomena affect transfer processes in metal-polymer systems have formed the basis for the discovery of some new ways to explain hydrogen absorption by a metal, which is known to cause embrittlement and dispersion of the frictionally deformed layer.[358] Hydrogen is evolved on frictional activation of dehydrogenation of the hydrocarbons in the plastic; the hydrogen may be supplied by products of the thermal degradation of the organic binder. The diffusional flux of hydrogen in steel is directed toward the region of maximum temperature, which lies at some depth below the friction surface,* where hydrogen dissolves readily.[359] Hydrogen absorbed by a metal is able to dissociate; then the uniquely small size of the proton ($1 \cdot 10^{-13}$ cm), and the fact that its charge is opposite to that of the metal, enable it to penetrate rather easily into the lattice.[358,360] Here in particular is where the role of frictional charging in hydrogen absorption by a steel surface is apparent.

6.4. EFFECT OF ELECTRICAL INSULATION OF THE TOOL IN THE MACHINING OF A DIELECTRIC

The following material has the principal aim of fixing the reader's closest attention on electrical phenomena during cutting in metal-dielectric systems. A kind of transitional variety of mechanical treatment, including elements of both cutting and "pure" friction (as distinct from, for example, friction at a tool rake face), is polishing. Lisov's dissertation dealt with the electrical phenomena that accompany the process of polishing.[361] Without going into specific detail on the conditions of contact and triboelectric charging in polishing, let us just remark that Lisov succeeded in using these phenomena. He was able to establish the link between the engineering parameters of polishing (rate of removal of metal, polishing-wheel lifetime) and the statistical characteristics of the charging of the workpiece being polished. (The potential of the workpiece changes continuously, because electric charges transfer across the regions of contact between it and the abrasive wheel; this transfer is random in nature). He naturally proposed using information on the state of the wheel cutting surface, which is contained in the electrical signals generated in the polishing process, to automate polishing operations. (Lisov's dissertation has much in common, in goals, problems, and even methods of solution, with the studies examined in Section 4.3.)

As we have already seen, one way to increase the lifetime of an electrically

---

*This displacement of the maximum-temperature zone, which is referred to in the Soviet literature as "temperature knife," was given a theoretical basis by V. A. Kudinov.

conductive tool in the cutting of metals is to break the tool-workpiece-machine thermoelectric circuit. Various devices, including those that insulate the tool electrically from the machine, are used to accomplish this (an electrically insulating layer can be placed anywhere in the circuit). The use of these devices in the cutting of dielectric materials has been considered pointless, since a dielectric is itself an insulator. It has become clear, however, that in a case where a new special holding device provides good insulation of the tool from the earth (from the grounded machine), it may also be desirable to use the device in the machining of dielectrics.[31]

The effect of "disconnecting the earth" was seen under plant conditions, in the production of a large batch of polyethylene parts on a model-6M82 horizontal milling machine. The polyethylene workpieces were machined with face-and-side slitting cutters with inserted blades of up to 160 mm in diameter; the blades were made of type-R18 high-speed steel. The feed rate was 300 mm/min; the depth of milling was 35 mm. The spindle turned at 300 rpm (without insulation) or 500 rpm (with insulation). The milling arbor with cutters was insulated at both ends with a cylindrical and a conical bushing, each made up of two sections assembled with type BF–4 bakelite-phenol adhesive.

Cutters operating without insulation had to be resharpened every two to three shifts, while insulated tools could be operated continuously for about 60 shifts (to the end of the batch of parts), although the milling speed was increased by a factor of more than one and a half. In addition, it was found that when the tool was electrically insulated from the earth, the machine operator was in no danger from the effects of electrical discharges.

According to the author and co-workers, electrostatic phenomena form the basis for the effect described.[30]

When a dielectric is cut with an electrically conductive tool on a grounded machine, the workpiece and chip become charged. This process can be represented schematically in the following way. As soon as the tool makes contact with the workpiece, the transfer of electrons (or ions) across the metal-dielectric interface begins; as a result, the contacting surfaces of the tool and the workpiece acquire charges of opposite signs. The charging of the workpiece, which moves relative to the tool, will include all new areas of the workpiece. The positive or negative charge transmitted to the dielectric is continuously made up, because of the large electrical capacitance of the tool-machine-earth system. As machining of the dielectric continues, electric charges build up on it, creating a field so strong that electrical erosion of the tool can take place.

It follows that extending the lifetime of the tool requires a reduction in the charging of the workpiece. This problem can be solved by electrical insulation of the tool from the earth, where the insulating layer lies between the tool and the machine. The tool and workpiece, once brought into contact, become charged as before. But as they move relative to each other, the tool ceases to give up charge to the workpiece. Since it is not connected to the earth, it behaves like a charged body whose field hinders the transfer of electrons (ions) across the metal-dielectric interface. Therefore, when the tool is insulated from the earth, it cannot communicate a significant quantity of charge to the dielectric, the workpiece becomes only weakly charged, and electrical-erosion wear of the tool (in our trial, the cutter) decreases.

Supporting the assumption that the electrical-erosion type of wear coexists with other types of wear is the fact that the friction of a metal against a dielectric is accompanied by electrical discharges. Because of these discharges, the rubbing couple turns into a source of electromagnetic radiation, which can be picked up by an antenna connected to an oscillograph (spikes on the screen) or to a radio receiver (distinctive clicks and crackling).

With the help of the circuit shown in Fig. 6-2, it is possible to trace how rapidly the tool becomes charged in the cutting of dielectrics. In the machining of workpieces made of plastic, wood, poly(methyl methacrylate), and other materials, capacitor C (connected to the tool, which is insulated from the machine by gaskets LG made of resin-dipped fabric laminate) becomes charged. The magnitude and sign of the charge are determined from the deflection of ballistic galvanometer G when key K is closed.

There is still no experimental confirmation that the electrical-erosion type of wear can predominate. Thus, it is still a question for discussion what mechanism brings about the effect seen in the milling of polyethylene. For example, the author does not think it impossible, in the light of the ideas set forth above, that the following takes place: the rupture of C–C bonds in polyethylene corresponds to ∼80 kcal/mole of work.[33] Under conditions when this occurs, a reduction in charging of the polyethylene surface by an artificial limitation on the quantity of charge transferred to the surface may greatly weaken the transfer of metal ions; it is the latter transfer that causes tool wear.

In conclusion, according to data available to the author, "electrostatic cooling" is used in the United States as an effective method of combating the wear

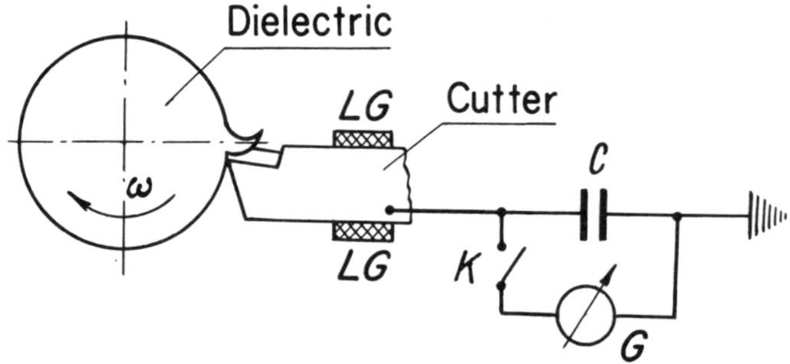

Fig. 6-2. Arrangement for measuring charge of cutter.

of a diamond tool. This method consists in lowering the temperature in the cutting zone by discharging static electricity on an auxiliary electrode, which is brought right up to the cutting edge and which moves along with the diamond tool. The physical essence of this phenomenon is evidently the same as in the effect that we have examined; however, in the new case, we see not a limited natural pumping of charges but an artificial, continuous suction of them.

7

Aspects of Magnetism in Cutting and Friction

7.1. MAGNETIC TREATMENT OF THE TOOL

The magnetic treatment of the tool is a question that goes somewhat beyond the bounds of phenomena in the region of contact interaction of solids, since magnetization of the tool in strong magnetic fields is accompanied by structural changes in the tool material. Nevertheless, as this chapter will show, the possibility of extending the life of a tool under the influence of a magnetic field is linked in the most intimate way with the problem of electrophysical processes in cutting.

From the literature, it is known (Table 7-1) that magnetic treatment of cutters and drills made of high-speed steel doubles their lives in the cutting of materials of construction. One of the main advantages of this method is its simplicity; thus, it has recently begun finding wider and wider application. At the same time, it should be emphasized that the effect of a magnetic field on phase transformations and mechanical properties of steels came under study as early as 1929.[366]

Bernshtein has reviewed investigations into thermomagnetic treatment.[367] Many of the studies he cites affirm in concert that hardening in a magnetic field with little or no tempering leads to an increase in the ultimate strength of steels. If there is moderate or high tempering, then thermomagnetic treatment leads to the same results as does conventional heat treatment. The application of a magnetic field in the tempering of a high-speed steel stimulates the transformation of residual austenite to martensite, with the simultaneous transformation of tetragonal to cubic martensite.

Table 7-1. Increase in cutting-tool life through magnetic treatment.

| Work Material | Tool Material | Type of Tool (diameter) | Cutting Conditions | | | Criterion of Efficiency (K_{em}) | Reference |
			v (meter/min)	s (mm/rev)	t (mm)		
40Kh	VK8	Cutter	120	0.11	1.0	1.3	362
40Kh	R12	Drill	26	0.28		1.6	362
		(22 mm)	47	0.22		2.0	362
40Kh	R18	Cutter	55–65	0.11	1.0	1.7–2.0	363
ShKh15	R18	Cutter	—	—	—	2.3	364
Type-30 steel	R6M3	Drill (25 mm)	—	—		3.7	364
Type-45 steel	R18	Drill (12 mm)	—	—		2.0	364
SCh18-36	R6M5	Drill (14 mm)	—	—		2.6	364
Type-45 steel	R18K5	Cutter	50–60	0.4	1.0	2.0–1.6	365
Type-45 steel	R18	Cutter	80	0.2	1.0	2.0	365
Type-45 steel	R9K5	Cutter	110	0.1	1.0	2.0	365
Type-45 steel	R9	Drill (12.5 mm)	20	0.16		2.0	365
Type-45 steel	T15K6	Cutter	150–230	0.4	1.0	1.0	365
S421–40	VK6	Cutter	125–200	0.1	0.5	1.0	365

On the basis of these same ideas, Dimitrov attempted to discover what role a magnetic field may play in increasing the wear resistance of drills and cutters made from types-R9, -R9K5, -R18, and -R18K5 high-speed steels.[365] He showed that, regardless of whether the magnetically treated tool was in the magnetized (any polarity of the cutting portion) or demagnetized state, its life was extended by a factor of 1.5–2 under various conditions of cutting. Magnetic treatment of cutters and drills in an alternating magnetic field with an intensity of up to $3.2 \cdot 10^5$ A/m yielded an increase in life of the same order as did magnetization in a steady field with an intensity of up to $5.6 \cdot 10^5$ A/m. Hence, Dimitrov concluded that the action of an intense magnetic field may result in changes in the structure of the tool material. Dimitrov justly remarks that, after conventional triple tempering, the austenite content in a high-speed steel has been reduced to a minimum, so that it is not very probable that any further austenite-martensite transformation in it under the action of the magnetic field would cause an improvement in the cutting properties right after heat treatment. The situation is different if the tool has been ground and a rehardened layer with a large content (40–80 percent) of residual austenite has formed on its working surface.[368] A system of this kind, on the contrary, may be rather sensitive to the action of a magnetic field in communicating to the system the activation energy needed for the $\gamma \longrightarrow \alpha$ transformation. As confirmation of this hypothesis, experimental results have been cited that indicate that ordinary and magnetically treated cutters have equal lives when the latter have been subjected to prolonged polishing after holding in a steady ($5.6 \cdot 10^5$ A/m) or alternating ($3.2 \cdot 10^5$ A/m) field.

As for Dimitrov's conclusion[365] on the character of structural changes under the action of a magnetic field, it will be shown below that his is not the only possible conclusion. Now, however, we turn to the question of how magnetic condition can affect tool life.

7.2. CUTTING WITH A MAGNETIZED TOOL

In the judgment of the researchers who first performed cutting with a magnetized tool,[369] the presence of a magnetic field causes a temperature change in the region of contact between the tool and work materials, on account of odd thermomagnetic effects (odd in the sense of "odd function"). From this standpoint, it would seem, it is fairly difficult to explain the fact that the tool life increased even after a magnetic treatment that resulted in the net magnetization remaining equal to zero. However, according to Yakunin and Molchanova, the steel's taking on a magnetic texture may have a favorable effect on tool life, because it makes it easier for magnetization of the tool to occur during cutting, under the action of local fields associated with internal thermoelectric circuits.[362]

As it did to Avakov and Markosyan,[370] it seemed rather doubtful to us too whether the Righi-Leduc thermomagnetic effect could be used to control the

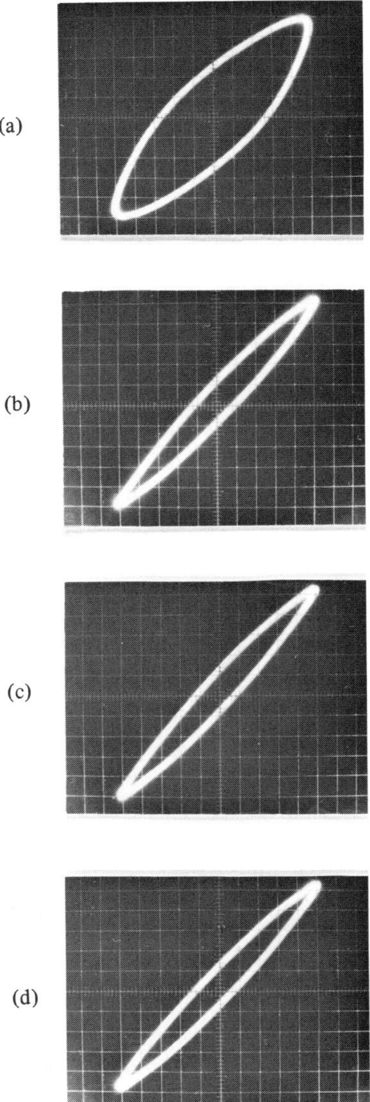

Fig. 7-1. Dynamic hysteresis loops taken under conditions far from saturation (H_m = 2,500 A/meter):[373] (a) type-U7A steel (B_m = 0.202 T); (b) type-R6M5 steel (B_m = 0.070 T); (c) type R9K5-steel (B_m = 0.048 T); (d) type-R18 steel (B_m = 0.54 T).

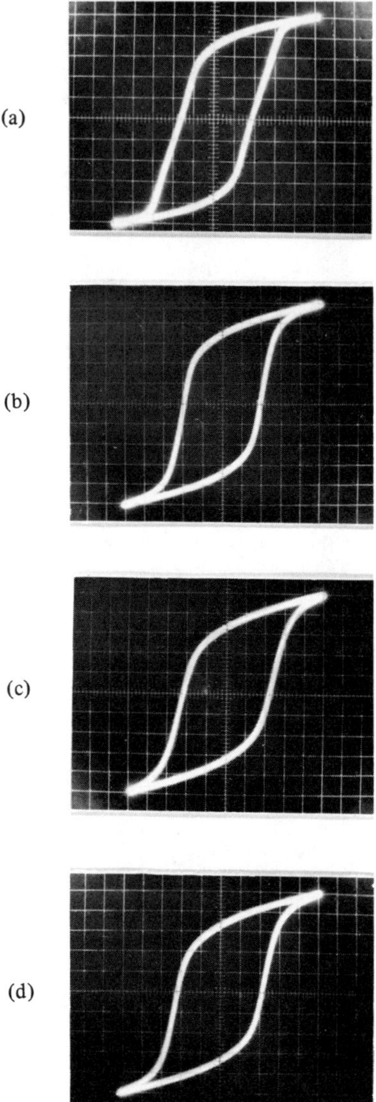

Fig. 7-2. Dynamic hysteresis loops taken under saturation conditions (H_m = 20,000 A/meter):[373] (a) type-U7A steel (B_m = 0.78 T); (b) type-R6M5 steel (B_m = 0.51 T); (c) type-R9K5 steel (B_m = 0.48 T); (d) type-R18 steel (B_m = 0.54 T).

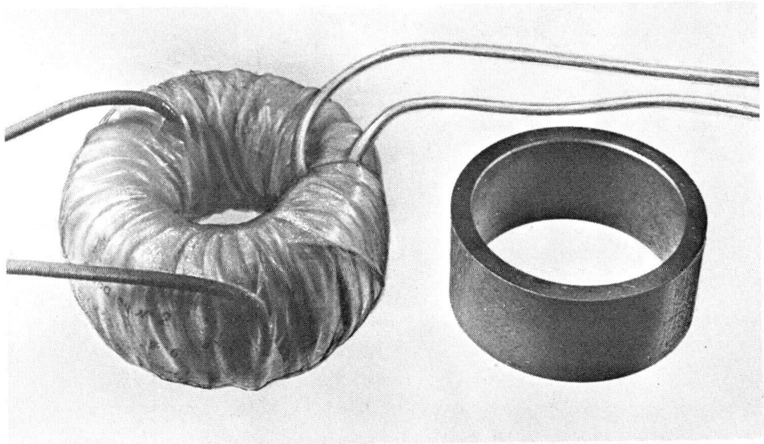

Fig. 7-3. Ring-shaped specimens for the taking of magnetic dynamic characteristics.

heat flux in the cutting zone, where the drift velocity of heat (charge) carriers is comparatively low.[32] Even with allowance for features of the electron-energy spectrum linked with the formation of semiconducting oxide films on solid surfaces it was supposed that the scale of the Righi-Leduc effect in the tool material would be too small to bring about a marked decrease in temperature at a site of intensive heat evolution. Dimitrov drew a conclusion analogous to this assumption, on the basis of measurements of the cutting-zone temperature.[365] However, according to model investigations of the change in direction of heat flux in type-R18 steel under the action of a magnetic field,[371] the transverse odd thermomagnetic effect in this steel may be significant because of semiconducting (carbide) inhomogeneities (with p-type conduction).

Another possible reason for the change in life of a magnetized tool is processes occurring in the cutting zone because of electromagnetic induction.[372] Among these processes, the additional liberation of heat by eddy currents deserves special interest. A comparison of the dynamic magnetization cycles shows that the rate of heat liberation by this means varies greatly from one to another combination of tool and work materials. According to one published report,[373] the area of the hysteresis loop taken in the dynamic regime is significantly greater for type-U7A high-carbon steel than for types-R18, -R9K5, and -R6M5 high-speed tungsten steels (see Figs. 7-1 and 7-2). The loops were obtained on ring specimens (Fig. 7-3) with the help of a type-U-54G ferrometer. Although the fraction of losses due to eddy currents in tool steels is sometimes nonnegligible, the principal objects of study, from the standpoint of magnetic dynamic characteristics, should remain those work materials in whose bulk most of the additional heat energy is liberated.

As for the static hysteresis cycles, these cannot yield ideas on the magnitude of energy losses due to Foucault currents. It should be noted, however, that there is a close connection between the microstructure and properties of a material and the static coercive force. For example, it is known from the reference literature that type-U7A steel treated to spheroidized pearlite has a coercive force of 4 oersteds; treated to lamellar pearlite, it has a coercive force of 12 oersteds. Annealed steels have coercive forces severalfold smaller than hardened steels (16 versus 80 oersteds for type R9K5, 10 versus 80 for type R6M5). This link makes it possible to consider the determination of coercive force as one method of nondestructive testing in the plant operations involved in manufacturing and heat-treating a cutting tool.[374]

If the quantity of heat liberated by eddy currents depends on the conductivity of the work material, then the attraction of the chip to the tool-magnet depends on the magnetic characteristics of the work material. In the cutting of a magnetic steel, say type-2Kh13, this attraction (Fig. 7-4) may have a detrimental effect on the tool life, despite other benefits gained by magnetization of the tool. That is, if the increase in life is due to changes in structure, not to the introduction of the magnetic field into the cutting zone, then a tool once subjected to magnetic treatment should preferably be used in the demagnetized condition. Incidentally, if this is the case, then it follows from an earlier report[31] that the behavior of the microchip in cutting may include manifestations of magnetism connected not only with the formation of ferromagnetic crystals of like polarity but also with the elastomagnetic interaction of the solids (see Section 7.5).

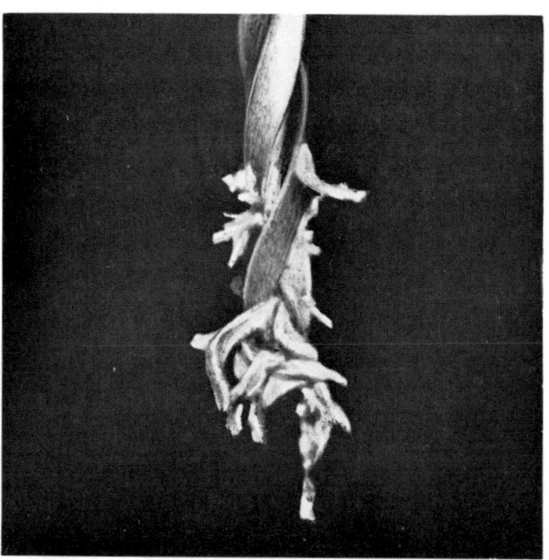

Fig. 7-4. Tip of a magnetized drill (type-R18 steel) with chip that has stuck to it, after withdrawal from a magnetic steel (type-2Kh13).

It should be emphasized, however, that the application of a magnetic field makes the $\alpha \longrightarrow \gamma$ reverse martensite transformation more difficult in the heating of a high-speed steel.[375] For this reason, the content of austenitic phase in the surface layers of a reground tool will depend on the magnetic condition of the tool. Therefore, if the ideas set forth by Dimitrov[365] are used as a basis, then to answer the question whether it is desirable to demagnetize the tool, it is evidently necessary to consider what kind of work material—ferromagnetic, paramagnetic, or diamagnetic—the tool is intended for.

The process of cutting with a magnetized tool can be considered as a natural mechanical-thermal-magnetic treatment. As the author remarked earlier,* the imposition of the external magnetic field of the tool on the phase transformations in the cutting zone may have a substantial effect on the character of the interaction of atom-electron systems. At the same time, the first cause of the long life of a magnetically treated tool probably lies in the change in the characteristic dislocational structure of the high-speed steel under the action of an external magnetic field. As for carbide-tipped cutters, these have proved practically insensitive to various types of magnetic treatment,[365] because of the relatively low content of ferromagnetic cobalt in the solid alloy.[362] Incidentally, the saturation of the alloy with carbides might result in a significant thermomagnetic effect in the presence of an external magnetic field;[371] it is easy to set up such a field with a solenoid surrounding the tool.

7.3. EXTENSION OF DRILL LIFE THROUGH PULSED MAGNETIC TREATMENT

In order to clarify the fundamental physical reason for the effect of magnetic treatment, it was necessary to achieve stable reproducibility of the effect. Drills made of type-R18 steel were selected as the object of laboratory and plant tests. These drills were magnetized with a pulsed magnetic field of intensity over 1.0 kilooersteds. Figure 7-5 gives an overall view of the device that set up the field. Operation of the device is based on accumulation of electrical energy in a charged capacitor for a long time and the discharge of this capacitor into an induction coil in a short time.

Drill life was seen to double when the current fed to the magnetizing coil had an amplitude of some 400 A ($H \sim 10^5$ A/m). Increasing the current in the coil (and simultaneously reducing the pulse repetition rate) resulted not only in the disappearance of the effect, but also in a tool life markedly shorter than the starting life; tool wear acquired traits characteristic of brittle fracture.

---

*In his contribution to the scientific and engineering seminar on "Modern technology for the production of instruments, hardware for automation, and control systems," Moscow, 1973.

Fig. 7-5. Experimental model of a device for magnetic treatment of the tool (type UMO-1).

Under thermally severe conditions of grinding, when type-St3 steel was being cut, the effect of magnetic treatment was reduced by 15 to 25 percent after two to three regrindings of the drills. The subsequent decline in tool life to the starting value—that is, to the life of drills not subjected to the action of a pulsed magnetic field—went rather quickly as a rule.

When type-Kh18N10T steel was drilled and grinding was done with an Elbor wheel and an oleic-acid–based lubricant compound, the effect of magnetic treatment was developed to practically the same degree, despite the large number of regrindings.[376] The trials were carried out on a model-2N118 vertical drilling machine; the cutting fluid was a 5 percent solution of self-emulsifying oil. Blind holes were drilled to a depth of 15 mm from both sides of workpieces that had been cut from a single blank. The drills used were 8 mm in diameter and had been checked with respect to geometric parameters (overall length, diameter, flute length, width of margin, length of cutting lips, helix angle), runout of margins, and hardness of the working portion. For the main series of tests, drills were selected with roughly equal wear rate up to a steady-state criterion of dullness on the flank face.

Figure 7-6 presents the results of the investigation, in the form of life curves, $T = f(v)$, taken at a feed of 0.14 mm/revolution. From a comparison of these curves, it follows that the life of a magnetically treated tool more than doubles in the range of cutting speeds recommended in practice. Borodkin and coworkers explain this behavior on the hypothesis advanced in the following concerning magnetostrictive strengthening of the tool material.[376] The decrease in the criterion of efficiency K_{em} upon a change to more drastic conditions is due, they say, to resoftening of the materials under the action of high temperatures in the cutting zone.[376]

In the course of plant tests, magnetized drills were ground with an electrocorundum wheel just before use in the cutting of type-40Kh steel. The lives of these drills were 1.5–2.5 times as long as the starting value depending on the process conditions.

Touching on the physical essence of the laws governing the wear of a magnet-

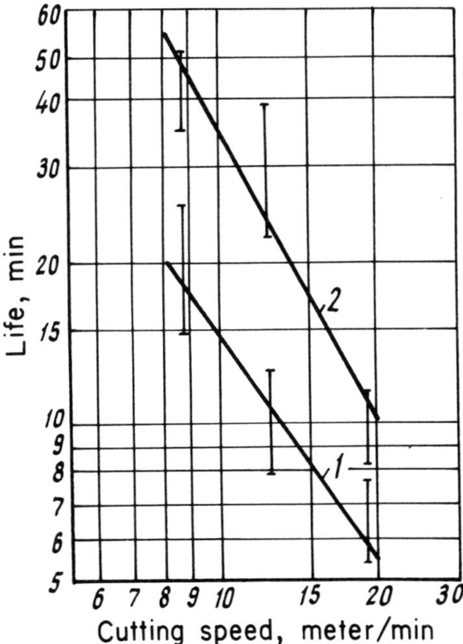

Fig. 7-6. Life of drills made from type-R18 steel versus cutting speed, before (1) and after (2) magnetic treatment.

ically treated tool, it is a question of no small importance whether the tool retains its increased life even after its net magnetization has vanished. Drills demagnetized in an alternating field may have the same life as magnetized drills;[376] this fact makes it possible to speak with great certainty of structural changes in high-speed steel as the most important result of magnetic treatment of the tool.

7.4. MAGNETOSTRICTIVE STRENGTHENING OF HIGH-SPEED STEELS IN PULSED MAGNETIC FIELDS

Analysis of the data obtained supports the author's hypothesis that the increase in life of a magnetically treated tool results from magnetostrictive strengthening and, less probably, from magnetodispersion hardening of the high-speed steel.[32,377] Let us examine the mechanisms of these processes, starting with some well-known concepts.

In the unmagnetized condition, the ferromagnetic matrix of the steel is divided into a great number of domains. When, under the influence of an external field, the domain boundaries are displaced and the spontaneous magnetizations rotate, the directions of the electron spin moments change; this change is accompanied by magnetostrictive deformation and, in accordance with Hooke's law, by elastic stresses. Interaction of the elastic field, due to magnetostriction of steel, with the elastic field of the real dislocational structure of the steel leads to the appearance of local overstresses. At these places, the probability of thermofluctuational breaking of stressed interatomic bonds rises sharply.[310] It is just here that the multiplication and migration of dislocations, and thus the formation of sites of plastic deformation, proceed most rapidly. As the dislocation density increases, when the growing dislocation forest makes it more and more difficult for the dislocations to migrate to other slip planes not belonging to the forest, the tool experiences something resembling cold work and becomes strain-hardened.[378] However, if the high-speed steel is "hammered" by impulses having too large current and power amplitudes, then the very great piling up of

Table 7.2. Values of the lattice parameter in the magnetostrictive strengthening of drills made of type-R18 steel.

State of Drill	Number of Drill	$a(\text{Å})$
Before magnetic treatment	1	3.5153
	2	3.5154
	3	3.5153
	4	3.5153
After magnetic treatment	5	3.5146
	6	3.5145
	7	3.5146
	8	3.5146

dislocations in intercommunicating grain boundaries brings about a reduction in the steel's resistance to brittle failure. It is not out of the question that the event responsible for the formation of microcracks is the magnetostriction of the paraprocess; systems having an $\alpha \rightleftharpoons \gamma$ transformation with large hysteresis are predisposed to this behavior.[379] In such systems, the energy of the exchange interaction (and it is many times greater than the energy of the magnetoelastic interaction) may depend rather strongly on the interatomic distances. At the same time, it was that steel whose cutting properties had improved after magnetization that exhibited a change in the martensite lattice parameter; this was easy to judge from X-ray structural analysis data on two groups of drills (see Table 7-2). These data were obtained by the ionization method, on a type-URS-50IM system (Cr radiation). The relative error of determination of the parameter was 10^{-4}. As we see, the influence of the pulsed magnetic field resulted in a decrease of roughly $8 \cdot 10^{-4}$ Å in the interatomic distance.

Figure 7-7 shows photomicrographs of steel sections typical for the indicated groups of drills; Fig. 7-8, of carbon replicas of the steel. The dark areas on the images of the replicas resulted from Bragg scattering of electrons on crystalline particles of a carbide phase that had adhered to the film; these were identified with the help of microdiffraction (Fig. 7-9). The photomicrographs bear witness that magnetostrictive deformation of high-speed steel is accompanied by size

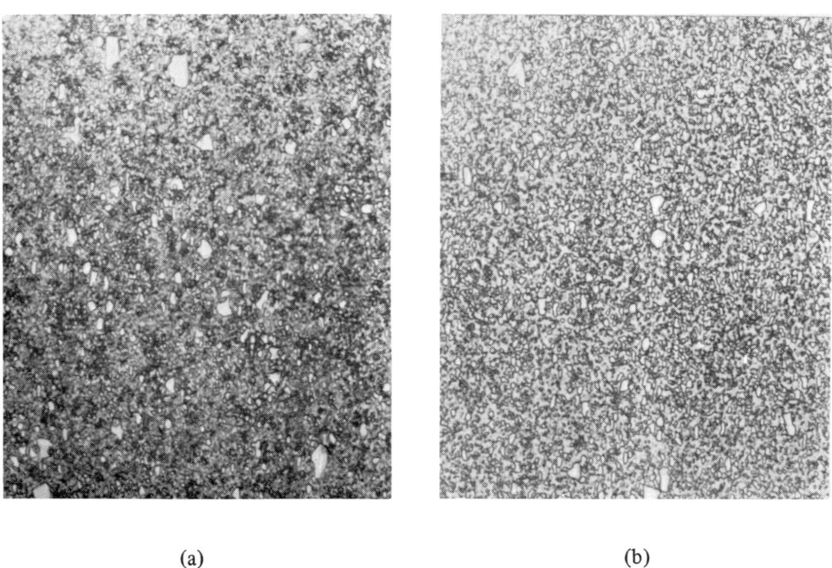

(a)   (b)

Fig. 7-7. Photomicrographs of ordinary (a) and magnetically treated (b) drills made from type-R18 steel, in cross section parallel to the axis (340×).

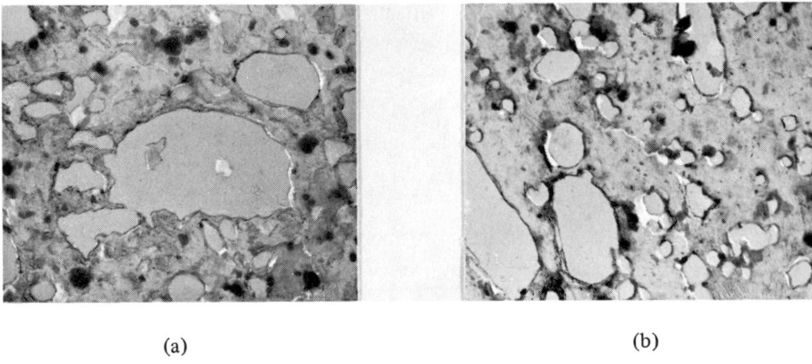

(a)   (b)

Fig. 7-8. Carbon replicas of type-R18 steel, sputtered at an angle of 30° onto etched surfaces of cross sections of ordinary (a) and magnetically treated (b) drills. Magnifications: (a) 5,000×, (b) 10,000×.

reduction of the carbides. This size reduction agrees with the decrease in size of the principal domains when the dislocation density increases.[380] But the structure still acquires great homogeneity from the large number of fine carbides precipitated from the principal phase, which becomes metastable in the process of strictive strengthening caused by shock waves. It is reported that segregation of impurity atoms, above all of carbon atoms, at grain and dislocation boundaries may occur under the action of local stresses.[381] With the formation of Cottrell atmospheres, the probability of carbon-concentration fluctuations along the dislocations rises sharply; this now leads to the nucleation of discrete particles of

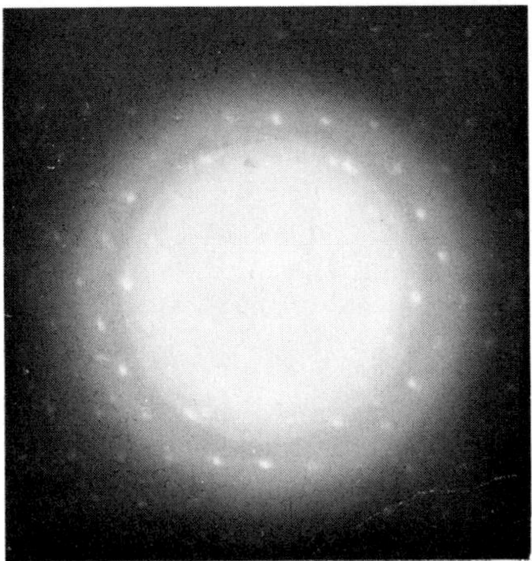

Fig. 7-9. Electron diffraction pattern of a carbide inclusion in a carbon replica of steel.

the carbide phase. It looks as if we have a picture of a kind of dispersion hardening; this idea has been confirmed in hardness and microhardness measurements on several batches of specimens before and after magnetization. At the same time, high-speed steel subjected to magnetic treatment exhibits not only a large content and a more uniform distribution of carbides. Another distinguishing feature of such a steel (see Fig. 7-7 [b]) is the more spherical shape of the carbides, which are known to play the role of stress concentrators. The sphericity acquired by the carbides signifies a decrease in the surface energy at the interface with the matrix, and consequently an increase in the activation energy for breaking of interatomic bonds in the crystal lattice of the matrix. Naturally, the strength characteristics of a tool steel are improved in this way.

Magnetostrictive cold work, in distinction to mechanical cold work, is more a bulk strengthening than a surface strengthening of the metal. It can of course be removed; that is, the metal can be resoftened, by heating, especially cyclic heating.[382] For this reason, high-speed steel can lose its enhanced strength properties fairly rapidly when the tool is reground, the more so when this is done under severe thermal conditions. The weakening of the effect of magnetic treatment is also aided by periodic, nonuniform heating of the tool in the drilling of holes one after another.

Following Dimitrov,[365] the author and coworkers suppose that the possibility of a decline in tool life after the first few grindings may be linked with a large percent content of residual austenite in the surface layer of the steel. In order to transform the paramagnetic austenite to ferromagnetic martensite, it is thought necessary to subject the tool to magnetic treatment after each regrinding. A measure of this kind may indeed prove useful, although the main task of magnetic treatment comes down, as was remarked above, to magnetostrictive strengthening of the steel, not to further decomposition of the residual austenite in it.

As a result of what has been said, it becomes clear that magnetic treatment of a high-speed tool may be considered one promising method of extending tool life, and a method that is easy to put into plant practice.

To complement the preceding material, let us now dwell on one of the experimental investigations mentioned in Section 7.2 that contributes to our understanding of the role played by electrophysical phenomena in friction and cutting.

7.5. ELASTOMAGNETIC INTERACTION OF SOLIDS
UNDER SEVERE CONDITIONS OF SLIDING

When a rotating cylinder of tool steel slides against a stationary cylinder crossed with it, the material of the stationary specimen has a significant effect on the behavior of the wear particles (see Fig. 7-10). Numerous experimental confirmations of this fact have been obtained under conditions of sliding that approached in their severity the conditions of the flow of a chip over the contact faces of a

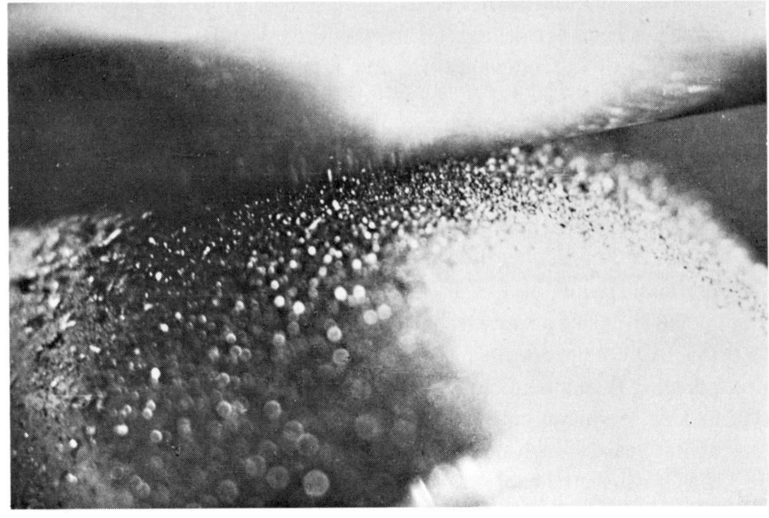

(a)

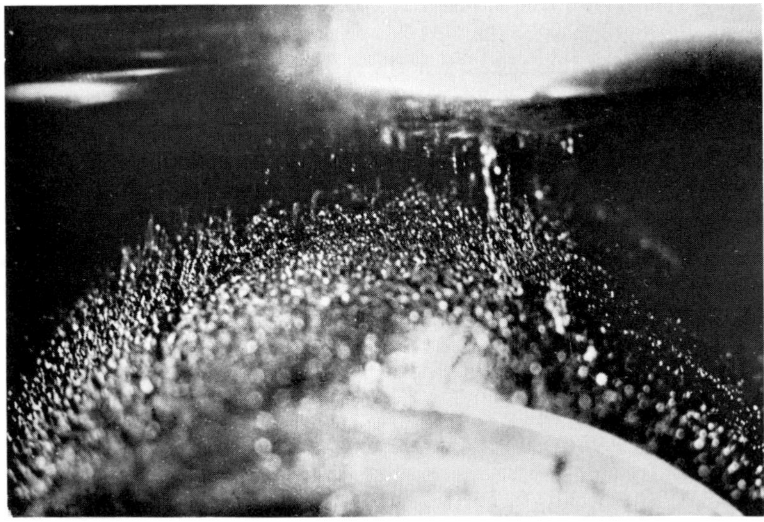

(b)

Fig. 7-10. Wear particles that have flown out of the zone of frictional interaction, (a) between type-R9K5 high-speed steel and type-Kh18N10T nonmagnetic steel and (b) between type-R9K5 and type-2Kh13 magnetic steel.

tool (specimen diameters 30 or 40 mm, applied load $\sim$ 80 N, sliding speed $\sim$ 1.5 meter/sec). Therefore, we will speak of phenomena associated with a heterogeneous process including, along with friction between physically and chemically clean oxidized surfaces, elements of microcutting.

In trials with specimens of a nonmagnetic steel, the trajectories of particles ejected from the sliding zone and the orientation of the particles on the surface of the stationary specimen are determined by the action of one force field only: the gravitational field (Fig. 7-10[a]). In the case where a specimen of magnetic material undergoes severe wear, the behavior of the microchip indicates that it is under the influence of not only the gravitational, but also a magnetic, field, the latter acting on wear particles with linear dimensions up to 1 mm (and even larger) as on natural ferromagnetic "filings" (Fig. 7-10[b]). Since both crossed specimens were demagnetized before the trial, it is natural to assume that the appearance of a magnetic field has to do with a mechanical-magnetic effect.

Let us attempt to picture what does happen in the sliding zone itself and in the air space in its vicinity. If the heat evolved is considerably greater than the heat withdrawn, the frictional couple begins to move gradually from the athermal regime of operation to the thermal (quasi-adiabatic) regime. The destruction of oxide films and an increase in the intensity of thermal oscillations of the atoms result in seizing and diffusional sintering of the materials. Cohesion between solids that are sliding relative to each other causes mechanical shear deformations to appear on both sides of the interface. The character of the magnetic interaction of atoms changes when the lattices of crystallites in the elastic-stress region become distorted. The magnetic moments of the regions of spontaneous magnetization become reoriented; the arrangement that proves energetically

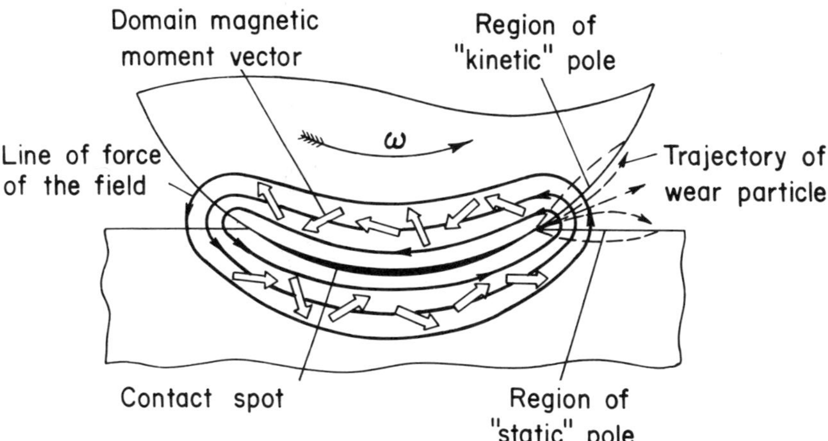

Fig. 7-11. Magnetic field within and outside the sliding zone in elastomagnetic interaction of rubbing solids.

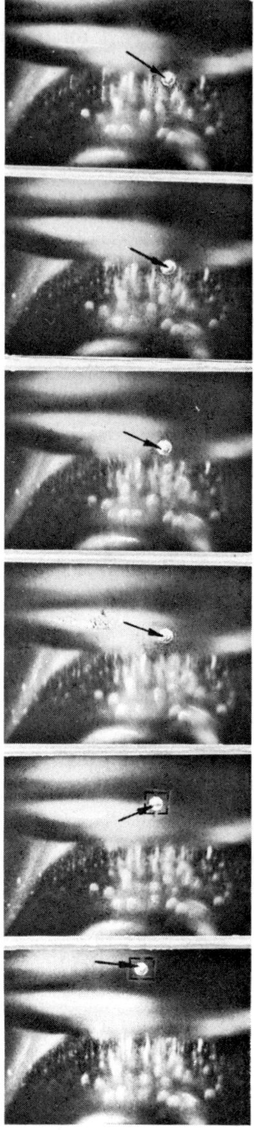

Fig. 7-12. Frames of high-speed motion-picture film demonstrating the travel of a particle entrained by the kinetic pole and its adaptation to the specimen surface. Times, top to bottom ($\tau \cdot 10^3$, sec): 0, 1, 5, 6, 9, 11.

most favorable is that in which the lines of magnetic induction "ring in" the contact spot and the layers adjacent to it (Fig. 7-11). Ferromagnetics are more heavily deformed, and thus more strongly magnetized, in the immediate vicinity of the interface; for this reason, the magnetic field intensity in the air gap between the specimens falls off especially rapidly with increasing distance from the contact spot.

With the help of such a picture of the field, it is not difficult to clarify how it affects the behavior of the wear particles. The microchip ejected from the sliding zone is attracted to the poles formed on the specimen surfaces. The microchip is strewn thickly over that part of the "static" pole that lies near the contact spot, aligning itself there in something like a "magnetic fence." When the rubbing surfaces are separated, this fence moves (migrates) in the direction toward the strong magnetization. Particles coming under the predominant influence of the "kinetic" pole move along rising trajectories; being attracted to the rotating specimen, they may be transported by it into the sliding zone, where they act as an abrasive. Figure 7-12 shows the travel of one such particle.

As the time of sliding increases, the dominance of the kinetic pole over the static pole becomes especially marked. Now the development of the tribomagnetic effect is expressed in the relatively quick thinning out of the magnetic fence on the surface of the stationary specimen: More and more particles are torn from the static pole and turn toward the kinetic pole. Probably the cause of this behavior of the microchip lies in the asymmetry of the sliding process. The temperature gradient in the stationary solid, one unchanging part of which is in contact with continuously changing parts of the rotating solid, is significantly higher. Here is why the decrease in elastic stresses due to the breaking of interatomic bonds (which is accomplished by thermal fluctuations) goes considerably more quickly in precisely those layers of the stationary specimen that lie adjacent to the contact spot. Under these conditions, the elastomagnetic effect and the resulting magnetic texture begin to disappear first; this involves a decrease in magnetization in the region of the static pole. What is more, the additional weakening of the tribomagnetic effect due to the sharp drop in specific load on the spot when its area increases should be taken into account.

It is interesting to note that phenomena similar to those described above are seen in the *cutting* of metals. According to Avakov, for example, only some of the very small particles of work material (that is, particles bordering on metal powder) settle on the tool rake face and fall to the machine as they build up. "Others of them, more or less in dependence on the conditions, stick in a thick 'brush' to the lower parts of the main and auxiliary flank faces, just as iron filings stick to a magnet."[383] It is logical to suppose that phenomena of this type, which are characteristic of friction and cutting processes, have as their basis (along with the formation of single-domain ferromagnetic crystals) the elastomagnetic interaction of solids.

Conclusion

From the incomplete review of research with which the reader has now become acquainted, it is clear that the current interest in electric phenomena in friction and cutting of solids reflects two fundamental questions. On the one hand, the ability to control these phenomena provides a basis for developing new methods and hardware for increasing the wear resistance of rubbing couples and cutting tools. This is the engineering question. At the same time, the discovery of the mechanism of electrophysical and electrochemical processes taking place in the contact interaction of solids would lift the veil that covers the complex of phenomena characteristic of friction and cutting processes. This is the scientific question.

How complex the first, and even more, the second of these questions are was shown by the extreme conflict of views on the role of electrical phenomena at the very initial stage of the study of these phenomena, when few believed that their role could be important against the background of intensive deformational and thermal processes. This is why the author's attempt to systematize and critically interpret the experimental and theoretical material in this field is justified, if only because it eliminates all doubt about the promise of investigations of triboelectricity (in the broad sense of the word).

In work on the problem of electrical phenomena, a number of new research methods have been singled out: combined measurement of work function and exoelectron-emission intensity during contact interaction; simultaneous recording of integral emf and contact resistance; harmonic analysis of the spectrum of fluctuations of the integral emf; calculation of equivalent electrical circuits for frictional systems; taking of current-voltage characteristics of boundary lubricant layers and oxide films; and so on. The use of these methods together with optical and electron microscopy, high-temperature X-ray diffractometry, neutron activation analysis and other procedures makes it possible to obtain information about the intimate aspects of the contact interaction of solids. The possibility of following the course of friction and cutting processes with the help of these methods means that every prerequisite exists for putting these processes under control or even making them self-governing.

As the reader has already seen, there are very many points at which our views on triboelectric phenomena can be developed and made more profound. Further development of the electromagnetic theory of friction, the electrochemical fundamentals of the selective-transfer phenomenon, the energy-fluctuation hypothesis of the formation of a cohesion joint, the theory of the electrical state of boundary lubricant layers and oxide films, methods of polarizing frictional systems, evaluation of the predominant type of wear from the spectrum of the ac component of emf, and so on—these are all urgent problems of tribophysics and tribochemistry, which apparently will be solved in the coming years. The question of whether practical application of some electrophysical or electrochemical phenomenon is desirable, particularly in the devising of frictional systems so as to provide optimal operating conditions, should be answered with allowance for new theoretical notions.

Appendix 1
Procedure for Determining the Criterion of Efficiency of Methods for Improving Tool Wear Resistance

The following procedure is suggested with the aim of significantly reducing expenditures of materials and time in laboratory tests of methods and hardware for extending cutting-tool life.*   The efficiency of any such method is determined, as a rule, by statistical treatment of experimental data. The quantity taken as the criterion of efficiency K_e is the ratio of life values, or of tool wear values, for operation under ordinary conditions and with the use of some method of increasing wear resistance. Selecting the tool for uniform wear rate makes it possible to narrow the interval of variation of this criterion; however, determining the most probable value of the criterion presupposes a comparison of the results obtained on the same drills, broaches, milling cutters, or whatever. This quantity will be nearer the actual value of the criterion of efficiency, the more statistical material is available to us for determining it.   Statistical material is accumulated through repeated determination of the life or wear values, new cutting conditions alternating with the starting conditions after each grinding of the tool.   The necessity of this alternation is due to the fact that the state of the tool

*Besides the author, Y. I. Zinkin and A. D. Ovchinnikov took part in the development of this procedure.

is modified, different from previous states, after each grinding. Thus, because the cutting conditions are poorly reproducible, the number of grindings sufficient for a comparison of averaged data inevitably becomes large. This is in fact the principal shortcoming of existing methods for the determination of K_e. This drawback is eliminated in the proposed variant, for the criterion of efficiency for a method or device used to extend tool life is determined while the tool is in operation, before it becomes dulled once. Together with a significant reduction in expenditure of materials and time, this procedure evidently makes it possible to obtain more reliable functional dependences for tool wear, since each dependence is recorded under a given set of cutting conditions, between two successive grindings.

The procedure is realized through repeated, sequential alternations of machining conditions, between the use of some method (device) for influencing the cutting process and the starting conditions, up to a single dulling of the tool. (Methods and devices used include opening the tool-workpiece-machine thermoelectric circuit, magnetizing the tool, using bonded holding devices for electrical insulation, using engineering cutting fluids, and so on.) The procedure in essence consists of three steps: (1) determining the changes, over equal time intervals, of quantities (instrument readings) that reflect the absolute variation of wear for each of a series of successive changeovers; (2) summing these changes for two sets of changeovers (first in one, then in the opposite direction); and (3) computing the criterion of efficiency of the method as the ratio of the sums obtained.

The schematic diagram of Fig. A.1-1 illustrates the basis of the procedure. We

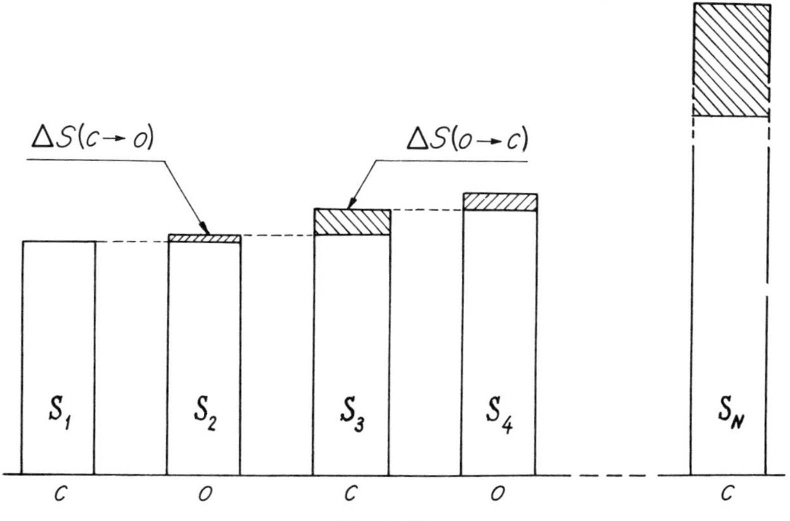

Fig. A.1-1

assume that, before becoming dull once (before grinding of the drill), the tool drills N holes, and that the use of some method of influencing the cutting process (e.g., opening the tool-workpiece-machine thermoelectric circuit) alternates with the starting conditions of machining (thermoelectric circuit closed) in the drilling of each subsequent hole. The instrument readings reflecting the degree of dullness of the tool are as follows:

$$S_{1(c)}, S_{2(o)}, S_{3(c)}, S_{4(o)}, \ldots, S_{N(c)}.$$

From these values, the criterion of efficiency of the method can be found as the ratio between the sum of changes in the instrument readings for $(N - 1)/2$ changes from open to closed thermoelectric circuit and the corresponding sum for the same number of changes in the opposite direction:

$$K_e = \frac{\Sigma \Delta S(o \longrightarrow c)}{\Sigma \Delta S(c \longrightarrow o)}.$$

Suppose, for example, we compare two cutting fluids ($CF1$ and $CF2$) and it has proved that, on changes from operation with a cutting fluid to dry machining and back again,

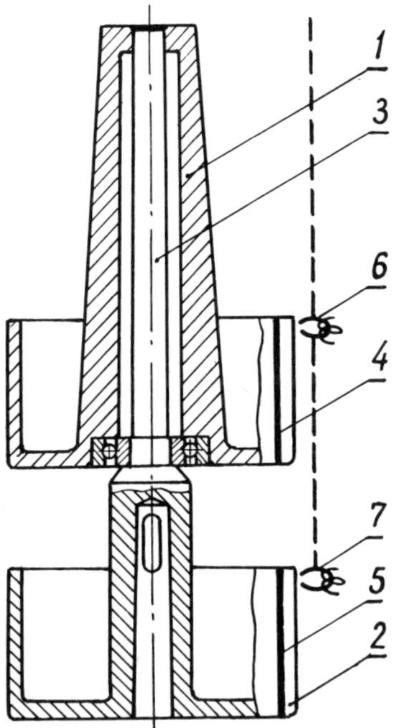

Fig. A.1-2

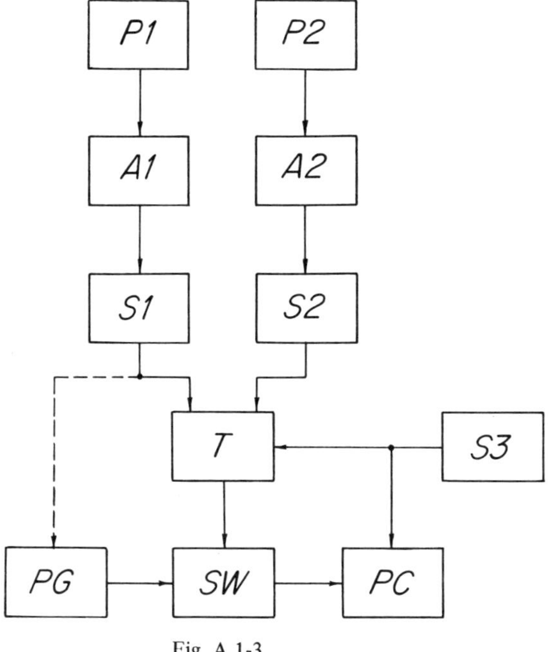

Fig. A.1-3

$$\frac{\Sigma\Delta S(CF1 \longrightarrow dry)}{\Sigma\Delta S(dry \longrightarrow CF1)} = \frac{\Sigma\Delta S(CF2 \longrightarrow dry)}{\Sigma\Delta S(dry \longrightarrow CF2)};$$

that is, $K_{e1} = K_{e2}$. This means that, from the standpoint of their effects on tool life, the selected cutting fluids are equivalent.

The idea of a rapid evaluation of the criterion of efficiency requires, of course, the development of knowledge most applicable to its practical implementation. The procedure can be realized, for example, with the help of a device consisting of two parts: the special arrangement of Fig. A.1-2 and the control system block-diagrammed in Fig. A.1-3.

The device through which torque is transmitted from the machine spindle to the tool (e.g., a drill) consists of two conical bushings 1 and 2 connected by an elastic shaft 3. During the cutting process, the shaft twists, so that the lower bushing rotates through some angle relative to the upper one. Narrow strips of magnetic tape 4 and 5, which are glued to the flanges of the bushings parallel to the axis of rotation, move apart when this rotation occurs; as the tool wears, the separation between the strips increases. Magnetic heads 6 and 7 are mounted to the headstock of the machine; electrical pulses are generated in these heads when they move past the bushing section holding the magnetic tape. The electrical signals from magnetic pickups *P1* and *P2* (see Fig. A.1-3) are amplified in *A1* and

A2 to the values necessary to actuate the transistor trigger. The signals go to shapers *S1* and *S2,* which are differentiating networks each with a diode switch. The output signal from each shaper is a differentiated pulse of definite polarity corresponding to the leading edge of the input signal. These pulses are fed to separate inputs of trigger *T.* The magnetic pickups are mounted in such a way that when the drill is running free the signals from *P1* and *P2* arrive simultaneously; in this case, the communications channel between the pulse generator *PG* and the pulse counter *PC* is blocked by switch *SW.* On the beginning of machining, the signals from the magnetic pickups are shifted in time by some amount τ; during this time, the pulse counter will count the pulses emitted by the generator. The number of pulses arriving at the counter over one revolution at the spindle is $n = \tau/t$, where t is the repetition period of the pulses from the generator. The time τ and the number n will correspond to the degree of deformation of the elastic shaft, that is, to the degree of dullness of the tool during a given revolution. The total number of pulses S_i $(i = 1, 2, 3, \ldots, N)$ arriving at the counter during one pass of the tool to a set drilling depth (one hole) will depend on the magnitude and rate of tool wear as it penetrates into the workpiece. A device *S3* for shaping reset pulses is needed to set counter *PC* to zero and to return trigger *T* to its starting state before the start of drilling on each hole. To assure stable operation of the whole device, the pulse generator operates in external-trigger mode; this is shown by the dashed line in the block diagram.

As we see, the device just described makes it possible to obtain the set of readings $S_1, S_2, S_3, \ldots, S_N$ needed for the determination of the criterion of efficiency by the proposed method.

Appendix 2
Method for Approximating the Response Function in the Multifactorial Experiment

When the mechanism of the process under study is incompletely known, the behavior of the response function in a given n-factor space is described by a regression equation. It would be incorrect to think that the polynomial of degree k approximating the response surface lacks any heuristic potential. A statistical connection between random quantities, if it is established for the first time, may suggest "shadowy" aspects of the process that have not been remarked on before. However, it is rather difficult to discover the character of the connection from the form of the polynomial equation (even when the effects are ranked). These same considerations justify the attempt to use the results of passive observation to obtain a well-fitting functional relation, which would reflect the connection in meaning between the random quantities and each of the probability relations that they replace.

Suppose that, in some system of random quantities $(X, Y, \ldots, W)$, the results of the experiment are given by a random response function

$$V(X, Y, Z, \ldots, P, Q, W) \qquad \text{(A.2-1)}$$

where the n variable factors comprising the arguments of this function and initially assumed independent are arranged in order of decreasing stochastic connection with V. A realization of function (A.2-1) approximating its mathematical expectation is the "almost nonrandom" function

$$v(x,y,z, \ldots ,w). \qquad \text{(A.2-2)}$$

It is required to find an analytical expression for the response function (A.2-2), where the numerical values of the random quantities $(v_i,x_i,y_i, \ldots ,w_i)$ are known for a sufficiently large number of observations $(i = 1, 2, 3, \ldots ,N)$.

We begin the treatment of the experiments by finding an estimate for the mathematical expectation of the random function $V(X)$. We make up a table of distribution, in which all the experimental material is divided into classes of equal class interval length. If it proves that some individual observations lie in intervals outside the limits of distribution of the given set, then these observations are not taken into account in the computation of estimates of the mathematical expectations of the corresponding sections of the function.*

It is natural to take the arithmetic mean of the observed values as an estimate of the mathematical expectation of a section of the function $V(X)$, which will henceforth be denoted $\tilde{m}_v$. Then, the estimate of $\tilde{m}_v$ will be consistent, unbiased, and, if V is normally distributed, efficient. However, since we have available only a limited amount of statistical material, it becomes impossible to compute the estimate $\tilde{m}_{vi}$ for any fixed value x_i. We find interval-by-interval estimates for the mathematical expectations of the quantities X and V (Fig. A.2-1), since $\tilde{m}_{xi}$ and $\tilde{m}_{vi}$ are equal to the abscissas and ordinates of the centers of gravity of the points lying in the i-th interval.

Using a number of values $(\tilde{m}_{xi},\tilde{m}_{vi})$, we construct an experimental curve of $\tilde{m}_v(\tilde{m}_x)$, which we take as the desired estimate of $\tilde{m}_v(x)$, as if it had been constructed from values $(x_i,\tilde{m}_{vi})$ on the condition that $x_i = \tilde{m}_{xi}$. To smooth out the resulting curve, we use the least-squares method; the function $\tilde{m}_v(x)$ is approximated by an expression $f(x)$. We can write

$$\tilde{m}_v = f(x). \qquad \text{(A.2-3)}$$

Using formula A.2-3, we refer all the numerical values of the quantity V to the same value of the quantity X, equal to x^*:†

$$v_{(x)i} = v_i \frac{f(x^*)}{f(x_i)}. \qquad \text{(A.2-4)}$$

This step accomplishes the conversion from function (A.2-1) to the "less random" function

$$V_{(x)} = V \frac{f(x^*)}{f(x)}, \qquad \text{(A.2-5)}$$

which has as arguments all the same random quantities except X:

---

*We remark, however, that anomalies in the distribution of statistical data may carry special information on the process, and for this reason require theoretical interpretation.
†The value of x^* is selected such that the greatest number of experimental points lie in its vicinity.

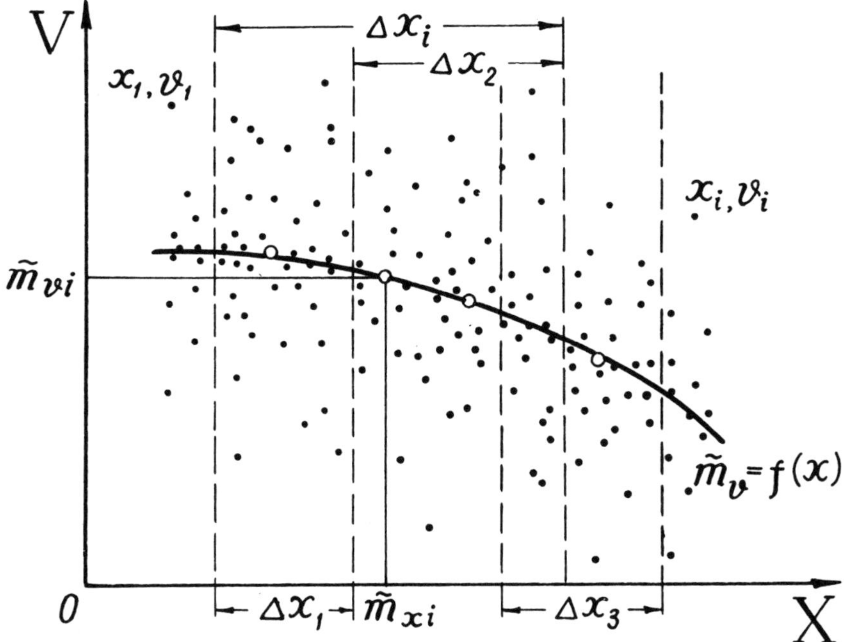

Fig. A.2-1

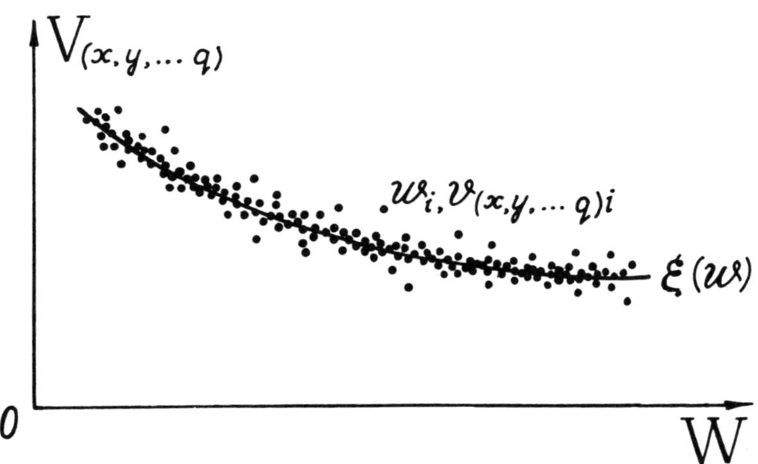

Fig. A.2-2

$$V_{(x)}(x^*,Y,Z,\ldots,P,Q,W). \tag{A.2-6}$$

In a completely similar way, we find the expression $\varphi(y)$ approximating the mathematical expectation of the random function $V_{(x)}(Y)$ (more precisely, its estimate), and obtain the function

$$V_{(x,y)} = V_{(x)}\,\frac{\varphi(y^*)}{\varphi(y)} \tag{A.2-7}$$

whose arguments do *not* include the two variable factors X and Y:

$$V_{(x,y)}(x^*,y^*,Z,\ldots,P,Q,W). \tag{A.2-8}$$

The value of y^* is determined from the condition $\varphi(y^*) = f(x^*)$.

In a similar way, we carry out other conversions in the direction of the further transformation of the random quantity V into a determinate quantity: $V_{(x,y)} \longrightarrow V_{(x,y,z)} \longrightarrow \cdots \longrightarrow V_{(x,y,\ldots,p)}$ and so on.

Now the transformation of the response function in which the variance of this function decreases step by step can be written as an arbitrary scheme:

$$V(X,Y,Z,\ldots,P,Q,W) \tag{A.2-1}$$
$$\downarrow$$
$$V_{(x)}(x^*,Y,Z,\ldots,P,Q,W) \tag{A.2-6}$$
$$\downarrow$$
$$V_{(x,y)}(x^*,y^*,Z,\ldots,P,Q,W) \tag{A.2-8}$$
$$\downarrow$$
$$\cdots\cdots\cdots\cdots\cdots\cdots\cdots$$
$$\downarrow$$
$$V_{(x,y,\ldots,p)}(x^*,y^*,z^*,\ldots,p^*,Q,W) \tag{A.2-9}$$
$$\downarrow$$
$$V_{(x,y,\ldots,q)}(x^*,y^*,z^*,\ldots,p^*,q^*,W), \tag{A.2-10}$$

with

$$V_{(x,y,\ldots,q)} = V_{(x,y,\ldots,p)}\,\frac{\chi(q^*)}{\chi(q)}. \tag{A.2-11}$$

The single argument of function A.2-10 is the last of the n factors selected. The principal assumption of the proposed method is just that not the mathematical expectation of the function $V_{(x,y,\ldots,q)}(W)$ but the function itself is considered nonrandom and is approximated by the expression $\xi(w)$ that completes the series of analytical expressions $f(x)$, $\varphi(y)$, $\ldots,\chi(q)$. The more factors are included (especially those that exert a dominant influence on the behavior of the response function), the smaller is the scatter of the empirical points (Fig. A.2-2) and the more accurate is the formula giving an approximate expression of the relation between $V_{(x,y,\ldots,q)}$ and W:

$$V_{(x,y,\ldots,q)}(W) \approx \xi(w). \tag{A.2-12}$$

Now, after making several substitutions one after the other (in reverse order), we can replace all the particular probability relations by a family of functional dependences:

$$\xi(w) \approx V_{(x,y,\ldots,p)} \frac{\chi(q^*)}{\chi(q)} = \cdots = V_{(x,y)} \frac{\psi(z^*)\ldots\gamma(p^*)\chi(q^*)}{\psi(z)\ldots\gamma(p)\chi(q)} = \cdots =$$

$$= V \frac{f(x^*)\varphi(y^*)\ldots\chi(q^*)}{f(x)\varphi(y)\ldots\chi(q)}. \tag{A.2-13}$$

Formula A.2-13 shows that the "randomness" of the quantity V has practically disappeared. Introducing the notation v for the determinate quantity and remembering that $f(x^*) = \varphi(y^*) = \cdots = \chi(q^*) = C$, we write

$$v \approx C^{1-n}f(x)\varphi(y)\ldots\xi(w). \tag{A.2-14}$$

This is now the analytical expression for the response function of interest to us. Incidentally, a similar formula in general form can readily be obtained through the use of the probability theory of regression.

It is not difficult to see that in the case of presentation by the formula

$$v_{(x,y,\ldots,w)i} = v_{(x,y,\ldots,q)i} \frac{\xi(w^*)}{\xi(w)} \tag{A.2-15}$$

not nearly all the numerical values of the quantity

$$V_{(x,y,\ldots,w)}(x^*,y^*,z^*,\ldots,p^*,q^*,w^*) \tag{A.2-16}$$

will coincide accurately with the modal value of the parameter C:

$$C = v(x^*,y^*,\ldots,w^*). \tag{A.2-17}$$

Random departures from C, as from the center of scatter, are linked with the influence of unaccounted-for factors. They may be used to identify these factors and to make formula A.2-14 more accurate.

Up to now the variables $X, Y, \ldots, W$ have been assumed independent, although some of them (and, in isolated cases, even all the arguments of the response function) are pairwise correlated.

Suppose, for example, that the factors X, Y, and W are among these interrelated values. Then, we represent the complete transformation of the function in the form of the following scheme, which is considered standard for other similar variants too:

$$V(X,Y,Z,\ldots,P,Q,W)$$
$$\downarrow$$

$$V_{(x)}(x^*,Y,Z,\ldots,P,Q,W);$$
$$D[V_{(x)}(Y)] < D[Y(X)];$$
$$D[V_{(x)}(W)] > D[W(X)]$$
$$\downarrow$$
$$V_{(x)}(x^*,Y,Z,\ldots,P,Q,W_{(x)})$$
$$\downarrow$$
$$V_{(x,y)}(x^*,y^*,Z,\ldots,P,Q,W_{(x)});$$
$$D[V_{(x,y)}(W_{(x)})] > D[W_{(x)}(Y)]$$
$$\downarrow$$
$$V_{(x,y)}(x^*,y^*,Z,\ldots,P,Q,W_{(x,y)})$$
$$\downarrow$$
$$\cdots\cdots\cdots\cdots\cdots\cdots\cdots\cdots\cdots\cdots\cdots\cdots\cdots$$
$$\downarrow$$
$$V_{(x,y,\ldots,q)}(x^*,y^*,z^*,\ldots,p^*,q^*,W_{(x,y)})$$
$$\downarrow$$
$$V_{(x,y,\ldots,q)}(W_{(x,y)}) \approx \xi(w_{(x,y)})$$
$$\downarrow$$
$$v \approx C^{1-n}f(x)\varphi(y)\ldots\xi(w_{(x,y)})$$

where D is the variance of the random function, and $C = v(x^*,y^*,\ldots,w^*_{(x,y)})$.

In conclusion, we remark that the use of this method makes it convenient to obtain the final result in graphical-analytical form; that is, to use the curves themselves instead of the unwieldy polynomials that express the complex experimental relationships.

References

1. *Proc. Conf. Electrical Phenomena in Friction and Cutting of Metals* (in Russian). Moscow: Nauka, 1969.
2. *Proc. Conf. Electrical Phenomena in Friction, Cutting, and Lubrication of Solids* (in Russian). Moscow: Nauka, 1973.
3. *Proc. Conf. Electrochemical Processes in Friction and Their Use Against Wear* (in Russian). Odessa: 1973.
4. *Proc. Conf. Theory of Friction and Wear* (in Russian). Tashkent: 1975.
5. Deryagin, B. V., Krotova, N. A., and Smilga, V. P. *Adhesion of Solids* (in Russian). Moscow: Nauka, 1973.
6. Deryagin, B. V., Abrikosova, I. I., and Lifshitz, E. M. *Usp. Fiz. Nauk* **64** (3) :493 (1958).
7. Akhmatov, A. S. In *Research in the Area of Surface Forces* (in Russian), p. 93. Moscow: Izd-vo AN SSSR, 1963.
8. Akhmatov, A. S. and Uchuvatkin, G. N. In reference 2, p. 7.
9. Bufeyev, V. A. In reference 3, p. 7.
10. Postnikov, S. N. In *Questions of the Electrophysics of Friction and Cutting* (in Russian), Tr. Gork. Politekh. Inst. **30** (4) :4 (1974).
11. Sidorov, V. P., Guslyakova, G. P., and Sokolov, L. D. In reference 9, p. 11.
12. Uchuvatkin, G. N. In reference 2, p. 35.
13. Yevdokimov, V. D. and Syomov, Yu. I. *Exoelectron Emission in Friction* (in Russian). Moscow: Nauka, 1973.
14. Postnikov, S. N. and Yashina, A. N. *Izv. Vyssh. Uchebn. Zaved., Fiz.* (7) :66 (1967).
15. Postnikov, S. N. *Tr. Gork. Politekh. Inst.* **22** (1) :117 (1966).
16. Melnichenko, I. M. In *Questions of the Electrophysics of Friction and Cutting* (in Russian), Tr. Gork. Politekh. Inst. **30** (4) :17 (1974).
17. Postnikov, S. N. In *Development and Application of Cutting Fluids in the Cutting of Metals* (in Russian), No. 2, p. 54. Moscow: Izd. Mosk. Dom Nauchno-Tekh. Propagandy im. F. E. Dzerzhinskogo, 1966.
18. Postnikov, S. N. *Wear* **10** (2) :142 (1967).
19. Postnikov, S. N. and Teplov, S. V. *Mashinostroitel* (5) :11 (1967).
20. Postnikov, S. N., and Starodubrovskaya, I. N. In *Friction, Lubrication, Wear* (in Russian). Tr. Gork. Politekh. Inst. **24** (11) :41 (1968).

21. Kretinin, O. V. In *Questions of the Technology of Machinery Construction* (in Russian), Tr. Gork. Politekh. Inst. 26 (4) :33 (1970).
22. Gordienko, P. L. and Gordienko, S. L. *Vestn. Mashinostr.* (7) :38 (1952).
23. Dubrov, Yu. S. and Nikolaeva, G. S. In reference 1, p. 56.
24. Opitz, H. In *New Works on Friction and Wear* (Russian translation). Moscow: IL, 1959.
25. Rizhkin, A. A. In reference 1, p. 70.
26. Postnikov, S. N. *Investigation of Electrical Phenomena in Friction and Cutting of Metals* (in Russian). Dissertation for candidate's degree, Gork. Politekh. Inst., 1966.
27. Postnikov, S. N. In reference 1, p. 35.
28. Hehenkamp, T. H. G. *Arch. Eisenhuettenwes.* 29 :4 (1958).
29. Bobrovsky, V. A. In reference 1, p. 7.
30. Postnikov, S. N., Borodkin, Yu. A., and Obidin, V. A. In reference 2, p. 61.
31. Postnikov, S. N., Avdonin, I. M., Gromyko, G. G., Zinkin, Yu. I., and Polunichev, A. I. In *Friction, Lubrication, Wear* (in Russian), Tr. Gork. Politekh. Inst. 29 (9): 29 (1973).
32. Postnikov, S. N., Godlina, A. F., and Tarakanov, V. N. In *Questions of the Electrophysics of Friction and Cutting* (in Russian), Tr. Gork. Politekh. Inst. 30 (4) :27 (1974).
33. Akhmatov, A. S. *Molecular Physics of Boundary Friction* (English translation). Jerusalem: Israel Program for Scientific Translations, 1966.
34. Smilga, V. P. *The Electron Theory of Adhesion* (in Russian). Dissertation for candidate's degree, Institut Fizicheskoi Khimii Akad. Nauk SSSR, 1961.
35. Deryagin, B. V. and Krotova, N. A. *Adhesion* (in Russian). Moscow: Izd-vo AN SSSR, 1949.
36. Mambetov, D. M. *Electrical Phenomena in the Adhesion and Cohesion Destruction of Solids* (in Russian). Frunze: Mektep, 1973.
37. Lifshitz, E. M. *Dokl. Akad. Nauk SSSR* 97 (4) :643 (1954).
38. Casimir, H. B. C. *J. Chim. Phys.* 46 :407 (1949).
39. Deryagin, B. V., Krotova, N. A., Knyazeva, N. P., and Toporov, Yu. P. In reference 3, p. 1.
40. Deryagin, B. V. *Zh. Fiz. Khim* 5 :1165 (1934).
41. Terzaghi, K. *Soil Mechanics* (in German). Vienna: 1925.
42. Bowden, F. P. and Tabor, D. *Friction and Lubrication of Solids,* Part II. Oxford: Clarendon Press, 1964.
43. Tabor, D. *Proc. R. Soc.* A251 :378 (1959).
44. Postnikov, S. N. *Contemp. Phys.* 6 (2) :100 (1964).
45. Brillouin, M. *Ann. Chim. Phys.* 16 :456 (1899).
46. Woog, P. *Contribution to the Study of Lubrication. Oiliness. Molecular Influences* (in French). Paris: 1926.
47. Adirovich, E. I. and Blokhintsev, D. I. *J. Phys. USSR* 7 (1) :29 (1943).
48. Postnikov, V. S. *Internal Friction in Metals* (in Russian). Moscow: Metallurgiya, 1974.
49. Gorelik, G. S. *Vibration and Waves* (in Russian). Moscow-Leningrad: GITTL, 1950.
50. Schulze, G. *Physics of Metals* (in Russian). Moscow: Mir, 1971.
51. Postnikov, S. N. and Navrotskaya, Ya. K. In reference 4, p. 22.
52. Bowden, F. P., and Freitag, E. H. *Proc. R. Soc.* A233 :429 (1958).
53. Granato, A. V. *ASTM STP* 378 :93 (1965).
54. Dukhovskoy, E. A., Onischenko, V. S., Ponomarev, A. N., Silin, A. A., and Talroze, V. L. *Dokl. Akad. Nauk SSSR* 189 (6) :1211 (1969).

55. Severdenko, V. P. and Gursky, L. I. *The Interior and Surface Structure of Rolled Materials* (in Russian). Minsk: Nauka I Tekhnika, 1972.

56. Kittel, C. *Elementary Solid State Physics* (Russian translation). Moscow: Nauka, 1965.

57. Breger, A. H. and Zhukhovitsky, A. A. *Zh. Fiz. Khim.* **20** (4–5) :355 (1946).

58. Nedorezov, S. S. *Zh. Eksp. Teor. Fiz.* **51** (3) :868 (1966).

59. Bekker, F. *Electron Theory* (in Russian). GITTL, 1941.

60. Kryuk, V. I. and Pavlov, V. V. In *Investigation of the Surface of Materials of Construction by the Exoelectron Emission Method* (in Russian). Tr. Ural. Politekh. Inst. (Sverdlovsk) **1969**, No. 177, 116.

61. Sokolov, L. D., Guslyakova, G. P., Shetulov, D. I., Myasnikov, A. M., Roganov, P. I., Sidorov, V. P., and Shibarov, V. V. *Proc. Conf. Physicochemical Mechanics of Contact Interaction and Fretting Corrosion* (in Russian), p. 155. Kiev: 1973.

62. Guslyakova, G. P., Sidorov, V. P., and Sokolov, L. D. *Proc. Conf. Modern Achievements in the Science and Practice of the Physics of Metals, Physical Metallurgy, and Pressure Treatment of Metals* (in Russian), p. 25. Krasnodar: 1973.

63. Cahn, R. (ed.). *Physical Metallurgy* (Russian translation). No. 1. Moscow: Mir, 1967.

64. Novikov, N. N. *Izv. Vyssh. Uchebn. Zaved., Fiz.* No. 7, 1972.

65. Mukha, I. M., Shtepan, Ya. G., and Valchuk, G. I. *Fiz.-Khim. Mekh. Mater.* (5): 111 (1971).

66. Stepanov, V. N. In *Strength of Metals under Cyclic Load* (in Russian), p. 181. Moscow: Nauka, 1970.

67. Alyabiev, A. Ya., Shevelya, V. V., and Anpilogov, V. N. *Proc. Conf. of Contact Interaction and Fretting Corrosion* (in Russian), p. 132. Kiev: 1973.

68. Bortnik, G. I., and Matyushenko, V. Ya. In reference 3, p. 132.

69. Matyushenko, V. Ya., Bortnik, G. I., Hanin, D. E., and Shpenkov, G. P. In reference 3, p. 157.

70. Demchenko, V. V. and Homutov, N. E. *Tr. Mosk. Khim.-Tekhnol. Inst.* (39) :115 (1962).

71. Avdentova, T. S., and Starodubrovskaya, I. N. In *Friction, Lubrication, Wear* (in Russian), Tr. Gork. Politekh. Inst. **29** (9) :24 (1973).

72. Markov, A. A. In reference 2, p. 28.

73. Kichkin, G. I., Markov, A. A., and Lashkhi, V. L. In reference 2, p. 37.

74. Markov, A. A., Kichkin, G. I., and Lashkhi, V. L. *Khim. Tekhnol. Topl. Masel,* No. 10, p. 44 (1970).

75. Garkunov, D. N., Markov, A. A., and Golikov, G. A. In *Theory of Lubricating Action and New Materials* (in Russian), p. 12. Moscow: Nauka, 1965.

76. Garkunov, D. N. and Kragelsky, I. V. *Dokl. Akad. Nauk SSSR* **113** (2) :326 (1957).

77. Conway, B. and Bockris, S. *J. Chem. Phys.* **26** :3 (1957).

78. Skorchelletti, V. V. *Theoretical Electrochemistry* (in Russian), p. 568. Leningrad: Goskhimizdat, 1963.

79. Balakin, V. A., Melnichenko, I. M., and Podalov, A. N. In reference 3, p. 129.

80. Gromyko, G. G., Volodin, G. F., Postnikov, S. N., and Tyurin, Yu. M. In *Friction, Lubrication, Wear* (in Russian), Tr. Gork. Politekh. Inst. **27** (13) :5 (1971).

81. Kubashevsky, O. and Gopkins, B. *Oxidation of Metals and Alloys* (in Russian). Moscow: Metallurgiya, 1965.

82. Cabrera, N. and Mott, N. E. Rep. Prog. Phys. **12** :163 (1948).

83. Cabrera, N. *Philos. Mag.* **40** (1) :175 (1949).

84. Dankov, P. D., Ignatov, D. V., and Shishakov, N. A. *Electron Diffraction Investigation of Oxide and Hydroxide Films on Metals* (in Russian). Moscow: Izd-vo AN SSSR, 1963.

85. Arslambekov, V. A. In *The Mechanism of the Reaction of Metals with Gases* (in Russian), p. 24. Moscow: Nauka, 1964.

86. Andreyev, L. A. and Paligeh, Ya. *Dokl. Akad. Nauk SSSR* 152 (5) :1086 (1963).

87. Jaeckel, B. and Wagner, B. *Vacuum* 13 (3) :509 (1963).

88. Volker, H. *Phys. Rev.* 138 (6A) :1689 (1965).

89. In *Acta Phys. Austriaca* 10 : (1957).

90. Kramer, J. In *Exoelectron Emission* (Russian translation), p. 91. Moscow: IL, 1962.

91. Haxel, O., Houtermanns, F., and Seeger, K. *Z. Phys.* 130 (1) :109 (1951).

92. Roykh, I. L. and Yarpovetsky, L. Ya. *Usp. Khim.* 28 (2) :168 (1959).

93. Bichevich, V. V. and Kyaehmbré, H. F. In *Exoelectron Processes in Alkali-Halide Crystals* (in Russian), Tr. Inst. Fiz. Astron., Akad. Nauk SSSR (Tartu) (38) :3 (1970).

94. Grunberg, L. and Wright, K. H. R. *Proc. R. Soc.* 232 (2) :403 (1955).

95. Von Voss, W. D. and Brotzen, F. R. *J. Appl. Phys.* 30 (11) :1639 (1959).

96. Mints, R. I. In *Investigation of the Surface of Materials of Construction by the Exoelectron Emission Method* (in Russian), Tr. Ural. Politekh. Inst. (177) :5 (1969).

97. Kortov, V. S. In reference 93, p. 18.

98. Yevdokimov, V. D. *Dokl. Akad. Nauk SSSR,* 180 (2) :338 (1968).

99. Gaprindashvili, A. I. and Yegolaev, V. F. In *Investigation of the Surface of Materials of Construction by the Exoelectron Emission Method* (in Russian), (177) :33 (1969).

100. Weiss, R. *Physics of Solids* (in Russian). Moscow: Atomizdat, 1968.

101. Van Buren. *Defects in Crystals* (Russian translation). Moscow: IL, 1962.

102. Trefilov, V. I. *Investigation of the Mechanism of Deformation and Destruction of Transition Metals with Body Centered Cubic Lattice* (in Russian). Author's abstract of doctor's dissertation. Kharkov: 1965.

103. Novikov, I. I. and Shishokin, D. P. *Dokl. Akad. Nauk SSSR* 164 (2) :307 (1965).

104. Kryuk, V. I. *Investigation of the Surface of Materials of Construction by the Exoelectron Emission Method* (in Russian), (177) :131 (1969).

105. Samsonov, G. V., Paderno, Yu. B., and Fomenko, V. S. *Zh. Tekh. Fiz.* 36 (8) :1436 (1966).

106. Mints, R. I. and Kortov, V. S. In *Atomic and Molecular Physics* (in Russian), Tr. Ural. Politekh. Inst. (143), 15 (1965).

107. Yevdokimov, V. D. *Dokl. Akad. Nauk SSSR* 175 (3) :563 (1967).

108. Yevdokimov, V. D. In *Investigation of the Surface of Materials of Construction by the Exoelectron Emission Method* (in Russian), Tr. Ural. Politekh. Inst. (177), 46 (1969).

109. Yevdokimov, V. D. In reference 3, p. 96.

110. Dubinin, A. D. *Energetics of Friction and Wear of Machine Parts* (in Russian). Moscow-Kiev: Mashgiz, 1963.

111. Bobrovsky, V. A. *Electrodiffusion Tool Wear* (in Russian). Moscow: Mashinostroenie, 1970.

112. Ioffe, A. F. *Semiconductor Thermocouples* (in Russian). Moscow-Leningrad: Izd-vo AN SSSR, 1960.

113. Postnikov, S. N. *Fiz.-Khim. Mekh. Mater.* 3 :313 (1947).

114. Lebedev, L. A. In reference 2, p. 21.

115. Frenkel, Ya. I. *Zh. Eksp. Teor. Fiz.* 16 (4) (1946).

116. Davis, R. M. *Stress Waves in Solids* (Russian translation). Moscow: IL, 1961.

117. Weinreich, G. *Phys. Rev.* 107 :317 (1957).

118. Kostetsky, B. I. and Zaporozhets, V. V. In *Theory of Friction and Wear* (in Russian), p. 125. Moscow: Nauka, 1965.

119. Holm, R. *Electrical contacts* (Russian translation). Moscow: IL, 1961.

120. Wilson, R. W. *Proc. Phys. Soc., London* **B58** :625 (1955).

121. Bowden, F. P. and Tabor, D. *Proc. R. Soc.* **A169** :391 (1939).

122. Kragelsky, I. V. and Schedrov, V. S. *The Development of Knowledge about Friction* (in Russian). Moscow: Izd-vo AN SSSR, 1956.

123. Kuznetsov, V. D. *Solid State Physics,* vol. 4 (in Russian). Tomsk: 1947.

124. Kragelsky, I. V. *Friction and Wear* (in Russian). Moscow: Mashgiz, 1962.

125. Kaidanovsky, N. L. and Haikin, S. Eh. *Zh. Tekh. Fiz.* **3** (1) :91 (1933).

126. Tolstoy, D. M. and Kaplan, R. L. In *Theory of Friction and Wear* (in Russian), p. 44. Moscow: Nauka, 1965.

127. Ioffe, A. F. *Physics of Semiconductors* (in Russian). Moscow: Izd-vo AN SSSR, 1957.

128. Bowden, F. P. and Williamson, J. B. P. *Res. Corresp.* (7) :A1 (1954).

129. Bowden, F. P. and Williamson, J. B. P. *Proc. R. Soc.* **A246** (1) (1958).

130. Postnikov, S. N. and Yashina, A. N. *Trudy Gork. Politekh. Inst.* **24** (4) :108 (1968).

131. *Thermoelectric Materials and Transformers* (in Russian). Moscow: Mir, 1964.

132. Bowden, F. P. and Tabor, D. *Friction and Lubrication* (in Russian). Moscow: Mashgiz, 1960.

133. Yakunin, G. I. and Mirbabaev, V. A. *Izv. Vyssh. Uchebn. Zaved., Mashinostr.* (6): 84 (1962).

134. Axer, H. In *Cost, Performance and Economy of Modern Machine Tools.* Aachen Machine Tool Colloquim 6, Essen, 1953.

135. Postnikov, S. N. *Trudy. Gork. Politekh Inst.* **24** (3) :156 (1968).

136. Bakradze, I. I., Vinogradov, G. V., Ivanov, M. K., Korepova, I. V., and Podolsky, Yu. Ya. In reference 2, p. 25.

137. Bakradze, I. I., Ivanov, M. K., Kolesnichenko, L. F., and Yuga, A. I. In reference 3, p. 4.

138. Pfail, B. *Proc. Conf. Physicochemical Mechanics of Contact Interaction and Fretting Corrosion* (in Russian), p. 38. Kiev: 1973.

139. Galey, M. T. In *Main Directions and Prospects for the Development of Instrument Construction Technology* (in Russian), p. 421. Moscow: ONTIPRIBOR, 1964.

140. Bowden, F. P. and Ridler, K. E. W. *Proc. R. Soc.* **A154** :640 (1936).

141. Shamshur, A. S. *Proc. Conf. Physicochemical Mechanics of Contact Interaction and Fretting Corrosion* (in Russian), p. 85. Kiev: 1973.

142. Shaw, M. C. and Yang, S. T. *Tool Eng.* **36** (4) (1956).

143. Latyshev, V. N. *Izv. Vyssh. Uchebn. Zaved., Mashinostr.* (5) :173 (1964).

144. Porter, A. I., Preis, G. A., and Sologub, N. A. *Proc. Conf. Physicochemical Mechanics of Contact Interaction and Fretting Corrosion* (in Russian), p. 63. Kiev: 1973.

145. Bowden, F. P. and Young, L. *Research* (3) :235 (1950).

146. Frumkin, A. N., Bagotsky, V. S., Iofa, Z. A., and Kabanov, B. N. *Kinetics of Electrode Processes* (in Russian). Moscow: Izd-vo MGU, 1952.

147. Tyurin, Yu. M. and Volodin, G. F. *Elektrokhimiya* **6** :1186 (1970); **7** :233 (1971).

148. Tyurin, Yu. M. and Volodin, G. F. *Elektrokhimiya* **5** :1203 (1969).

149. Afonshin, G. N., Volodin, G. F., and Tyurin, Yu. M. *Elektrokhimiya* **7** :1338 (1971).

150. Fleischmann, M., Mansfield, J. R., and Wynne-Jones, W. F. K. *J. Electroanal. Chem.* **10** :511 (1965).

151. Volodin, G. F. and Tyurin, Yu. M. In *The Double Layer and Adsorption on Solid Electrodes* (in Russian), vol. 2, p. 124. Tartu: Izd TGU, 1970.

152. Balej, J. and Spalek, O. *CITCE, Extended Abstracts, 21 Meeting,* p. 114. Prague: 1970.

153. Porter, A. I., Preis, G. A., and Sologub, N. A. In reference 3, p. 118.

154. Preis, G. A., Sologub, N. A., and Porter, A. I. *Proc. Conf. Physicochemical Mechanics of Contact Interaction and Fretting Corrosion* (in Russian), p. 117. Kiev: 1973.

155. Kuznetsov, V. A., Korobov, Y. M., Kotlov, Y. G., Preis, G. A., Moiseyenko, A. A., and Novitsky, A. E. In reference 3, p. 19.

156. Korobov, Y. M., Kuznetsov, V. A., Preis, G. A., Kotlov, Y. G., Novitsky, A. E., and Moiseyenko, A. A. In reference 3, p. 64.

157. Movseyev, G. E. In reference 3, p. 112.

158. Garkunov, D. N. and Kragelsky, I. V. *Effect of Selective Transfer in Friction (Effect of Wearlessness)* (in Russian). Dipl. 41, *Byull. Izobr.*, No. 17, p. 5 (1965).

159. Garkunov, D. N., Kragelsky, I. V., and Polyakov, A. A. *Selective Transfer at Friction Points* (in Russian). Moscow: Transport, 1969.

160. Polyakov, A. A. In reference 3, p. 122.

161. Postnikov, S. N. and Yashina, A. N. In *Friction, Lubrication, Wear* (in Russian), Tr. Gork. Politekh. Inst. **24** (11) :33 (1968).

162. Shpenkov, G. P. and Podalov, A. N. In *Selective Transfer at Friction Points* (in Russian), p. 44. Moscow: Izd. Mosk. Dom Nauchno-Tekh. Propagandy im. F. E. Dzerzhinskogo, 1971.

163. Dyomkin, N. B. *Actual Contact Area of Solid Surfaces* (in Russian). Moscow: Izd-vo AN SSSR, 1962.

164. Diachenko, P. E., Tolkachyova, N. N., Andreyev, G. A., and Karpova, T. M. *Area of Actual Contact of Joined Surfaces* (in Russian). Moscow: Izd-vo AN SSSR, 1963.

165. Dyomkin, N. B. In *Theory of Friction and Wear* (in Russian), p. 26. Moscow: Nauka, 1965.

166. Dyomkin, N. B., and Lankov, A. A. *Zavod. Lab.*, No. 6 (1965).

167. Kitamura, S. *Denki Gakkai Zasshi* **86** (6) :1022 (1966).

168. Korshunov, L. G., and Mints, R. I. *Fiz. Khim. Obrab. Mater.* (2) :86 (1971).

169. Polyakov, A. A. *Dokl. Akad. Nauk SSSR* **205** (2) :332 (1972).

170. Yevdokimov, Yu. A., Dobrinsky, G. K., and Domb, S. S. In reference 3, p. 139.

171. Lyashenko, A. B. and Snitkovsky, M. M. In reference 3, p. 157.

172. Shimanovsky, V. G., Hovrin, E. V., and Afanasiev, V. A. In reference 3, p. 183.

173. Melnichenko, I. M. In reference 3, p. 159.

174. Onsager, R. *Amer. Chem. Soc.* **58** :1486 (1936).

175. Akhmatov, A. S. *Proc. Conf. Friction and Wear in Machines* (in Russian), vol. 2, p. 119. Moscow: Izd-vo AN SSSR, 1960.

176. Bowden, F. P. and Tabor, D. *Friction and Lubrication of Solids,* Part I. Oxford: Clarendon Press, 1954.

177. Matveyevsky, R. M. *The Temperature Method of Estimating the Maximal Lubricating Ability of Machine Oils* (in Russian). Moscow: Izd-vo AN SSSR, 1956.

178. Matveyevsky, R. M., Markov, A. A., and Buyanovsky, I. A. *Proc. Conf. Physicochemical Mechanics of Contact Interaction and Fretting Corrosion* (in Russian), p. 46. Kiev: 1973.

179. Lashkhi, V. L., Vipper, A. B., Sanin, P. I., Markov, A. A., Shepelyova, E. S., Lozovoy, Yu. A., and Yermolov, F. N. In reference 3, p. 27.

180. Shor, G. I. and Lapin, V. P. In reference 1, p. 108.

181. Shor, G. I., Lapin, V. P., and Yevstingneyev, E. V. In reference 2, p. 41.

182. Lapin, V. P., and Shor, G. I. In reference 3, p. 24.

183. Salomon, T. *J. Inst. Pet.* **45** :423 (1959).

184. Lazarenko, B. R., and Lazarenko, N. I. In *Electromachining of Metals* (in Russian), Tr. Tsentr. Nauchno-Issled. Lab. Electr. Obrab. Mater., Akad. Nauk SSSR, No. 1, p. 9, Moscow, 1957.

185. Raiko, M. V. and Pavlov, V. N. In reference 3, p. 30.

186. Ventsel, S. V., Lobkin, A. M., and Scherbinin, A. I. In reference 3, p. 10.

187. Deinega, Y. F., Dumansky, A. V., and Vinogradov, G. V. *Kolloidn. Zh.* **23** (1) :25 (1961).

188. Deinega, Y. F., Vinogradov, G. V. *Kolloidn. Zh.* **25** (3) :379 (1963).

189. Ventsel, V. S., Garbuz, V. T., Yemets, B. G., and Tsiganok, A. A. In reference 3, p. 87.

190. Vieweg, V. and Kluge, J. *Arch. Eisenhuettenwes.* **2** :805 (1929).

191. Frumkin, A. *Z. Phys. Chem.* **109** :34 (1924).

192. Guyot, M. I. *Ann. Phys.* **2** :506 (1924).

193. Deryagin, B. V., Strakhovsky, G. M., and Malisheva, D. S. *Zh. Eksp. Teor. Fiz.* **16**: 171 (1946).

194. Deryagin, B. V. and Pichugin, E. F. *Proc. Conf. Friction and Wear in Machines (in Russian)*, vol. 1, p. 103, 1947; vol. 3, p. 101, 1949; Moscow-Leningrad: Izd-vo AN SSSR.

195. Deryagin, B. V. and Pichugin, E. F. *Dokl. Akad. Nauk SSSR* **63** :53 (1948).

196. Snitkovsky, M. M. In reference 3, p. 32.

197. Deryagin, B. V., Snitkovsky, M. M., and Lemza, V. D. In reference 3, p. 188.

198. Snitkovsky, M. M. *Proc. Conf. Physicochemical Mechanics of Contact Interaction and Fretting Corrosion* (in Russian), p. 68. Kiev: 1973.

199. Yuriev, V. N. In reference 3, p. 41.

200. Migal, V. D. and Pavlov, V. N. *Proc. Conf. Physicochemical Mechanics of Contact Interaction and Fretting Corrosion* (in Russian), p. 92. Kiev: 1973.

201. Hardy, W. *Philos. Mag.* **40** :201 (1920).

202. Lunn, B. *VDI-Berichte* **20** (1957).

203. Khanmamedov, S. A., Popovsky, Y. M., and Kilimnik, I. M. In reference 3, p. 35.

204. Gebda, M. and Vakhal, A. *Proc. Conf. Physicochemical Mechanics of Contact Interaction and Fretting Corrosion* (in Russian), p. 65. Kiev: 1973.

205. Hardy, W. *J. Chem. Soc.* **127** :1207 (1925).

206. Zisman, W. In Davies, R. (ed.), *Friction and Wear*, p. 110. Edited by R. Davies. Amsterdam: Elsevier, 1959.

207. Fuks, G. I. In *Research in the Area of Surface Forces* (in Russian), p. 99. Moscow: Izd-vo AN SSSR, 1961.

208. Fein, R. S., Rowe, C. N., and Kreuz, K. L. *Trans. ASLE* **2** :50 (1959).

209. Bowden, F. P., Gregory, I. N., and Tabor, D. *Nature* **156** :97 (1945).

210. St. Pierre, L. E., Owens, R. S., and Klint, R. V. *Nature* **202** :1204 (1964).

211. Fein, R. S. and Kreuz, K. L. *Trans ASLE* **8** :29 (1965).

212. Haikin, S. W. *Wear* **10** :49 (1967).

213. Tabor, D. *Proc. Conf. Lubrication and Wear, Sept. 25-29, 1967.* In *Proc. Inst. Mech. Eng., London* **182**, Part 3A, 262 (1967–68).

214. Moskalyov, V. I. In reference 3, p. 161.

215. Avakov, A. A. *Physical Grounds for the Theory of Cutting Tool Life* (in Russian). Moscow: Mashgiz, 1960.

216. Danielyan, A. M. and Bobrovsky, V. A. *Tr. Voen. Akad. BTV*, No. 11–12 (1952).

217. Semko, M. F. *"Cutting Heat and Tool Life"* (in Russian). Leningrad: 1937. Also in *Materials for the Conference on Cutting of Metals* (in Russian). Len. Otd. Vses. Nauchn. Inzh.-Tekh. Obshchestva Mashinostroitelei, 1940.

218. Yakovlev, G. M. *Some Questions of High-Speed Milling and Turning* (in Russian). Minsk: Gosizdat, 1960.

219. Krivets, D. V., Rizhkin, A. A., and Dmitriev, V. S. In reference 3, p. 67.

220. Dmitriev, V. S., Rizhkin, A. A., and Krivets, D. V. In reference 3, p. 57.

221. Bobrovsky, V. A. and Soloviov, Y. A. In reference 1, p. 97.

222. Afanasiev, F. Z. and Bobrovsky, V. A. In reference 2, p. 54.

223. Pisarev, V. S. *Thermoelectric Currents in Drilling and the Life of Special Small-Diameter Drills* (in Russian). Author's abstract of dissertation for candidate's degree, Moscow Aviation Technology Institute. Moscow: 1969.

224. Bobrovsky, V. A. *Vestn. Mashinostr.* (8) :65 (1966).

225. Barrow, G. and Spenser, R. M. *Ann. CIRP* 18 (2) :199 (1970).

226. Markosyan, R. G. *Sb. Nauchn. Tr. Leninakansk. Fil. Erevan. Politekh. Inst. im. K. Marksa* 1 (4) :57 (1971); in reference 3, p. 73.

227. Kulikovsky, L. F., Melik-Shakhnazarov, A. M., Rabinovitch, S. G., and Seliber, B. A. *Galvanometric Compensators* (in Russian). Moscow-Leningrad: Energiya, 1964.

228. Markosyan, R. G. *Investigation of the Effect of Thermoelectric and Thermomagnetic Phenomena on Cutting Tool Life* (in Russian). Author's abstract of dissertation for candidate's degree, Gruzinskii Politekh. Inst. im V. I. Lenina. Tbilisi: 1973.

229. Dubrov, Yu. S., Nikolaeva, G. S., and Filonenko, V. S. In reference 2, p. 70.

230. Yakunin, G. I., Umarov, Eh. A., and Antsupov, A. A. In reference 2, p. 132.

231. Kasahara, H. and Kinoshita, N. *Rikagaku Kenkyusho Hokoku* 37 (3) :177 (1961).

232. Bobrovsky, V. A. and Pisarev, V. S. In reference 1, p. 83.

233. Bobrovsky, V. A. and Pisarev, V. S. *Proc. Semin. New Cutting Tools and Progressive Cutting Processes* (in Russian), p. 42. Kharkov: 1968.

234. Postnikov, S. N., Borodkin, Y. A., Volovskaya, N. D., and Obidin, V. A. In *Friction, Lubrication, Wear* (in Russian), Tr. Gork. Politekh. Inst. 24 (11) :44 (1968).

235. Andzhus, P. A. *Investigation and Selection of High-Speed Steel and Heat Treatment of Drills* (in Russian). Author's abstract of dissertation for candidate's degree, Moscow Machine Tool Institute. Moscow: 1972.

236. Postnikov, S. N., Borodkin, Yu. A., and Obidin, V. A. *Inf. Listok Volgo-Vyatskogo TsNTI*, No. 170 (1969).

237. Postnikov, S. N. and Polunichev, A. I. In *Electrical Phenomena in Cutting and Friction of Metals* (in Russian), p. 121. Izd. Leninakansk. Fil. Erevan. Politekh. Inst. im. K. Marksa, 1970.

238. Markosyan, R. G. *Sb. Nauchn. Tr. Leninakansk. Fil. Erevan. Politekh. Inst. im. K. Marksa* 1 (4) :417 (1971).

239. Stepanenko, A. V. In reference 2, p. 102.

240. Shulga, V. A. In reference 2, p. 85.

241. Shulga, V. A. In reference 3, p. 81.

242. Loladze, T. N. *Cutting Tool Wear* (in Russian). Moscow: Mashgiz, 1958.

243. Yakunin, G. I. and Yakubov, F. Ya. *Tr. Tashk. Politekh. Inst.* (40) :13 (1966).

244. Rizhkin, A. A. and Dmitriev, V. S. In reference 2, p. 116.

245. Korobov, Yu. M. *Investigation of the Process of Finish Turning and Accompanying Thermoelectric Phenomena* (in Russian). Author's abstract of dissertation for candidate's degree, Leningrad. Politekh. Inst. im. M. I. Kalinina. Leningrad: 1965.

246. Korobov, Yu. M. In reference 2, p. 109.

247. Landberg, P. and Blankevoort, P. *Met. Working* 20 :16 (1954).

248. Engstrand, G. *Trans. R. Inst. Technol., Stockholm* 144 (1959).

249. Engstrand, G. *Eng. Dig.* 20 :12 (1959).

250. Ishibashi, A. and Katsuki, A. *Kisyu Daigaku Kogaku Syuho*, No. 42 (1970).

251. Fyodorov, R. *Nauka i Zhizn* 1962 (6) :68.

252. Yegorov, S. V. and Rudnev, A. V. *Stanki Instrum.* 1963 (5) :27.

253. Levin, V. I. *Proc. Semin. New Cutting Tools and Progressive Cutting Processes* (in Russian), p. 32. Kharkov, 1968.

254. Dubrov, Y. S., Nikolaeva, G. S., and Ter-Minosian, S. M. *Stanki Instrum.* 1968 (8) :28.
255. Bobrovsky, V. A. *Stanki Instrum.* (12) :20 (1966).
256. Gulyaev, A. P. *Mechanical Engineering* (in Russian). Moscow: Metallurgiya, 1966.
257. Korobov, Yu. M. *Stanki Instrum.* (3) :25 (1968).
258. Andreyevsky, V. M. *Investigation of the Forces of Friction and Wear of Steel Couples During Vibration* (in Russian). Author's abstract of dissertation for candidate's degree, Moscow Machine Tool Institute. Moscow: 1968.
259. Dikushin, V. I. In reference 1, p. 6.
260. Kretinin, O. V. *Proc. Conf. on the Treatment of Metals by Cutting* (in Russian). Gorky: 1970.
261. Dwyer, J. J. *Am. Mach.* 108 (6) :105 (1964).
262. Likhtman, V. I., Rebinder, P. A., and Karpenko, G. V. *Effect of Surfactant Medium on Metal Deformation Processes* (in Russian). Moscow: Izd-vo AN SSSR, 1954.
263. Yepifanov, G. I., Pletneva, N. A., and Rebinder, P. A. *Dokl. Akad. Nauk SSSR* 97 (2) :277 (1954).
264. Rebinder, P. A. *Izv. Akad. Nauk SSSR, Otd. Khim. Nauk.* (11) :1284 (1957).
265. Cassin, C. and Boothroyd, G. *J. Mech. Eng. Sci.* 7 (1) :67 (1965).
266. Macklin, E. S. and Jankee, W. R. *J. Appl. Phys.* 25 :5 (1954).
267. Zorev, N. N. *Questions of the Mechanics of the Process of Cutting of Metals* (in Russian). Moscow: Mashgiz, 1956.
268. Arzt, P. R. and Stewart, I. J. *Lubr. Eng.* 19 :7 (1963).
269. Deryagin, B. V. and Obukhov, E. V. *Kolloidn. Zh.* 1 :385 (1935).
270. Deryagin, B. V. *Kolloidn. Zh.* 17 (3) :207 (1955).
271. Isaev, A. I. *Process of the Formation of a Surface Layer in Cutting of Metals* (in Russian). Moscow: Mashgiz, 1950.
272. Timofeyev, V. P. *Cutting Fluids* (in Russian). Moscow: Mashgiz, 1960.
273. Bibikov, E. S. and Turik, V. F. *Priroda* 1966 (12) :59.
274. Filonenko, V. S. and Dubrov, Yu. S. In reference 2, p. 81.
275. Ladakina, E. P., Nikolaeva, G. S., and Dubrov, Yu. S. In reference 2, p. 95.
276. Makarov, A. D. and Kolenchenko, V. M. In *Electrical Phenomena in Cutting and Friction of Metals* (in Russian), p. 107. Izd. Leninakansk. Fil. Erevan. Politekh. Inst. im. K. Marksa, 1970.
277. Bobrovsky, V. A. *Proc. Semin. New Cutting Tools and Progressive Cutting Processes* (in Russian), p. 23. Kharkov: 1968.
278. Rizhkin, A. A. *Investigation of the Process of Drilling High-Temperature Steels with High-Speed and Carbide-Tipped Small-Diameter Drills* (in Russian). Author's abstract of dissertation for candidate's degree. Rostov-na-Donu, 1966.
279. Rizhkin, A. A., Dmitriev, V. S., Krivets, D. V., and Bineyev, R. E. In reference 3, p. 78.
280. Chertavskikh, A. K. and Kan, K. N. *Zh. Teor. Fiz.* 14 :9 (1944).
281. Kostetsky, B. I., Topekha, P. K., and Nesterovsky, S. E. In *Advanced Technology of Machinery Construction* (in Russian), p. 461. Moscow: Izd-vo AN SSSR, 1955.
282. Furuichi, R. and Tamamura, K. *Mem. Fac. Eng., Osaka City Univ.* 2 (1960).
283. Benar, Zh. (ed.). *Oxidation of Metals* (in Russian), vol. 1. Moscow: Metallurigya, 1968.
284. Zhilin, V. A. In reference 3, p. 62.
285. Germanchuk, F. K., Pyatnitsky, I. E., and Dokuchaev, V. G. *Proc. Conf. Physico-chemical Mechanics of Contact Interaction and Fretting Corrosion* (in Russian), p. 79. Kiev: 1973.

286. Frantsevitch, I. N. and Kalinovich, D. *Vopr. Prochn. Metall. Prochn. Mater.* 1956 (3) :45.
287. Kovensky, I. I. *Fiz. Tverd. Tela* 5 (5) :1423 (1963).
288. Lebedev, T. and Guterman, V. *Corrosion, Corrosion Protection and Electrolysis* (in Russian). GNTI [State Scientific and Technical Publishing House], 1948.
289. Kalinovich, D. F., Kovensky, I. I., Smolin, M. D., and Frantsevitch, I. N. *Fiz. Met. Metalloved.* 10 (1) :42 (1960).
290. Frantsevitch, I. N., Kalinovich, I. I., Kovensky, I. I., and Smolin, M. D. *Fiz. Tverd. Tela* 5 (5) :1238 (1963).
291. Kalinovich, D. F., Kovensky, I. I., Smolin, M. D., and Frantsevitch, I. N. *Izv. Akad. Nauk SSSR, Otd. Tekh. Nauk, Metall. Topl.* 1959 (1) :71.
292. Galey, M. T. *Priborostroenie* 1962 (10) :15.
293. Galey, M. T. In reference 1, p. 27.
294. Shulga, V. A. *Investigation of the Life of Gear-Cutting Tools and Roughness of Working Surfaces of the Teeth of Wheels as a Function of Thermoelectric Phenomena* (in Russian). Author's abstract of dissertation for candidate's degree, Kuibyshev. Politekh. Inst. im. V. V. Kuibysheva. Kuibyshev: 1970.
295. Tyushev, V. S. In reference 2, p. 48; in reference 3, p. 206.
296. Budzynski, A. F. *Mechanik,* No. 7, 1971.
297. Ladakina, E. P. and Gufan, R. M. In reference 3, p. 69.
298. Usmanov, K. B. and Yakunin, G. I. *Izv. Akad. Nauk Uzb. SSR, Ser. Tekh. Nauk,* No. 5, 1968.
299. Hashimov, A. N. and Yakunin, G. I. In reference 2, p. 137.
300. Makarov, A. D. *Wear and the Life of Cutting Tools* (in Russian). Moscow: Mashinostroenie, 1966.
301. Semyonov, A. P. *Seizing of Metals* (in Russian). Moscow: Mashgiz, 1958.
302. Semyonov, A. P. In *Theory of Friction and Wear* (in Russian), p. 164. Moscow: Nauka, 1965.
303. Semyonov, A. P. In *The Nature of Seizing of Solids* (in Russian), p. 44. Moscow: Nauka, 1968.
304. Semyonov, A. P. In *The Nature of Friction of Solids* (in Russian), p. 109. Minsk: Nauka I Tekhnika, 1971.
305. Semyonov, A. P. *Friction and Adhesion Interaction of Refractory Materials at High Temperatures* (in Russian). Moscow: Nauka, 1972.
306. Smit, M. K. *Fundamentals of the Physics of Metals* (in Russian). Moscow: Metallurgizdat, 1962.
307. Likhtman, V. I. In *The Nature of Seizing of Solids* (in Russian), p. 30. Moscow: Nauka, 1968.
308. Ainbinder, S. B. *Cold Welding of Metals* (in Russian). Riga: Izd-vo AN Latv. SSR, 1957.
309. Zhurkov, S. N. *Vestn. Akad. Nauk SSSR* 38 (3) :46 (1968).
310. Regel, V. R., Slutsker, A. I., and Tomashevsky, Eh. E. *Usp. Fiz. Nauk* 106 (2) :193 (1972).
311. Indenbom, V. L. and Orlov, A. N. In "Thermally Activated Processes in Crystals" (introductory article), *News of Solid State Physics* (in Russian), No. 2, Moscow: Mir, 1973.
312. Larin, M. N. and Martinov, G. A. *Proc. Semin. New Cutting Tools and Progressive Cutting Processes* (in Russian), p. 8. Kharkov: 1968.
313. Kurochkina, L. F. and Kolev, N. S. In reference 3, p. 110.
314. Ainbinder, S. B. In *The Nature of Seizing of Solids* (in Russian), p. 35. Moscow: Nauka, 1968.

315. Moiseyev, S. I. *Effect of Mechanical, Physical and Chemical Properties of Metals on Their Cutting* (in Russian). Author's abstract of dissertation for candidate's degree, Gork. Politekh. Inst. im. A. A. Zhdanov. Gorky: 1971.
316. Rikalin, N. N., Shorshorov, M. H., and Krasulin, Y. *Izv. Akad. Nauk SSSR, Neorg. Mater.* 1 (1) :29 (1965).
317. Landau, L. D. and Lifshitz, E. M. *Statistical Physics* (in Russian). Moscow: Nauka, 1964.
318. Rozenberg, Y. A. *Effect of Lubricating Oils on the Life and Dependability of Machine Parts* (in Russian). Moscow: Mashinostroenie, 1970.
319. Levin, V. I. and Nikitina, S. V. In reference 3, p. 72.
320. Grigoriev, N. A. In reference 3, p. 126.
321. Gegel, V. R. In *Some Questions of the Physics of Crystal Plasticity* (in Russian), p. 12. Itogi Nauki, No. 3. Moscow: Izd-vo AN SSSR, 1960.
322. Lovosky, V. N. In *The Nature of Seizing of Solids* (in Russian), p. 33. Moscow: Nauka, 1968.
323. Danielyan, A. M. *Vestn. Mashinostr.* **1957** (1) :39.
324. Postnikov, S. N. In *Friction, Lubrication, Wear* (in Russian), Tr. Gork. Politekh. Inst. **29** (9) :39 (1973).
325. Reznikov, N. I. *Science of Metal Cutting* (in Russian). Moscow: Mashgiz, 1947.
326. Reichenbach, G. S. *Trans. ASME* **80**, 3 (1958).
327. Reznikov, A. N. *Heat Exchange in Cutting and Cooling of Tools* (in Russian). Moscow: Mashgiz, 1963.
328. Telegin, A. A. *Investigation of the Thermal State of a Cutting Tool* (in Russian). Author's abstract of dissertation for candidate's degree, Moscow Aviation Technol. Inst. Moscow: 1968.
329. Reznikov, A. N. *Thermophysics of Cutting* (in Russian). Moscow: Mashinostroenie, 1969.
330. Kostetsky, N. I. *Zavod. Lab.*, No. 11–12, 1946.
331. Tashlitsky, N. I. *Effect of the Thermal Conductivity of Steel on the Cutter Temperature and the Machinability* (in Russian). Moscow: Mashgiz, 1948.
332. Tashlitsky, N. I. *Effect of Mechanical Properties and Thermal Conductivity of Steels on Their Machinabilities* (in Russian), p. 56. Moscow: Mashgiz, 1952.
333. Kazakov, N. F. *Diffusion Welding in a Vacuum* (in Russian). Moscow: Mashinostroenie, 1968.
334. Postnikov, S. N., Gromyko, G. G., Zinkin, Y. I., and Polunichev, A. I. In reference 3, p. 75.
335. Drozdov, N. G. *Static Electricity in Industry* (in Russian). Gosenergoizdat, 1949.
336. Lyob, L. *Static Charging* (in Russian). Moscow-Leningrad: GEI (State Scientific and Technical Power Engineering Publishing House), 1963.
337. Zolotikh, B. N. In *Electromachining of Metals,* Tr. Tsentr. Nauchno-Issled. Lab. Elektr. Obrab. Mater., Akad. Nauk SSSR 1957, No. 1, p. 38.
338. Zingerman, A. S. *Izv. Vyssh. Uchebn. Zaved., Fiz.* **1963** (1) :20.
339. Vizhilin, N. N., Mizeri, A. A., and Molodenskaya, K. V. In reference 3, p. 13.
340. Yevdokimov, Yu. A., and Sanches, S. S. In reference 3, p. 147.
341. Arbit, S. Eh., Gromyko, G. G., Postnikov, S. N., Shteinberg, M. I., Ivanov, A. A., and Likhachov, Yu. N. *Elektron. Tekh., Ser. 7: Tekhnologiya, Organizatsiya Proizvodstva i Oborudovanie* 4 (56) :65 (1973).
342. Golisheva, G. I., Gromyko, G. G., Arbit, S. Eh., Shteinberg, M. I., and Postnikov, S. N. In *Friction, Lubrication, Wear* (in Russian), Tr. Gork. Politekh. Inst. **29** (2) :4 (1973).
343. Harper, W. R. *Adv. Phys.* **6** :365 (1957).
344. Kornfeld, M. I. *Fiz. Tverd. Tela* **11** :1611 (1969).

345. Kornfeld, M. I. *Fiz. Tverd. Tela* **10** :2422 (1968).

346. Kornfeld, M. I. *Fiz. Tverd. Tela* **13** :474 (1971).

347. Coehn, A. and Lotz, A. *Z. Phys.* **5** :242 (1921).

348. Davies, D. K. *Br. J. Appl. Phys.* **2** (11) :1533 (1969).

349. Böning, P. *Z. Angew. Phys.* **8** :516 (1956).

350. Grassi, N. *Chemistry of Processes of Polymer Degradation* (Russian translation). Moscow: IL, 1959.

351. Butyagin, P. Y., Dubinskaya, A. M., and Radtsig, V. A. *Usp. Khim.* **38** :593 (1969).

352. Baramboim, N. K. *Mechanochemistry of High-Molecular-Weight Compounds* (in Russian). Moscow: Nauka, 1971.

353. Vinogradov, G. V. *Lubricating Effects of Hydrocarbon Synthetic Liquids and Solid Polymers* (in Russian), p. 9. Izd. Instituta Neftekhimicheskogo Sinteza AN SSSR, 1962.

354. Bilik, Sh. M. and Tsurkan, V. P. In *Theory of Lubrication and New Materials* (in Russian), p. 222. Moscow: Nauka, 1965.

355. Georgiyevsky, G. A., Lebedev, L. A., and Borozdinsky, E. M. In reference 2, p. 12.

356. Haritonov, V. V., Shpenkov, G. P., and Matyushenko, V. Ya. In reference 3, p. 208.

357. Georgiyevsky, G. A. and Lebedev, L. A. In reference 3, p. 15.

358. Polyakov, A. A. and Garkunov, D. N. *Proc. Conf. Physicochemical Mechanics of Contact Interaction and Fretting Corrosion* (in Russian), p. 11. Kiev: 1973.

359. Garkunov, D. N., Matyushenko, V. Y., Haritonov, V. V., and Shpenkov, G. P. In reference 3, p. 88.

360. Georgiyevsky, G. A., Lebedev, L. A., Matyushenko, V. Ya., and Polyakov, A. A. In reference 3, p. 138.

361. Lisov, B. N. *Investigation of Electrical Phenomena Accompanying the Process of Polishing and Methods for Their Use* (in Russian). Author's abstract of dissertation for candidate's degree, Chelyabinsk. Politekh. Inst. im. Leninskogo Komsomola. Chelyabinsk: 1972.

362. Yakunin, G. I. and Molchanova, N. G. In reference 2, p. 128.

363. Hashimov, A. N. *Effect of the Rake Angle of the Cutting Tool on Its Life with the Use of Cutting Fluids, Magnetization and Passage of Thermoelectric Currents through the Cutting Zone* (in Russian). Author's abstract of dissertation for candidate's degree, Tashkent. Politekh. Inst. im. A. R. Biruni. Tashkent: 1969.

364. Galey, M. T. *Stanki Instrum.* 1973 (5) :31.

365. Dimitrov, L. P. *Sb. Dokl. na VIMMESS—Ruse* **13** (2) :62 (1971).

366. Herbert, E. G. *J. Iron Steel Inst.* **120** :2 (1929).

367. Bernshtein, M. L. *Thermomagnetic Treatment of Steel* (in Russian). Moscow: Metallurgiya, 1968.

368. Landa, V. A. In *Physical Methods of Investigating and Monitoring the Structure of Tool Steels* (in Russian). Moscow: Mashgiz, 1963.

369. Yakunin, G. I. and Balabekov, M. T. *Yangi Tekhnika*, Tashkent, No. 4 (1965).

370. Avakov, A. A. and Markosyan, R. G. *Sb. Nauchn. Tr. Leninakansk. Fil. Erevan. Politekh. Inst. im. K. Marksa* **1** (4) :119 (1971).

371. Gavrilov, G. M., Skidanenko, V. I., Budyak, V. M., and Ionichev, A. I. In *Thermophysics of Engineering Processes* (in Russian), p. 135. Izd. Saratovskogo Gos. Universiteta, 1973.

372. Balabekov, M. T., Chernoglazov, M. I., Usmanov, A. I., and Bashkevich, S. V. *Sb. Mater. Itogam NIR Mechanich. Fak. Tashkent. Politekh. Inst. im A. R. Biruni* **1972** (83) :4.

373. Arkhangelskaya, G. A., Dikan, A. I., and Dobrokhotova, V. B. In *Questions of Elec-*

trophysics of Friction and Cutting (in Russian), Tr. Gork. Politekh. Inst. 30 (4) :40 (1974).

374. Galey, M. T. In reference 2, p. 125.

375. Sadovsky, V. D., Smirnov, L. V., Fokina, E. A., Malinen, P. A., and Sorokin, I. P. *Fiz. Met. Metalloved.* 24 (5) :918 (1967).

376. Borodkin, Yu. A., Kurbatov, V. A., Siluanov, A. E., and Tkachuk, V. N. In *Questions of Electrophysics of Friction and Cutting* (in Russian), Tr. Gork. Politekh. Inst. 30 (4) :36 (1974).

377. Postnikov, S. N., Kurbatov, V. A., Borodkin, Yu. A., and Ovchinnikov, A. D. *Vopr. Sudostr. Tekhnolog. Organ. Proizvod. Sud. Mashinostr.* (3) :92 (1975).

378. Oding, I. A. *Izv. Akad. Nauk SSSR, Otd. Tekh. Nauk,* No. 12 (1948).

379. Belov, K. P. *Elastic, Thermal and Electrical Phenomena in Ferromagnetic Metals* (in Russian). Moscow-Leningrad: GITTL, 1951.

380. Mishin, D. D. *Effect of Dislocational Structure on the Susceptibility and Coercivity of Ferrosilicon* (in Russian). Author's abstract of dissertation for doctor's degree, Tomsk. Gos. Univ. im. V. V. Kuibysheva. Tomsk, 1970.

381. Cahn, R. (ed.). *Physical Metallurgy* (Russian translation), No. 2. Moscow: Mir, 1968.

382. Gorelik, S. S. *Recrystallization of Metals and Alloys* (in Russian). Moscow: Metall-urgiya, 1967.

383. Avakov, A. A. *Sb. Tr. TBIIZhT im. V. I. Lenina* 1948 (27–28) :288.

Author Index

Subject Index